Biomass Gasification and Pyrolysis

Practical Design and Theory

Prabir Basu

ELSEVIER

AMSTERDAM • BOSTON • HEIDELBERG • LONDON
NEW YORK • OXFORD • PARIS • SAN DIEGO
SAN FRANCISCO • SINGAPORE • SYDNEY • TOKYO

Academic Press is an imprint of Elsevier

Academic Press is an imprint of Elsevier
30 Corporate Drive, Suite 400
Burlington, MA 01803, USA

Elsevier, The Boulevard, Langford Lane
Kidlington, Oxford, OX5 1GB, UK

Library of Congress Cataloging-in-Publication Data

Basu, Prabir

Biomass gasification and pyrolysis : practical design and theory / Prabir Basu.

p. cm.

Includes bibliographical references and index.

ISBN 978-0-12-374988-8 (alk. paper)

1. Biomass gasification. 2. Biomass—Combustion. 3. Pyrolysis. 4. Gas manufacture and works. I. Title.

TP339.B355 2010

662′.88—dc22

2010010068

British Library Cataloguing-in-Publication Data

A catalogue record for this book is available from the British Library.

For information on all Academic Press publications visit our Web site at *www.elsevierdirect.com*

Printed in the United States

10 11 12 13 14 10 9 8 7 6 5 4 3 2 1

To Mother.

Contents

Preface

The art of gasification and pyrolysis of biomass is as old as our natural habitat. Both processes have been at work since the early days of vegetation on this planet. Flame leaping from forest fires is an example of "flaming pyrolysis." Blue hallow in a swamp is an example of methane gas formation through decomposition of biomass and its subsequent combustion in contact with air. Human beings, however, learned to harness these processes much later.

First, large-scale application of gasification for industry and society concentrated on coal as the fuel. It was primarily for lighting of city streets and affluent people's houses. Use of gasification, though nearly as ancient as combustion technologies, did not rise with industrialization the same as combustion because of the abundant supply and low prices of oil and natural gas. Only in the recent past has there been an upsurge in interest in gasification, fueled by several factors:

- Interest in the reduction in greenhouse gas emissions as a result of energy production
- Push for independence from the less reliable supply and fluctuating prices of oil and gas
- Interest in renewable and locally available energy sources
- Rise in the price of oil and natural gas

Several excellent books on coal gasification are available, but a limited few are available about biomass gasification and pyrolysis. A large body of peer-reviewed literature on biomass gasification and pyrolysis is available; some recent books on energy also include brief discussions on biomass gasification. However, there is a dearth of comprehensive publications specifically on gasification and pyrolysis; this is especially true for biomass.

Engineers, scientists, and operating personnel of biomass gasification plants clearly need such information from a single easy-to-access source. Better comprehension of the main aspects of gasification technology could help an operator understand the workings of the gasification plant, a design engineer to size the gasifier, and a planner to evaluate different options. The present book was written to fill this important need. It attempts to mold available research results in an easy-to-use design methodology whenever possible. Additionally, it brings into focus new advanced processes such as supercritical water gasification and torrefaction of biomass.

This book is comprised of nine chapters and three appendices, which include several tables that could be useful for the design of biomass gasifiers and their components. Chapter 1 introduces readers to the art of gasification and its present state of art. It also discusses the relevance of gasification to the current energy scenario around the world. Chapter 2 presents the properties of biomass with specific relevance to gasification and pyrolysis of biomass. The basics of pyrolysis are discussed in Chapter 3, which also covers torrefaction. In addition, it introduces readers to the design of a pyrolyzer and elements of the torrefaction process.

Chapter 4 deals with an important practical aspect of biomass gasification—the tar issue. This chapter provides information on the limits of tar content in product gas for specific applications. It also discusses several means of reduction in tar in the product gas. Chapter 5 concerns the basics of the gasification of biomass. It explains the gasification process and important chemical reactions that guide pyrolysis and gasification. Chapter 6 discusses design methodologies for gasifiers and presents some worked-out examples on design problems. Chapter 7 introduces the new field of hydrothermal gasification, with specific reference to gasification of biomass in supercritical water. It covers the basics of this relatively new field.

One of the common, but often neglected, problems in the design of a gasification plant is the handling of biomass. Chapter 8 discusses issues related to this and provides guidelines for the design and selection of handling equipment. The production of chemicals and synthetic fuels is gaining importance, so Chapter 9 provides a brief outline of how some important chemicals and fuels are produced from biomass through gasification. Production of diesel and biogasoline is also discussed briefly here. Appendix A contains definitions of biomass and Appendix B lists physical constants. Appendix C includes several tables containing design data. The Glossary presents definitions of some terms used commonly in the chemical and gasification industries.

ACKNOWLEDGMENTS

The author is greatly indebted to a large number students, professional colleagues, and institutions who helped revise numerous drafts of this book and provided permission for the use of published materials. Drs. D. Groulux, I. Ugursal, N. Mahinpey, N. Bakshi, A. Dutta, A. M. Leon, and P. Kaushal read parts of the manuscript and provided valuable suggestions. Many students worked tirelessly to support the work on this book. Special efforts were made by M. Greencorn, S. Rao, V. Mettanant, B. Acharya, A. Dhungana, and A. Basu. My hope is that is that what is here will benefit at least some students and/or practicing professionals in making the world around us a little greener and more habitable.

Finally, this book would not have materialized without the constant encouragement of my wife, Rama Basu.

About the Author

Dr. Prabir Basu is an active researcher and designer of gasifiers with a specific interest in fluidized-bed gasification of biomass. His current research interests include frontier areas, such as supercritical gasification, as well as applied research dealing with biomass co-firing. He is the founder of the prestigious triennial International Conference series on Circulating Fluidized Beds, and founder of Greenfield Research Incorporated, a private research and development company based in Canada that specializes in fluidized-bed boilers and gasification.

Professor Basu has been working in the field of energy conversion and the environment for more than 30 years. Prior to joining the engineering faculty at Dalhousie University (formerly known as the Technical University of Nova Scotia), he worked with both a government research laboratory and a private boiler manufacturing company.

Dr. Basu's passion for the transformation of research results into industrial practice is well known, as is his ongoing commitment to spreading advanced knowledge around the world. He has authored more than 200 research papers and six monographs in emerging areas of energy and environment, some of which have been translated into Chinese and Korean. He teaches short courses and seminars in industries and at universities across the globe. Presently, he leads the Circulating Fluidized-Bed Research Laboratory at Dalhousie University in Canada.

Chapter 1

Introduction

Gasification is a chemical process that converts carbonaceous materials like biomass into useful convenient gaseous fuels or chemical feedstock. Pyrolysis, partial oxidation, and hydrogenation are related processes. Combustion also converts carbonaceous materials into product gases, but there are some important differences. For example, combustion product gas does not have useful heating value, but product gas from gasification does. Gasification packs energy into chemical bonds while combustion releases it. Gasification takes place in reducing (oxygen-deficient) environments requiring heat; combustion takes place in an oxidizing environment giving off heat.

The purpose of gasification or pyrolysis is not just energy conversion; production of chemical feedstock is also an important application. In fact, the first application of pyrolysis of wood into charcoal around 4000 B.C.E. was not for heating but for iron ore reduction. In modern days, gasification is not restricted to solid hydrocarbons. Its feedstock includes liquid or even gases to produce more useful fuels. Partial oxidation of methane gas is widely used in production of synthetic gas, or *syngas*, which is a mixture of H_2 and CO.

Pyrolysis (see Chapter 3), the pioneer in the production of charcoal and the first transportable clean liquid fuel *kerosene*, produces liquid fuels from biomass. In recent times, gasification of heavy oil residues into syngas has gained popularity for the production of lighter hydrocarbons. Many large gasification plants are now dedicated to production of chemical feedstock from coal or other hydrocarbons. Hydrogenation, or hydrogasification, which involves adding hydrogen to carbon to produce fuel with a higher hydrogen-to-carbon (H/C) ratio, is also gaining popularity. Supercritical gasification (see Chapter 7), a new option for gasification of very wet biomass, is drawing growing interest.

This chapter introduces the subject of biomass gasification with a short description of its historical developments, its motivation, and its products. It also gives a brief introduction to the chemical reactions that are involved in gasification.

Biomass Gasification and Pyrolysis. DOI: 10.1016/B978-0-12-374988-8.00001-5

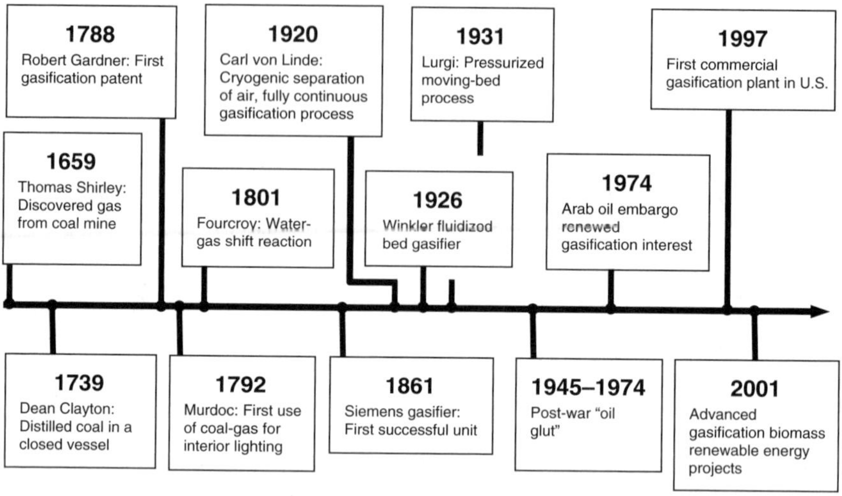

FIGURE 1.1 Milestones in gasification development.

1.1 HISTORICAL BACKGROUND

The earliest known investigation into gasification was carried out by Thomas Shirley, who in 1659 experimented with "carbureted hydrogen" (now called methane). Figure 1.1 shows some of the important milestones in the progression of gasification.

The pyrolysis of biomass to produce charcoal was perhaps the first large-scale application of a gasification-related process. When wood, owing to its overuse, became scarce toward the beginning of the eighteenth century, coke was produced from coal through pyrolysis, but the use of by-product gas from pyrolysis received little attention. Early developments were inspired primarily by the need for town gas for street lighting. The salient features of town gas from coal were demonstrated to the British Royal Society in 1733, but the scientists of the time saw no use for it. In 1798, William Murdoch used coal-gas (also known as town gas) to light the main building of the Soho Foundry, and in 1802 he presented a public display of gas lighting, astonishing the local population. Friedrich Winzer of Germany patented coal-gas lighting in 1804 (*www.absoluteastronomy.com/topics/coal gas*).

By 1823 numerous towns and cities throughout Britain were gas-lit. At the time, the cost of gas light was 75% less than that for oil lamps or candles, and this helped accelerate its development and deployment. By 1859, gas lighting had spread throughout Britain. It came to the United States probably in 1816, with Baltimore the first city to use it.

The history of gasification may be divided into four periods, as described in the following:

1850–1940: During this early stage, the gas made from coal was used mainly for lighting homes and streets and for heating. Lighting helped along the Industrial Revolution by extending working hours in factories, especially on short winter days. The invention of the electric bulb circa 1900 reduced the need for gas for lighting, but its use for heating and cooking continued. With the discovery of natural gas, the need for gasification of coal or biomass decreased. All major commercial gasification technologies (Winkler's fluidized-bed gasifier in 1926, Lurgi's pressurized moving-bed gasifier in 1931, and Koppers-Totzek's entrained-flow gasifier) made their debut during this period.

1940–1975: The period 1940–1975 saw gasification enter two fields of application as synthetic fuels: internal combustion and chemical synthesis into oil and other process chemicals. In the Second World War, Allied bombing of Nazi oil refineries and oil supply routes greatly diminished the crude oil supply that fueled Germany's massive war machinery. This forced Germany to synthesize oil from coal-gas using the Fischer-Tropsch (see Eq. 1.13) and Bergius processes ($n\text{C} + (n + 1)\text{H}_2 \rightarrow \text{C}_n\text{H}_{2n+2}$). Chemicals and aviation fuels were also produced from coal.

A large number of cars and trucks in Europe operated on coal or biomass gasified in onboard gasifiers. During this period over a million small gasifiers were built primarily for transportation (see Figure 1.2). The end of the Second World War and the availability of abundant oil from the Middle East eliminated the need for gasification for transportation and chemical production.

FIGURE 1.2 Bus with an onboard gasifier during the Second World War. *(Source: http://www.woodgas.com/history.htm.)*

The advent of plentiful natural gas in the 1950s dampened the development of coal or biomass gasification, but syngas production from natural gas and naphtha by steam reforming increased, especially to meet the growing demand for fertilizer.

1975–2000: The third phase in the history of gasification began after the Yom Kippur War, which triggered the 1973 oil embargo. On October 15, 1973, members of the Organization of Arab Petroleum Exporting Countries (OPEC) banned oil exports to the United States and other western countries, which were at that time heavily reliant on oil from the Middle East. This shocked the western economy and gave a strong impetus to the development of alternative technologies like gasification in order to reduce dependence on imported oil.

Besides providing gas for heating, gasification found major commercial use in chemical feedstock production, which traditionally came from petroleum. The subsequent drop in oil price, however, dampened this push for gasification, but some governments, recognizing the need for a cleaner environment, gave support to large-scale development of integrated gasification combined cycle (IGCC) power plants.

Post-2000: Global warming and political instability in some oil-producing countries gave a fresh momentum to gasification. The threat of climate change stressed the need for moving away from carbon-rich fossil fuels. Gasification came out as a natural choice for conversion of renewable carbon-neutral biomass into gas.

The quest for energy independence and the rapid increase in crude oil prices prompted some countries to recognize the need for development of IGCC plants. The attractiveness of gasification for extraction of valuable feedstock from refinery residue was rediscovered, leading to the development of some major gasification plants in oil refineries. In fact, chemical feedstock preparation took a larger share of the gasification market than energy production.

1.2 BIOMASS AND ITS PRODUCTS

Biomass is formed from living species like plants and animals—that is, anything that is now alive or was a short time ago. It is formed as soon as a seed sprouts or an organism is born. Unlike fossil fuel, biomass does not take millions of years to develop. Plants use sunlight through photosynthesis to metabolize atmospheric carbon dioxide and grow. Animals grow by taking in food from biomass. Fossil fuels do not reproduce whereas biomass does, and, for that reason, is considered renewable. This is one of its major attractions as a source of energy or chemicals.

Every year, a vast amount of biomass grows through photosynthesis by absorbing CO_2 from the atmosphere. When it burns it releases carbon dioxide that the plants had absorbed from the atmosphere only recently (a few

TABLE 1.1 Sources of Biomass

Farm products	Corn, sugar cane, sugar beet, wheat, etc.	Produces ethanol
	Rape seed, soybean, palm sunflower seed, Jatropha, etc.	Produces biodiesel
Ligno-cellulosic materials	Straw or cereal plants, husk, wood, scrap, slash, etc.	Can produce ethanol, bioliquid, and gas

years to a few hours). Thus, any burning of biomass does not add to the Earth's carbon dioxide inventory. For this reason biomass is considered a "carbon-neutral" fuel.

Of the vast amount of biomass, only 5% (13.5 billion metric tons) can be potentially mobilized to produce energy. This quantity is still large enough to provide about 26% of the world's energy consumption, which is equivalent to 6 billion tons of oil (IFP, 2007).

Biomass covers a wide spectrum—from tiny grass to massive trees, from small insects to large animal wastes, and the products derived from these. The principal types of harvested biomass are *cellulosic* (noncereal) and *starch* and *sugar* (cereal).

All parts of a harvested crop like corn plant are biomass, but its fruit (corn) is a starch while the rest of it is ligno-cellulose. The crop (corn) can produce ethanol through fermentation, but the ligno-cellulosic part of the corn plant requires a more involved process through gasification or hydrolysis.

Table 1.1 lists the two types of harvested biomass in food and nonfood categories, and indicates the potential conversion products from them. The division is important because the production of transport fuel (ethanol) from cereal, which is relatively easy and more established, is already being pursued commercially on a large scale. The use of such food stock for energy production, however, may not be sustainable as it diverts cereal from the traditional grain market to the energy market, with economic, social, and political consequences. Efforts are thus being made to produce more ethanol from nonfood resources like ligno-cellulosic materials such that the world's food supply is not strained by its energy hunger.

1.2.1 Products of Biomass

Three types of primary fuel are produced from biomass:

- *Liquid* (ethanol, biodiesel, methanol, vegetable oil, and pyrolysis oil)
- *Gaseous* (biogas (CH_4, CO_2), producer gas (CO, H_2, CH_4, CO_2 , H_2), syngas (CO, H_2), substitute natural gas (CH_4))
- *Solid* (charcoal, torrefied biomass)

From these come four major categories of product:

- Chemicals such as methanol, fertilizer, and synthetic fiber
- Energy such as heat
- Electricity
- Transportation fuel such as gasoline and diesel

The use of ethanol and biodiesel as transport fuels reduces the emission of CO_2 per unit of energy production. It also lessens dependence on fossil fuel. Thus, biomass-based energy not only is renewable but is also clean from a greenhouse gas (GHG) emission standpoint, and so it can take the center stage on the global energy scene. This move is not new. Civilization began its energy use by burning biomass. Fossil fuels came much later, around 1600 A.D. Before the twentieth century, wood (a biomass) was the primary source of the world's energy supply. Its large-scale use during the early Industrial Revolution caused so much deforestation in England that it affected industrial growth. As a result, from 1620 to 1720 iron production decreased from 180,000 to 80,000 tons per year (Higman and van der Burgt, 2008, p. 2). This situation was rectified by the discovery of coal, which began displacing wood for energy as well as for metallurgy.

Chemicals

Most chemicals produced from petroleum or natural gas can be produced from biomass. The two principal platforms for chemical production are sugar based and syngas based. The former involves sugars like glucose, fructose, xylose, arabinose, lactose, sucrose, and starch.

The syngas platform synthesizes the hydrogen and carbon monoxide constituent of syngas into chemical building blocks. Intermediate building blocks for different chemicals are numerous in this route. They include hydrogen, methanol, glycerol (C3), fumaric acid (C4), xylitol (C5), glucaric acid (C6), and gallic acid (Ar), to name a few (Werpy and Petersen, 2004). These intermediates are synthesized to produce large numbers of chemicals for industries involving transportation, textiles, food, the environment, communications, health, housing, and recreation. Werpy and Petersen (2004) identified 12 intermediate chemical building blocks having the highest potential for commercial products.

Energy

Biomass was probably the first on-demand source of energy that humans exploited. However, less than 22% of our primary energy demand is currently met by biomass or biomass-derived fuels. The position of biomass as a primary source of energy varies widely depending on the geographical and socioeconomic conditions. For example, it constitutes 90% of the primary energy source in Nepal but only 0.1% in the Middle East. Cooking, although highly

FIGURE 1.3 Cooking stove using fire logs.

FIGURE 1.4 A biomass fired bubbling fluidized bed in Canada. *(Source: Photo by author.)*

inefficient, is one of the most extensive uses of biomass in less developed countries. Figure 1.3 shows a cooking stove still employed by millions in the rural areas using twigs or logs as fuel. A more efficient modern commercial use of biomass is in the production of steam for process heat and electricity generation like the facility shown in Figure 1.4.

Heat and electricity are two forms of primary energy derived from biomass. The use of biomass for efficient energy production is presently on the rise in developed countries because of its carbon-neutral feature while its use for cooking is declining because of a shortage of biomass in less-developed countries.

Transport Fuel

Diesel and gasoline from crude petro-oil are widely used in modern transportation industries. Biomass can help substitute these petro-derived transport fuels. Ethanol, produced generally from sugarcane and corn, is used in gasoline (spark-ignition) engines, while biodiesel, produced from vegetable oils such as rape seed, is used in diesel (compression-ignition) engines.

Pyrolysis, fermentation, and mechanical extraction are three major ways to produce transport fuel from biomass. Of these, commercially the most widely used method is fermentation, where sugar (sugarcane, etc.) or starch (corn, etc.) produces ethanol. It involves a relatively simple process where yeast helps ferment sugar or starch into ethanol and carbon dioxide. The production and refining of marketable ethanol takes a large amount of energy.

Extraction of vegetable oil from seeds, like rape seed, through mechanical means has been practiced for thousands of years. Presently, oils like canola oil are refined with alcohol (transesterification) to produce methyl ester or biodiesel.

Pyrolysis involves heating biomass in the absence of air to produce gas, char, and liquid. The liquid is a precursor of bio-oil, which may be hydrotreated to produce "green diesel" or "green gasoline." At this time, ethanol and biodiesel dominate the world's biofuels market.

Gasification and anaerobic digestion can produce methane gas from biomass. The methane gas can then be used directly in some spark-ignition engines for transportation, or converted into gasoline through methanol.

1.3 BIOMASS CONVERSION

The bulky and inconvenient form of biomass is a major barrier to a rapid shift from fossil to biomass fuels. Unlike gas or liquid, biomass cannot be handled, stored, or transported easily, especially in its use for transportation. This provides a major motivation for the conversion of solid biomass into liquid and gaseous fuels, which can be achieved through one of two major paths (Figure 1.5): (1) biochemical (fermentation) and (2) thermochemical (pyrolysis, gasification).

Biochemical conversion is perhaps the most ancient means of biomass gasification. India and China produced methane gas for local energy needs by anaerobic microbial digestion of animal wastes. In modern times, most of the ethanol for automotive fuels is produced from corn using fermentation. Thermochemical conversion of biomass into gases came much later. Large-scale use of small biomass gasifiers began during the Second World War, when more than a million units were in use. Figure 1.5 shows that the two broad routes of

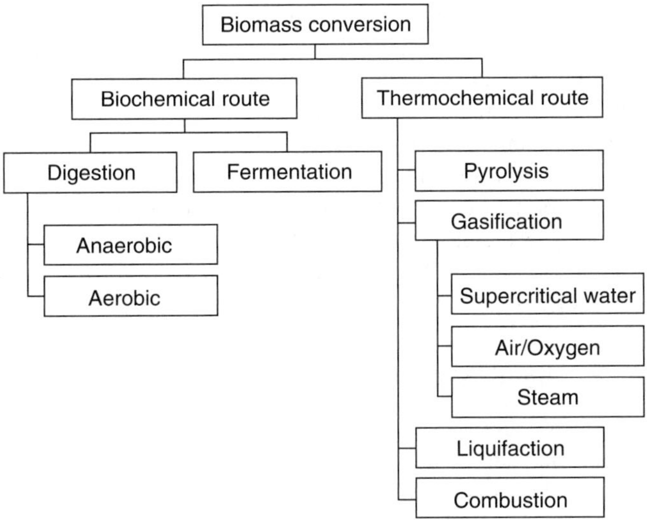

FIGURE 1.5 Two paths, biological and chemical, for conversion of biomass into fuel, gases, or chemicals.

conversion are subdivided into several categories. A brief description of these follows.

1.3.1 Biochemical Conversion

In biochemical conversion, biomass molecules are broken down into smaller molecules by bacteria or enzymes. This process is much slower than thermochemical conversion, but does not require much external energy. The three principal routes for biochemical conversion are:

- Digestion (anaerobic and aerobic)
- Fermentation
- Enzymatic or acid hydrolysis

The main products of anaerobic digestion are methane and carbon dioxide in addition to a solid residue. Bacteria access oxygen from the biomass itself instead of from ambient air.

Aerobic digestion, or composting, is also a biochemical breakdown of biomass, except that it takes place in the presence of oxygen. It uses different types of microorganisms that access oxygen from the air, producing carbon dioxide, heat, and a solid digestate.

In fermentation, part of the biomass is converted into sugars using acid or enzymes. The sugar is then converted into ethanol or other chemicals with the help of yeasts. The lignin is not converted and is left either for combustion or for thermochemical conversion into chemicals. Unlike in anaerobic digestion, the product of fermentation is liquid.

Fermentation of starch and sugar-based feedstock (i.e., corn and sugarcane) into ethanol is fully commercial, but this is not the case with cellulosic biomass because of the expense and difficulty in breaking down (hydrolyzing) the materials into fermentable sugars. Ligno-cellulosic feedstock, like wood, requires hydrolysis pretreatment (acid, enzymatic, or hydrothermal) to break down the cellulose and hemicellulose into simple sugars needed by the yeast and bacteria for the fermentation process. Acid hydrolysis technology is more mature than enzymatic hydrolysis technology, though the latter could have a significant cost advantage. Figure 1.6 shows the schemes for fermentation (of sugar) and acid hydrolysis (of cellulose) routes.

1.3.2 Thermochemical Conversion

In thermochemical conversion, the entire biomass is converted into gases, which are then synthesized into the desired chemicals or used directly (Figure 1.7). The Fischer-Tropsch synthesis of syngas into liquid transport fuels is an example of thermochemical conversion. Production of thermal energy is the main driver for this conversion route that has four broad pathways:

- Combustion
- Pyrolysis
- Gasification
- Liquefaction

Table 1.2 compares these four thermochemical paths for biomass conversion. It also shows the typical range of their reaction temperatures.

Combustion involves high-temperature conversion of biomass in excess air into carbon dioxide and steam. Gasification, on the other hand, involves a chemical reaction in an oxygen-deficient environment. Pyrolysis takes place at a relatively low temperature in the total absence of oxygen. In liquefaction, the large feedstock molecules are decomposed into liquids having smaller molecules. This occurs in the presence of a catalyst and at a still lower temperature.

Table 1.3 presents a comparison of the thermochemical and biochemical routes for biomass conversion (see page 13). It shows that the biochemical route for ethanol production is more commercially developed than the thermochemical route, but the former requires sugar or starch for feedstock; it cannot use ligno-cellulosic stuff. As a result, a larger fraction of the available biomass is not converted into ethanol.

For example, in a corn plant only the kernel is used for production. The stover, stalk, roots, and leaves, which are ligno-cellulosic, are left as wastes. Even though the enzymatic or biochemical route is more developed, this is a batch process and takes an order of magnitude longer to complete than the thermochemical process. In the thermochemical route, the biomass is first converted into syngas, which is then converted into ethanol through synthesis or some other means.

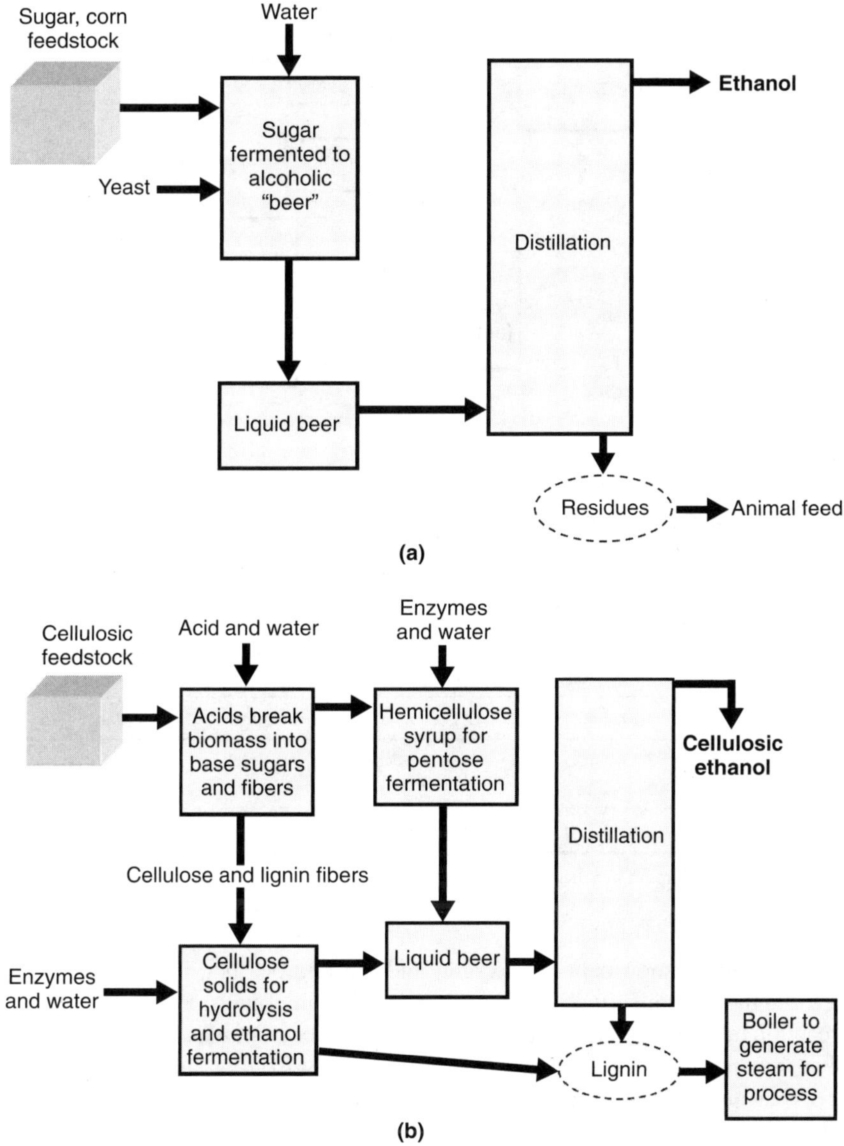

FIGURE 1.6 Two biochemical routes for production of ethanol from (noncellulosic) sugar (a) and (cellulosic) biomass (b).

Combustion

Combustion represents perhaps the oldest utilization of biomass, given that civilization began with the discovery of fire. The burning of forest wood taught humans how to cook and how to be warm. Chemically, combustion is an

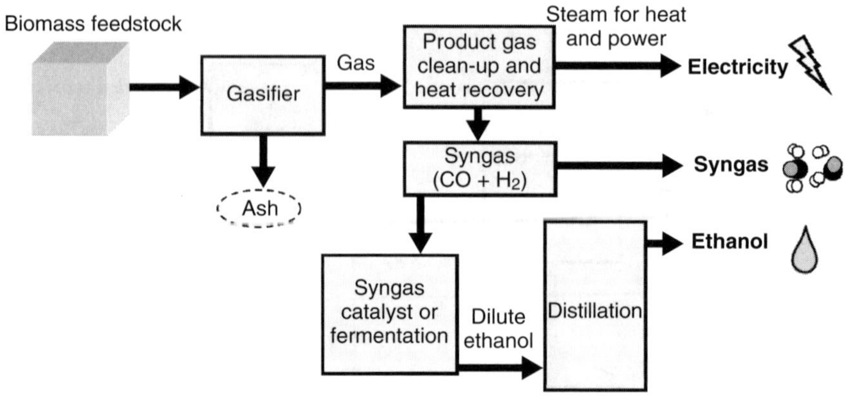

FIGURE 1.7 Thermochemical route for production of energy, gas, and ethanol.

TABLE 1.2 Comparison of Four Major Thermochemical Conversion Processes

Process	Temperature (°C)	Pressure (MPa)	Catalyst	Drying
Liquefaction	250–330	5–20	Essential	Not required
Pyrolysis	380–530	0.1–0.5	Not required	Necessary
Combustion	700–1400	>0.1	Not required	Not essential, but may help
Gasification	500–1300	>0.1	Not essential	Necessary

Source: Adapted from Demirbas, 2009.

exothermic reaction between oxygen and the hydrocarbon in biomass. Here, the biomass is converted into two major stable compounds: H_2O and CO_2. The reaction heat released is presently the largest source of human energy consumption, accounting for more than 90% of the energy from biomass.

Heat and electricity are two principal forms of energy derived from biomass. Biomass still provides heat for cooking and warmth, especially in rural areas. District or industrial heating is also produced by steam generated in biomass-fired boilers. Pellet stoves and log-fired fireplaces are as well a direct source of warmth in many cold-climate countries. Electricity, the foundation of all modern economic activities, may be generated from biomass combustion. The most common practice involves the generation of steam by burning biomass in a boiler and the generation of electricity through a steam turbine. In some places, electricity is produced by burning combustible gas derived from biomass through gasification.

TABLE 1.3 Comparison of Biochemical and Thermochemical Routes for Biomass Conversion into Ethanol

	Biochemical (sugar fermentation)	Thermochemical
Feedstock	Sugarcane, starch, corn	Cellulosic stock, wood, municipal solid waste
Reactor type	Batch	Continuous
Reaction time	2 days	7 minutes
Water usage	3.5–170 liter/liter ethanol	<1 liter/liter ethanol
By-products	Distiller's dried grain	Syngas/electricity
Yield	450 liter/ton	265–492 liter/ton
Technology maturity	>100 in U.S. plants	Pilot plant

Biomass is used either as a standalone fuel or as a supplement to fossil fuels in a boiler. The latter option is becoming increasingly common as the fastest and least-expensive means for decreasing the emission of carbon dioxide from an existing fossil fuel plant (Basu et al., 2009). This option is called co-combustion or co-firing.

Pyrolysis

Unlike combustion, pyrolysis takes place in the total absence of oxygen, except in cases where partial combustion is allowed to provide the thermal energy needed for this process. Pyrolysis is a thermal decomposition of the biomass into gas, liquid, and solid. It has three variations:

- Torrefaction, or mild pyrolysis
- Slow pyrolysis
- Fast pyrolysis

In pyrolysis, large hydrocarbon molecules of biomass are broken down into smaller hydrocarbon molecules. Fast pyrolysis produces mainly liquid fuel, known as bio-oil; slow pyrolysis produces some gas and solid charcoal (one of the most ancient fuels, used for heating and metal extraction before the discovery of coal). Pyrolysis is promising for conversion of waste biomass into useful liquid fuels. Unlike combustion, it is not exothermic.

Torrefaction, which is currently being considered for effective biomass utilization, is also a form of pyrolysis. In this process (named for the French word for roasting), the biomass is heated to 230 to 300 °C without contact with oxygen. The chemical structure of the wood is altered, which produces carbon dioxide, carbon monoxide, water, acetic acid, and methanol.

Torrefaction increases the energy density of the biomass. It also greatly reduces its weight as well as its hygroscopic nature, thus enhancing the commercial use of wood for energy production by reducing its transportation cost.

Gasification

Gasification converts fossil or nonfossil fuels (solid, liquid, or gaseous) into useful gases and chemicals. It requires a medium for reaction, which can be gas or supercritical water (not to be confused with ordinary water at subcritical condition). Gaseous mediums include air, oxygen, subcritical steam, or a mixture of these.

Presently, gasification of fossil fuels is more common than that of nonfossil fuels like biomass for production of synthetic gases. It essentially converts a potential fuel from one form to another. There are three major motivations for such a transformation:

- To increase the heating value of the fuel by rejecting noncombustible components like nitrogen and water.
- To remove sulfur and nitrogen such that when burnt the gasified fuel does not release them into the atmosphere.
- To reduce the carbon-to-hydrogen (C/H) mass ratio in the fuel.

In general, the higher the hydrogen content of a fuel, the lower the vaporization temperature and the higher the probability of the fuel being in a gaseous state. Gasification or pyrolysis increases the relative hydrogen content (H/C ratio) in the product through one the followings means:

Direct: Direct exposure to hydrogen at high pressure.
Indirect: Exposure to steam at high temperature and pressure, where hydrogen, an intermediate product, is added to the product. This process also includes steam reforming.
Pyrolysis: Reduction of carbon by rejecting it through solid char or CO_2 gas.

Gasification of biomass also involves removal of oxygen from the fuel to increase its energy density. For example, a typical biomass has about 40 to 60% oxygen by weight, but a useful fuel gas contains only a small percentage of oxygen (Table 1.4). The oxygen is removed from the biomass by either dehydration or decarboxylation. The latter process, which rejects the oxygen through CO_2, increases the H/C ratio of the fuel so that it emits less greenhouse gas when combusted:

$$\text{Dehydration:}\quad C_mH_nO_q \rightarrow C_mH_{n-2q} + qH_2O \quad O_2 \text{ removal through } H_2O \qquad (1.1)$$

$$\text{Decarboxylation:}\quad C_mH_nO_q \rightarrow C_{m-q/2}H_n + qCO_2 \quad O_2 \text{ removal through } CO_2 \qquad (1.2)$$

Hydrogen, when required in bulk for the production of ammonia, is produced from natural gas (~CH_4) through steam reforming, which produces

TABLE 1.4 Carbon to Hydrogen Ratio of Different Fuels

Fuel	C/H mass ratio	Oxygen %	Energy density, GJ/t
Anthracite	~44	~2.3	~27.6
Bituminous coal	~15	~7.8	~29
Lignite	~10	~11	~9
Peat	~10	35	~7
Crude oil	~9		42 (mineral oil)
Biomass/Cedar	7.6	40	20
Gasoline	6	0	46.8
Natural gas (~CH_4)	3	0	56
Syngas (CO:H_2 in 1:3 ratio)	2	0	24

syngas (a mixture of H_2 and CO). The CO in syngas is indirectly hydrogenated by steam to produce methanol (CH_3OH), an important feedstock for a large number of chemicals. These processes, however, use fossil fuels, which are not only nonrenewable but are responsible for the net addition of carbon dioxide (a major greenhouse gas) in the atmosphere.

Biomass can deliver nearly everything that fossil fuels provide, whether fuel or chemical feedstock. Additionally, it provides two important benefits that make it a viable feedstock for syngas production. First, it does not make any net contribution to the atmosphere when burnt; second, its use reduces dependence on nonrenewable and often imported fossil fuel.

For these reasons, biomass gasification into CO and H_2 provides a good base for production of liquid transportation fuels, such as gasoline, and synthetic chemicals, such as methanol. It also produces methane, which can be burned directly for energy production. Gasification is carried out generally in one of the three major types of gasifiers:

- Moving bed (also called fixed bed)
- Fluidized bed
- Entrained flow

Downdraft and updraft are two common types of moving-bed gasifier. A survey of gasifiers in Europe, the United States, and Canada shows that downdraft gasifiers are the most common (Knoef, 2000). It shows that 75% are downdraft, 20% are fluidized beds, 2.5% are updraft, and 2.5% are of various other designs.

Liquefaction

Liquefaction of solid biomass into liquid fuel can be done through pyrolysis, gasification as well as through hydrothermal process. In the latter process, biomass is converted into an oily liquid by contacting the biomass with water at elevated temperatures (300–350 °C) with high (12–20 MPa) for a period of time. There are several other means including the supercritical water process (Chapter 7) for direct liquefaction of biomass. Behrendt et al. (2008) present a review of these processes.

1.4 MOTIVATION FOR BIOMASS CONVERSION

Gasification is almost as ancient as combustion, but it is less developed because commercial interest in it has not been as strong as in combustion. However, there has been a recent surge of interest in conversion of biomass into gas or liquid due to three motivating factors:

- Renewability benefits
- Environmental benefits
- Sociopolitical benefits

1.4.1 Renewability

Fossil fuels like coal, oil, and gas are good and convenient sources of energy, and they meet the energy demands of society very effectively. However, there is one major problem: Fossil fuel resources are finite and not renewable. Biomass, on the other hand, grows and is renewable. A crop cut this year will grow again next year; a tree cut today may grow up within a decade. Unlike fossil fuels, then, biomass is not likely to be depleted with consumption. For this reason, its use, especially for energy production, is rising fast.

We may argue against cutting trees for energy because they serve as a CO_2 sink. This is true, but a tree stops absorbing CO_2 after it dies. On the other hand, if left alone in the forest it can release CO_2 in a forest fire or release more harmful CH_4 when it decomposes in water. The use of a tree as fuel after its life provides carbon-neutral energy as well as avoids greenhouse gas release from deadwood. The best option is new planting following cutting, as is done by some pulp industries. Fast-growing plants like switch grass and *Miscanthus* are being considered as fuel for new energy projects. These plants have very short growing periods that can be counted in months.

1.4.2 Environmental Benefits

With growing evidence of global warming, the need to reduce human-made greenhouse gas emissions is being recognized. Emission of other air pollutants, such as NO_2, SO_2, and Hg, is no longer acceptable, as it was in the past. In elementary schools and in corporate boardrooms, the environment is a major issue, and it has been a major driver for gasification for energy production. Biomass has a special appeal in this regard, as it makes no net contribution to carbon dioxide emission to the atmosphere. Regulations for making biomass economically viable are in place in many countries. For example, if biomass replaces fossil fuel in a plant, that plant earns credits for CO_2 reduction equivalent to what the fossil fuel was emitting. These credits can be sold on the market for additional revenue in countries where such trades are in practice.

Carbon Dioxide Emission

When burned, biomass releases the CO_2 it absorbed from the atmosphere in the recent past, not millions of years ago, as with fossil fuel. The net addition of CO_2 to the atmosphere through biomass combustion is thus considered to be zero.

Even if the fuel is not carbon-neutral biomass, CO_2 emissions from the gasification of the fuel are slightly less than those from its combustion on a unit heat release basis. For example, emission from an IGCC plant is 745 g/kWh compared to 770 g/kWh from a combustion-based subcritical pulverized-coal (PC) plant (Termuehlen and Emsperger, 2003, p. 23). Sequestration of CO_2 is becoming an important requirement for new power plants. On that note, a gasification-based power plant has an advantage over a conventional combustion-based PC power plant. In an IGCC plant, CO_2 is more concentrated in the flue gas, making it easier to sequestrate than it is in a conventional PC plant. Table 1.5 compares the emissions from different electricity-generation technologies.

TABLE 1.5 A Comparison of Emissions from Electricity-Generation Technologies

Emission	Pulverized-Coal Combustion	Gasification	Combined Natural-Gas Combustion
CO_2 (kg/1000 MWh)	0.77	0.68	0.36
Water use (L/1000 MWh)	4.62	2.84	2.16
SO_2 (kg/MWh)	0.68	0.045	0
NO_x (kg/MWh)	0.61	0.082	0.09
Total solids (kg/100 MWh)	0.98	0.34	~0

Source: Recompiled from graphs by Stiegel, 2005.

Sulfur Removal

Most virgin or fresh biomass contains little to no sulfur. Biomass-derived feedstock such as municipal solid waste (MSW) or sewage sludge does contain sulfur, which requires limestone for the capture of it. Interestingly, such derived feedstock also contains small amounts of calcium, which intrinsically aids sulfur capture.

Gasification from coal or oil has an edge over combustion in certain situations. In combustion systems, sulfur in the fuel appears as SO_2, which is relatively difficult to remove from the flue gas without adding an external sorbent. In a typical gasification process 93 to 96% of the sulfur appears as H_2S with the remaining as COS (Higman and van der Burgt, 2008, p. 351). We can easily extract sulfur from H_2S by absorption. Furthermore, in a gasification plant we can extract it as elemental sulfur, thus adding a valuable by-product for the plant.

Nitrogen Removal

A combustion system firing fossil fuels can oxidize the nitrogen in fuel and in air into NO, the acid rain precursor, or into N_2O, a greenhouse gas. Both are difficult to remove. In a gasification system, nitrogen appears as either N_2 or NH_3, which is removed relatively easily in the syngas-cleaning stage.

Nitrous oxide emission results from the oxidation of fuel nitrogen alone. Measurement in a biomass combustion system showed a very low level of N_2O emission (Van Loo and Koppejan, 2008, p. 295).

Dust and Other Hazardous Gases

Highly toxic pollutants like dioxin and furan, which can be released in a combustion system, are not likely to form in an oxygen-starved gasifier. (This observation is disputed by some.) Particulate in the syngas is also reduced significantly by multiple gas cleanup systems, including a primary cyclone, scrubbers, gas cooling, and acid gas–removal units. These reduce the particulate emissions by one to two orders of magnitude (Rezaiyan and Cheremisinoff, 2005, p. 15).

1.4.3 Sociopolitical Benefits

The sociopolitical benefits of biomass are substantial. For one, biomass is a locally grown resource. For a biomass-based power plant to be economically viable, the biomass needs to come from within a certain distance from it. This means that every biomass plant can prompt the development of associated industries for biomass growing, collecting, and transporting. Some believe that a biomass fuel plant could create up to 20 times more employment than that created by a coal- or oil-based plant (Van Loo and Koppejan, 2008, p. 1). The biomass industry thus has a positive impact on the local economy.

Another very important aspect of biomass-based energy, fuel, or chemicals is that they reduce reliance on imported fossil fuels. The volatile global political landscape has shown that supply and price can change dramatically within a short time, with a sharp rise in the price of feedstock. Locally grown biomass is relatively free from such uncertainties.

1.5 COMMERCIAL ATTRACTION OF GASIFICATION

A major attraction of gasification is that it can convert waste or low-priced fuels, such as biomass, coal, and petcoke, into high-value chemicals like methanol. Biomass holds great appeal for industries and businesses, especially in the energy sector. For example:

- Downstream flue-gas cleaning in a gasification plant is less expensive than that in a coal-fired plant with flue-gas desulphurization, selective catalytic reducers (SCRs), and electrostatic precipitators.
- Polygeneration is a unique feature of a gasifier plant. It can deliver steam for process, electricity for grid, and gas for synthesis, thereby providing a good product mix. Additionally, for high-sulfur fuel a gasifier plant produces elemental sulfur as a by-product; for high-ash fuel, slag or fly ash is obtained, which can be used for cement manufacture.
- For power generation, an IGCC plant can achieve a higher overall efficiency (38–41%) than can a combustion plant with a steam turbine alone. Gasification therefore offers lower power production costs.
- Carbon dioxide capture and sequestration (CCS) may become mandatory for power plants. An IGCC plant can capture and store CO_2 at one-half of what it costs a traditional PC plant (*www.gasification.org*). Other applications of gasification that produce transport fuel or chemicals may have even lower CCS costs. Established technologies are available to capture carbon dioxide from a gasification plant, but that is not so for a combustion plant.
- A process plant that uses natural gas as feedstock can use locally available biomass or organic waste instead, and thereby reduce dependence on imported natural gas, which is not only rising sharply in price but is also experiencing supply volatility.
- Gasification provides significant environmental benefits, as described in Section 1.4.2.
- Total water consumption in a gasification-based power plant is much less than that in a conventional power plant (Table 1.5). Furthermore, a plant can be designed to recycle its process water. For this reason, all zero-emission plants use gasification technology.
- Gasification plants produce significantly lower quantities of major air pollutants like SO_2, NOx, and particulates. Figure 1.8 compares the emission from a coal-based IGCC plant with that from a combustion-based coal-fired

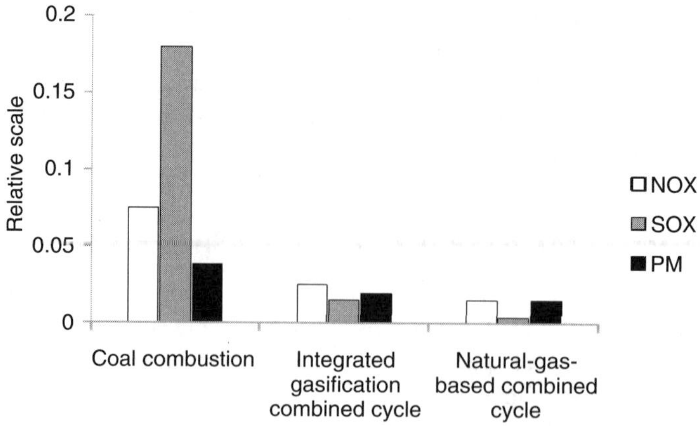

FIGURE 1.8 Comparison of pollutant emissions from a coal-based steam plant, an IGCC plant, and a natural-gas-fired combined-cycle plant.

steam power plant and a natural-gas-fired plant. It shows emissions from the gasification plant are similar to those from a natural-gas-fired plant.

- An IGCC plant produces lower CO_2 per MWh than a combustion-based steam power plant.

1.5.1 Comparison of Gasification and Combustion

With heat or power production, the obvious question is why a solid fuel should be gasified and then the gas burned for heat, losing some part of its energy content in the process. Does it not make more sense to directly burn the fuel to produce heat? Example 1.1 may provide an answer to this question by comparing two energy conversion options. The comparison is based, where applicable, on an IGCC plant and a PC-fired plant, both generating electricity with coal as the fuel.

- For a given throughput of fuel processed, the volume of gas obtained from gasification is much less compared to that obtained from a direct combustion system. The lower volume of gas requires smaller equipment and hence results in lower overall costs.
- A gasified fuel can be used in a wider range of application than can its precursor solid fuel. For example, sensitive industrial processes such as glass blowing and drying cannot use dirty flue gas from combustion of coal or biomass, but they can use heat from the cleaner and more controllable combustion of gas produced through gasification.
- Gas can be more easily carried and distributed than a solid fuel among industrial and domestic customers. Transportation of synthetic gas, or the liquid fuel produced from it, is both less expensive and less energy intensive than transportation of solid fuel for combustion.

- The concentration of CO_2 in the product of gasification is considerably higher than that of combustion, so it is less expensive to separate and sequestrate the CO_2 in IGCC.
- SO_2 emissions are generally lower in an IGCC plant (Table 1.5). Sulfur in a gasification plant appears as H_2S and COS, which can be easily converted into saleable elemental sulfur or H_2SO_4. In a combustion system sulfur appears as SO_2, which needs a scrubber producing ash-mixed $CaSO_4$, which has less market potential.
- Gasification produces much less NO_x than a combustion system (Table 1.5). In gasification, nitrogen can appear as NH_3, which washes out with water and as such does not need a SCR to meet statutory limits. A PC system, on the other hand, requires SCR for this purpose.
- The total solid waste generated in an IGCC plant is much lower than that generated in a comparable combustion system (Table 1.5). Furthermore, the ash in a slagging entrained-flow gasifier appears as glassy melt, which is much easier to dispose of than the dry fly ash of a PC system.
- For mechanical work or electricity in a remote location, a power pack comprising a gasifier and a compression ignition engine can be employed. For a combustion system, a boiler, a steam engine, and a condenser might be needed, making the power pack considerably more bulky and expensive.
- The producer gas from a gasifier can be used as a feedstock for the production of fertilizer, methanol, and gasoline. A gasification-based energy system has the option of producing value-added chemicals as a side stream. This polygeneration feature is not available in direct combustion.
- A gas produced in a central gasification plant can be distributed to individual houses or units in a medium-size to large community.
- If heat is the only product that is desired, combustion seems preferable, especially in small-scale plants. Even for a medium-capacity unit such as for district heating, central heating, and power, combustion may be more economical.

Example 1.1

Compare the theoretical thermodynamic efficiency of electricity generation from biomass through the following two routes:

1. Biomass is combusted in a boiler with 95% efficiency (on lower heating value [LHV] basis) to generate steam, which expands in a steam turbine from 600 °C to 100 °C driving a generator.
2. Biomass is gasified with 80% efficiency; the product gas is burnt into hot gas at 1200 °C. It expands in a gas turbine to 600 °C. Waste gas from the gas turbine enters a heat recovery steam generator to produce steam at 400 °C. This steam expands to 100 °C in a steam turbine.

Both turbines are connected to electricity generators. Neglect losses in the generators.

Solution

For a steam power plant, given:

Combustion efficiency: $\eta_c = 0.95$
Inlet steam temperature: $T_1 = 600\ °C = 873K$
Exhaust steam temperature: $T_2 = 100\ °C = 373K$

We assume the turbine to be an ideal heat engine, operating on a Carnot cycle. This makes the efficiency simple to calculate.

$$\text{Steam turbine efficiency: } \eta_s = 1-(T_2/T_1) = 1-(373/873) = 0.573$$

The overall efficiency of the first route is the combination of the two separate efficiencies.

$$\text{Overall efficiency: } \eta_a = \eta_c \times \eta_s = 0.95 \times 0.573 = \mathbf{0.544},\ \eta_a = \mathbf{54.4\%}$$

For an IGCC plant, given:

Gasification efficiency: $\eta_g = 0.8$
Energy supplied to steam: 20%
Inlet steam temperature: $T_1 = 400\ °C = 673K$
Exhaust steam temperature: $T_2 = 100\ °C = 373K$

We assume that all of the waste energy from the gas turbine is used to heat the steam. This means that 20% of the energy input to the gas turbine is used for steam heating; the remaining is used to generate electricity.

$$\text{Gas turbine efficiency: } \eta_g = 1-(T_{g2}/T_{g1}) = 1-(873/1473) = 0.407$$
$$\eta_b = \eta_g \times \eta_{cc} = 0.407 \times 0.85 = \mathbf{0.346}$$

The steam turbine is again considered to be a Carnot heat engine and the efficiency can easily be calculated. From basic thermodynamics the ideal cycle efficiency is written as:

$$\text{Steam turbine efficiency: } \eta_s = 1-(T_2/T_1) = 1-(373/673) = 0.446$$

Both the steam and gas turbines have been assumed to be ideal, so the ideal efficiency of the combined cycle can be calculated using the expression for combined cycle efficiency given in basic thermodynamics.

$$\text{Combined efficiency: } \eta_{cc} = \eta_g + \eta_s - \eta_g \times \eta_s = 0.346 + 0.446 - (0.346 \times 0.446) = \mathbf{0.638}$$

Thus, the IGCC plant has an overall efficiency of 63.8% compared to 54.4% for a PC boiler steam power plant.

1.6 BRIEF DESCRIPTION OF GASIFICATION AND RELATED PROCESSES

When a biomass or other carbonaceous material is heated in a restricted oxygen supply, it is first pyrolyzed or decomposed into solid carbon and condensable and noncondensable gases.

1.6.1 Pyrolysis

The solid carbon as well as the condensed liquid enters the gasification reaction with carbon dioxide, oxygen, or steam to produce combustible or synthetic gas. To illustrate the different reactions we take simple carbon as the feedstock.

$$C_nH_mO_p + Heat \Rightarrow \sum_{liquid} C_aH_bO_c + \sum_{gas} C_xH_yO_z + \sum_{solid} C \qquad (1.3)$$

1.6.2 Combustion of Carbon

When 1 kmol of carbon is burnt completely in adequate air or oxygen, it produces 394 MJ heat and carbon dioxide. This is a combustion reaction. The positive sign on the right side (+*Q* kJ/kmol) implies that heat is absorbed in the reaction. A negative sign (−*Q* kJ/kmol) means that heat is absorbed in the reaction.

$$C + O_2 \rightarrow CO_2 - 393{,}770 \text{ kJ/kmol} \qquad (1.4)$$

1.6.3 Gasification of Carbon

Instead of burning it entirely, we can gasify the carbon by restricting the oxygen supply. The carbon then produces 72% less heat than that in combustion, but the partial gasification reaction shown here produces a combustible gas, CO.

$$C + 1/2\, O_2 \rightarrow CO - 110{,}530 \text{ kJ/kmol} \qquad (1.5)$$

When the gasification product, CO, subsequently burns in adequate oxygen, it produces the remaining 72% (283 MJ) of the heat. Thus, the CO retains only 72% of the energy of the carbon, but in complete gasification the energy recovery is 75 to 88% owing to the presence of hydrogen and other hydrocarbons.

The producer gas reaction is an endothermic gasification reaction, which produces hydrogen and carbon monoxide from carbon. This product gas mixture is also known as synthesis gas, or *syngas*.

$$\text{Producer gas reaction:} \quad C + H_2O \rightarrow CO + H_2 + 131{,}000 \text{ kJ/kmol} \qquad (1.6)$$

Production of heavy oil residue in oil refineries is an important application of gasification. Low-hydrogen residues are gasified into hydrogen.

$$\text{Heavy oil gasification:} \quad C_nH_m + (n/2)O_2 \rightarrow nCO + (m/2)H_2 \qquad (1.7)$$

This hydrogen can be used for hydrocracking of other heavy oil fractions into lighter oils.

The reaction between steam and carbon monoxide is also used for maximization of hydrogen production in the gasification process at the expense of CO.

$$\text{Shift reaction:} \quad CO + H_2O \rightarrow H_2 + CO_2 - 41{,}000 \text{ kJ/kmol} \qquad (1.8)$$

1.6.4 Syngas Production

Syngas is also produced from natural gas (>80% CH_4), using a steam-methane-reforming reaction, instead of from solid carbonaceous fuel alone. The reforming reaction is, however, not gasification but a molecular rearrangement.

$$\text{Steam reforming reaction:} \quad CH_4 + H_2O\,(\text{catalyst}) \rightarrow CO + 3H_2 + 206{,}000\ \text{kJ/kmol} \tag{1.9}$$

Partial oxidation of natural gas or methane is an alternative route for production of syngas. In contrast to the reforming reaction, partial oxidation is exothermic. Partial oxidation of fuel oil also produces syngas.

$$\text{Partial oxidation reforming:} \quad CH_4 + 1/2\,O_2 \rightarrow CO + 2H_2 - 36{,}000\ \text{kJ/kmol} \tag{1.10}$$

The hydrogen may be used as fuel in fuel cells or in production of chemical feedstocks like methanol and ammonia.

1.6.5 Methanol Synthesis

Syngas provides the feedstock for many chemical reactions, including methanol synthesis (Eq 1.11). Methanol (CH_3OH) is a basic building block of many products, including gasoline.

$$CO + 2H_2(\text{catalysts}) \rightarrow CH_3OH \tag{1.11}$$

1.6.6 Ammonia Synthesis

Ammonia (NH_3) is an important feedstock for fertilizer production. It is produced from pure hydrogen and nitrogen from air.

$$3H_2 + N_2\,(\text{catalysts}) \rightarrow 2NH_3 - 92{,}000\ \text{kJ/kmol} \tag{1.12}$$

1.6.7 Fischer-Tropsch Reaction

The Fischer-Tropsch synthesis reaction can synthesize a mixture of CO and H_2 into a range of hydrocarbons, including diesel oil.

$$(2n+1)\,H_2 + nCO\,(\text{catalyst}) \rightarrow C_nH_{(2n+2)} + nH_2O \tag{1.13}$$

Here, $C_nH_{(2n+2)}$ represents a mixture of hydrocarbons ranging from methane and gasoline to wax. Its relative distribution depends on the catalyst, the temperature, and the pressure chosen for the reaction.

1.6.8 Methanation Reaction

Methane (CH_4), an important ingredient in the chemical and petrochemical industries, can come from natural gas as well as from solid hydrocarbons like biomass or coal. For the latter source, the hydrocarbon is hydrogenated to produce synthetic gas, or *substitute natural gas* (SNG). The overall reaction for SNG production may be expressed as

$$C + 2H_2 \rightarrow CH_4 - 74{,}800\ \mathrm{kJ/kmol} \tag{1.14}$$

More details on these reactions are given in Chapters 5 and 9.

Symbols and Nomenclature

LHV = lower heating value of gas (kJ/mol)
η_c = combustion efficiency (–)
η_s = steam turbine efficiency (–)
η_a = efficiency of combustion and steam turbine (–)
η_g = gas turbine efficiency (–)
η_{cc} = combined efficiency of steam turbine and gas turbine (–)
η_b = overall efficiency, including gasification efficiency (–)
T_1 = inlet steam temperature in the steam turbine (K)
T_2 = exhaust steam temperature in the steam turbine (K)
T_{g1} = inlet temperature in the gas turbine (K)
T_{g2} = exhaust temperature in the gas turbine (K)

Chapter 2

Biomass Characteristics

2.1 INTRODUCTION

The characteristics of biomass greatly influence the performance of a biomass gasifier. A proper understanding of the physical and the chemical properties of biomass feedstock is essential for the design of a biomass gasifier to be reliable. This chapter discusses some important properties of biomass that are relevant to gasification and related processes.

2.2 WHAT IS BIOMASS?

Biomass refers to any organic materials that are derived from plants or animals (Loppinet-Serani et al., 2008). A generally accepted definition is difficult to find. However, the one used by the United Nations Framework Convention on Climate Change (UNFCCC, 2005) is relevant here:

[A] non-fossilized and biodegradable organic material originating from plants, animals and micro-organisms. This shall also include products, by-products, residues and waste from agriculture, forestry and related industries as well as the non-fossilized and biodegradable organic fractions of industrial and municipal wastes.

Biomass also includes gases and liquids recovered from the decomposition of nonfossilized and biodegradable organic materials. In the United States, there has been much debate on a legal definition. Appendix A gives a recent legal interpretation of renewable biomass.

As a sustainable and renewable energy resource, biomass is constantly being formed by the interaction of CO_2, air, water, soil, and sunlight with plants and animals. After an organism dies, microorganisms break down biomass into elementary constituent parts like H_2O, CO_2, and its potential energy. Because the carbon dioxide, a biomass releases through the action of microorganisms or combustion was absorbed by it in the recent past, biomass combustion does not increase the total CO_2 inventory of the Earth. It is thus called *greenhouse gas* neutral or *GHG neutral*.

Biomass Gasification and Pyrolysis. DOI: 10.1016/B978-0-12-374988-8.00002-7

Biomass includes only living and recently dead biological species that can be used as fuel or in chemical production. It does not include organic materials that over many millions of years have been transformed by geological processes into substances such as coal or petroleum. Biomass comes from botanical (plant species) or biological (animal waste or carcass) sources, or from a combination of these. Common sources of biomass are:

- *Agricultural:* food grain, bagasse (crushed sugarcane), corn stalks, straw, seed hulls, nutshells, and manure from cattle, poultry, and hogs
- *Forest:* trees, wood waste, wood or bark, sawdust, timber slash, and mill scrap
- *Municipal:* sewage sludge, refuse-derived fuel (RDF), food waste, waste paper, and yard clippings
- *Energy:* poplars, willows, switchgrass, alfalfa, prairie bluestem, corn, and soybean, canola, and other plant oils
- *Biological:* animal waste, aquatic species, biological waste

2.2.1 Biomass Formation

Botanical biomass is formed through conversion of carbon dioxide (CO_2) in the atmosphere into carbohydrate by the sun's energy in the presence of chlorophyll and water. Biological species grow by consuming botanical or other biological species.

Plants absorb solar energy in a process called *photosynthesis* (Figure 2.1). In the presence of sunlight of particular wavelengths, green plants break down water to obtain electrons and protons and use them to turn CO_2 into glucose

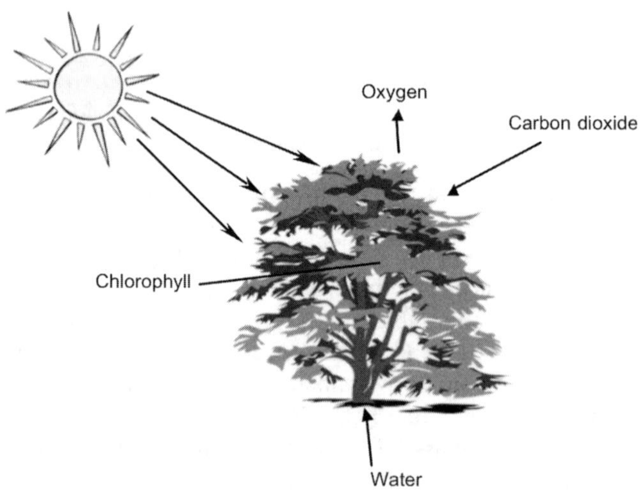

FIGURE 2.1 Biomass grows by absorbing solar energy, carbon dioxide, and water through photosynthesis.

(represented by CH_mO_n), releasing O_2 as a waste product. The process may be represented by this equation (Hodge, 2010, p. 297):

$$\text{Living plant} + CO_2 + H_2O + \text{Sunlight} \xrightarrow{\text{Chlorophyll}} (CH_mO_n) + O_2 - 480\ \text{kJ/mol} \tag{2.1}$$

For every mole of CO_2 absorbed into carbohydrate or glucose in biomass, 1 mole of oxygen is released. This oxygen comes from water the plant takes from the ground or the atmosphere (Klass, 1998, p. 30). The chlorophyll promotes the absorption of carbon dioxide from the atmosphere, adding to the growth of the plant. Important ingredients for the growth of biomass are:

- Living plant
- Visible spectrum of solar radiation
- Carbon dioxide
- Chlorophyll (serving as catalyst)
- Water

The chemical energy stored in plants is then passed on to the animals and to the human that take the plants as food. Animal and human waste also contribute to biomass.

2.2.2 Types of Biomass

Biomass comes from a variety of sources as shown in Table 2.1. Loosely speaking, these sources include all plants and plant-derived materials, including

TABLE 2.1 Two Major Groups of Biomass and Their Subclassifications

Virgin	Terrestrial biomass	Forest biomass Grasses Energy crops Cultivated crops
	Aquatic biomass	Algae Water plant
Waste	Municipal waste	Municipal solid waste Biosolids, sewage Landfill gas
	Agricultural solid waste	Livestock and manures Agricultural crop residue
	Forestry residues	Bark, leaves, floor residues
	Industrial wastes	Demolition wood, sawdust Waste oil or fat

livestock manures. Primary or virgin biomass comes directly from plants or animals. Waste or derived biomass comes from different biomass-derived products. Table 2.1 lists a range of biomass types, grouping them as virgin or waste. Biomass may also be divided into two broad groups:

- Virgin biomass includes wood, plants, and leaves (ligno-cellulose); and crops and vegetables (carbohydrates).
- Waste includes solid and liquid wastes (municipal solid waste (MSW)); sewage, animal, and human waste; gases derived from landfilling (mainly methane); and agricultural wastes.

Ligno-Cellulosic Biomass

A major part of biomass is ligno-cellulose, so this type is described in some detail. Ligno-cellulosic material is the nonstarch, fibrous part of plant materials. Cellulose, hemicellulose, and lignin are its three major constituents. Unlike carbohydrate or starch, ligno-cellulose is not easily digestible by humans. For example, we can eat the rice, which is a carbohydrate, but we cannot digest the husk or the straw, which are ligno-cellulose. Ligno-cellulosic biomass is not part of the human food chain, and therefore its use for biogas or bio-oil does not threaten the world's food supply.

A good example of ligno-cellulosic biomass is a woody plant—that is, any vascular plant that has a perennial stem above the ground and is covered by a layer of thickened bark. Such biomass is primarily composed of structures of cellulose and lignin. A detailed description of wood structure is given in Section 2.3.1. Woody plants include trees, shrubs, cactus, and perennial vines. They can be of two types: (1) herbaceous and (2) nonherbaceous.

An herbaceous plant is one with leaves and stems that die annually at the end of the growing season. Wheat and rice are examples of herbaceous plants that develop hard stems with vascular bundles. Herbaceous plants do not have the thick bark that covers nonherbaceous biomass like trees.

Nonherbaceous plants are not seasonal; they live year-round with their stems above the ground. Large trees fall in this category. Nonherbaceous perennials like woody plants have stems above ground that remain alive during the dormant season, and grow shoots the next year from their above-ground parts. These include trees, shrubs, and vines.

The trunk and leaves of tree plants form the largest group of available biomass. These are classified as ligno-cellulosic, as their dominant constituents are cellulose, hemicellulose, and lignin. Table 2.2 shows the distribution of these components in some plants. Section 2.3.2 presents further discussions of ligno-cellulose components.

There is a growing interest in the cultivation of plants exclusively for production of energy. These crops are ligno-cellulosic. They typically have a short growing period and high yields, and require little or no fertilizer, so they provide quick return on investment. Energy crops are densely planted. For

TABLE 2.2 Composition of Some Ligno-Cellulosic Wood

Plant	Lignin (%)	Cellulose (%)	Hemicellulose (%)
Deciduous	18–25	40–44	15–35
Coniferous	25–35	40–44	20–32
Willow	25	50	19
Larch	35	26	27

Source: Adapted from Bergman et al., 2005, p. 15.

energy production, woody crops such as *Miscanthus*, willow, switchgrass, and poplar are widely utilized. These plants have high energy yield per unit of land area and require much less energy for cultivation.

Crops and Vegetables

While the body of a plant or tree (trunk, branches, leaves, etc.) is ligno-cellulosic, the fruit (cereal, vegetable) is a source of carbohydrate, starch, and sugar. Some plants like canola also provide fat. The fruit is digestible by humans, but the ligno-cellulosic body is not. (Because of special chemicals in their stomach, some animals can digest ligno-cellulosic biomass.) Because they serve as human food, the use of crops or vegetables for the production of chemicals and energy must be weighed carefully as it might affect food supplies.

Compared to ligno-cellulosic compounds, carbohydrates are easier to dissolve, so it is relatively easy to derive liquid fuels from them through fermentation or other processes. For this reason, most commercial ethanol plants use crops as feedstock. There are two types of crop biomass: (1) the regularly harvested agricultural crops for food production and (2) the energy crops for energy production.

Natural crops and vegetables are a good source of starch and sugars and can be hydrated. Some vegetables and crops (coconut, sunflower, mustard, canola, etc.) contain fat, providing a good source of vegetable oil. Animal waste (from land and marine mammals) also provides fat that can be transformed into bio-oil. If carbohydrate is desired for the production of biogas, whole crops, such as maize, Sudan grass, millet, and white sweet clover, can be made into silage and then converted into biogas.

Wastes

Wastes are secondary biomass, as they are derived from primary biomass (trees, vegetables, meat) during different stages of their production or use. MSW is an important source of waste biomass, and much of it comes from renewables like food scraps, lawn clippings, leaves, and papers. Nonrenewable components of MSW like plastics, glass, and metals are not considered biomass. The combustible part of MSW is at times separated and sold as refuse-derived fuel (RDF). Sewage sludge that contains human excreta, fat, grease, and food wastes is an important biomass source. Another waste is sawdust, produced in sawmills during the production of lumber from wood. Table 2.3 lists the composition and heating values of some waste biomass products.

Landfills have traditionally been an important means of disposing of garbage. A designated area is filled with waste, which decomposes, producing methane gas. Modern landfilling involves careful lining of the containment cell (Figure 2.2) so that leached liquids can be collected and treated instead of leaking into groundwater. The containment cells are covered with clay or earth to avoid exposure to wind and rain.

TABLE 2.3 Typical Composition of Some Waste Biomass

Biomass	Moisture (wt. %)	Organic Matter (dry wt. %)	Ash (dry wt. %)	Higher Heating Value (MJ/dry kg)
Cattle manure	20–70	76.5	23.5	13.4
Sewage	90–98	73.5	26.5	19.9
RDF	15–30	86.1	13.9	12.7
Sawdust	15–60	99.0	1.0	20.5

Source: Adapted from Klass, 1998, p. 73.

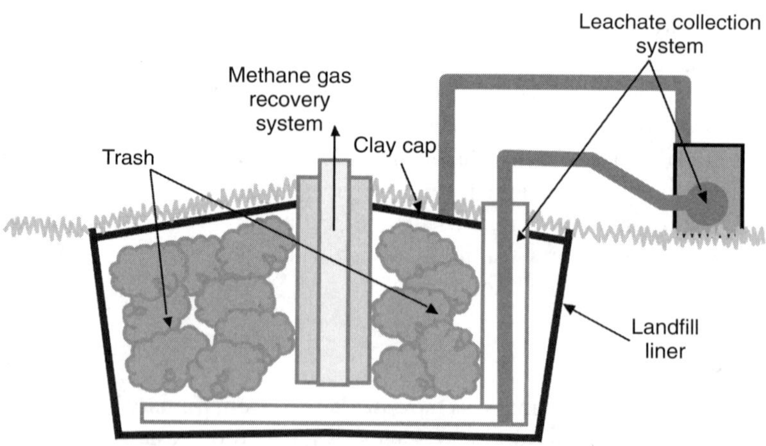

FIGURE 2.2 Anaerobic digestion of biodegradable waste.

An increasing number of municipalities are separating biodegradable wastes and subjecting them to digestion for degradation. This avoids disposal of leachate and reduces the volume of waste. Two types of waste degradation are used: aerobic digestion and anaerobic digestion.

Aerobic digestion: This process takes place in the presence of air and so does not produce fuel gas. Here, the leachate is removed from the bottom layer of the landfill and pumped back into the landfill, where it flows over the waste repeatedly. Air added to the landfill enables microorganisms to work faster to degrade the wastes into compost, carbon dioxide, and water. Since it does not produce methane, aerobic digestion is most widely used where there is no additional need for landfill gas.

Anaerobic digestion: This process does not use air and hence produces the fuel gas methane. Here, the land-filled solids are sealed against contact with the atmosphere oxygen. The leachate is collected and pumped back into the landfill as in aerobic digestion (Figure 2.2). Additional liquids may be added to the leachate to help biodegradation of the waste. In the absence of oxygen, the waste is broken down into methane, carbon dioxide, and digestate (or solid residues). Methanogenesis bacteria like thermophiles (45–65 °C), mesophiles (20–45 °C), and psychophiles (0–20 °C) facilitate this process (Probstein and Hicks, 2006). These biodegradation reactions are mildly exothermic. The process is represented by Eq. (2.2):

$$\begin{aligned} C_6H_{12}O_6 \text{ (representing wastes)} + \text{bacteria} \rightarrow 3CO_2 \\ + 3CH_4 + \text{digestate} \end{aligned} \tag{2.2}$$

Methane is an important constituent of landfill gas. A powerful greenhouse gas (~ 21 times stronger than CO_2), it is often burnt in a flare or utilized in a gas engine or in similar energy applications. Anaerobic digestion is very popular in farming communities, where animal excreta are collected and stored because the gas produced can be collected in a gas holder for use in cooking and heating while the residual solid can be used as fertilizer.

2.3 STRUCTURE OF BIOMASS

Biomass is a complex mixture of organic materials such as carbohydrates, fats, and proteins, along with small amounts of minerals such as sodium, phosphorus, calcium, and iron. The main components of plant biomass are extractives, fiber or cell wall components, and ash (Figure 2.3).

Extractives: Substances present in vegetable or animal tissue that can be separated by successive treatment with solvents and recovered by evaporation of the solution. They include protein, oil, starch, sugar, and so on.

Cell wall: Provides structural strength to the plant, allowing it to stand tall above the ground without support. A typical cell wall is made of

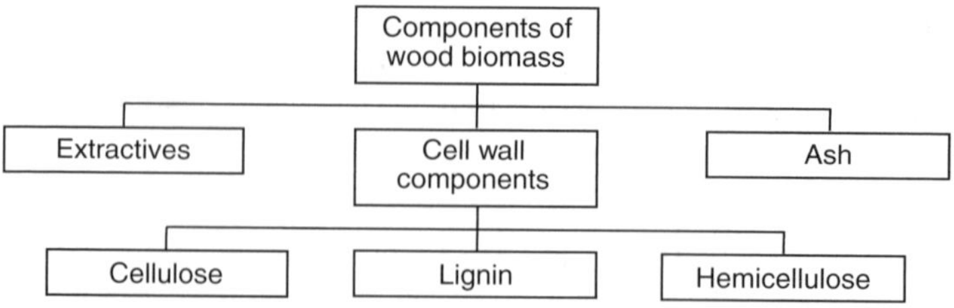

FIGURE 2.3 Major constituents of a woody biomass.

carbohydrates and lignin. Carbohydrates are mainly cellulose or hemicellulose fibers, which impart strength to the plant structure; the lignin holds the fibers together. These constituents vary with plant type. Some plants, such as corn, soybeans, and potatoes, also store starch (another carbohydrate polymer) and fats as sources of energy, mainly in seeds and roots.

Ash: The inorganic component of the biomass.

Wood and its residues are the dominant form of the biomass resource base. A detailed discussion of wood-derived biomass is presented next.

2.3.1 Structure of Wood

Wood is typically made of hollow, elongated, spindle-shaped cells arranged parallel to each other. Figure 2.4 is a photograph of the cross-section of a tree trunk showing the overall structure of a mature tree wood.

FIGURE 2.4 Cross-section of a tree trunk showing outer dead bark, inner live bark, sapwood, heartwood, and wood rays. *(Source: Photograph by author.)*

Bark is the outermost layer of a tree trunk or branch. It comprises an outer dead portion and an inner live portion. The inner live layer carries food from the leaves to the growing parts of the tree. It is made up of another layer known as *sapwood*, which carries sap from the roots to the leaves. Beyond this layer lies the inactive heartwood. In any cut wood we easily note a large number of radial marks. These radial cells (wood rays) carry food across the wood layers.

Wood cells that carry fluids are also known as fibers or *tracheids*. They are hollow and contain extractives and air. The cells vary in shape but are generally short and pointed. The length of an average tracheid is about 1000 microns (μm) for hardwood and typically 3000 to 8000 μm for softwood (Miller, R. B., 1999).

Tracheids are narrow. For example, the average diameter of the tracheid of softwood is 33 μm. These cells are the main conduits for the movement of sap along the length of the tree trunk. They are mostly aligned longitudinally, but there are some radial tracheids (G) that carry sap across layers. Lateral channels, called *pith*, transport water between adjacent cells across the cell layers. Softwood can have cells or channels for carrying resins. A hardwood, on the other hand, contains large numbers of pores or open vessels.

The tracheids or cells typically form an outer primary and an inner secondary wall. A layer called the *middle lamella*, joins or glues together adjacent cells. The middle lamella is predominantly made of lignin. The secondary wall (inside the primary layer) is made up of three layers: S1, S2, and S3 (Figure 2.5). The thickest layer, S2, is made of *macrofibrils*, which consist of long cellulose molecules with embedded hemicellulose. The construction of cell walls in wood is similar to that of steel-reinforced concrete, with the cellulose fibers

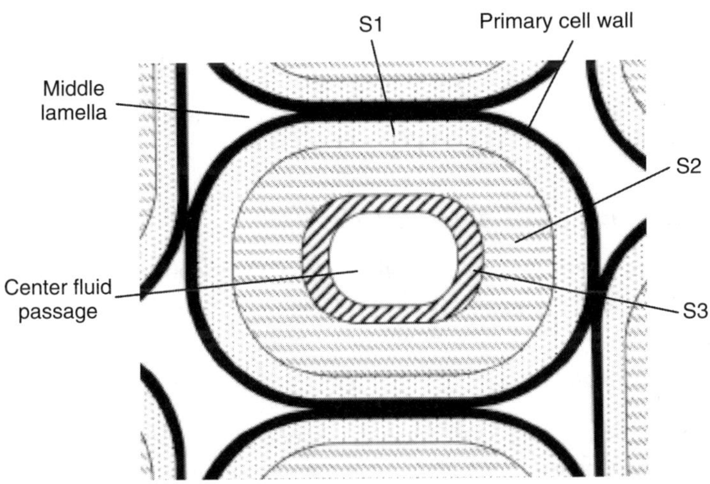

FIGURE 2.5 Layers of a wood cell. The actual shape of the cell cross-section is not necessarily as shown.

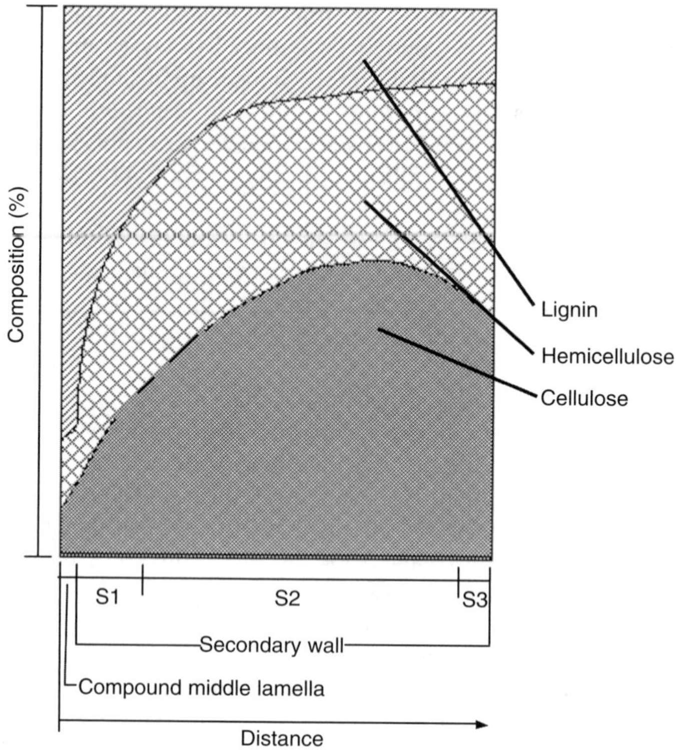

FIGURE 2.6 Distribution of cellulose, hemicellulose, and lignin in cell walls and their layers.

acting as the reinforcing steel rods and hemicellulose surrounding the cellulose microfibrils acting as the cement-concrete. The S2 layer has the highest concentration of cellulose. The highest concentration of hemicellulose is in layer S3. The distribution of these components in the cell wall is shown in Figure 2.6.

2.3.2 Constituents of Biomass Cells

The polymeric composition of the cell walls and other constituents of a biomass vary widely (Bergman et al., 2005a), but they are essentially made of three major polymers: cellulose, hemicellulose, and lignin.

Cellulose

Cellulose, the most common organic compound on Earth, is the primary structural component of cell walls in biomass. Its amount varies from 90% (by weight) in cotton to 33% for most other plants. Represented by the generic formula $(C_6H_{10}O_5)_n$, cellulose is a long chain polymer with a high degree of

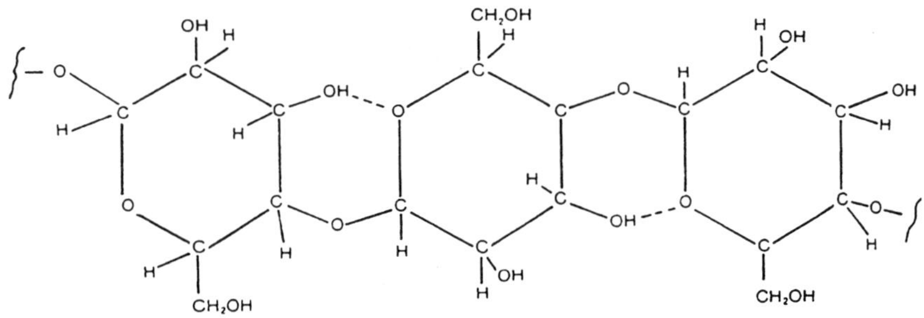

FIGURE 2.7 Molecular structure of cellulose.

polymerization (~10,000) and a large molecular weight (~500,000). It has a crystalline structure of thousands of units, which are made up of many glucose molecules. This structure gives it high strength, permitting it to provide the skeletal structure of most terrestrial biomass (Klass, 1998, p. 82). Cellulose is primarily composed of d-glucose, which is made of six carbons or hexose sugars (Figure 2.7).

Cellulose is highly insoluble and, though a carbohydrate, is not digestible by humans. It is a dominant component of wood, making up about 40 to 44% by dry weight.

Hemicellulose

Hemicellulose is another constituent of the cell walls of a plant. While cellulose is of a crystalline, strong structure that is resistant to hydrolysis, hemicellulose has a random, amorphous structure with little strength (Figure 2.8). It is a group of carbohydrates with a branched chain structure and a lower degree of polymerization (~100–200), and may be represented by the generic formula $(C_5H_8O_4)_n$ (Klass, 1998, p. 84). Figure 2.8 shows the molecular arrangement of a typical hemicellulose molecule, xylan.

There is significant variation in the composition and structure of hemicellulose among different biomass. Most hemicelluloses, however, contain some simple sugar residues like d-xylose (the most common), d-glucose, d-galactose, l-ababinose, d-glucurnoic acid, and d-mannose. These typically contain 50 to 200 units in their branched structures.

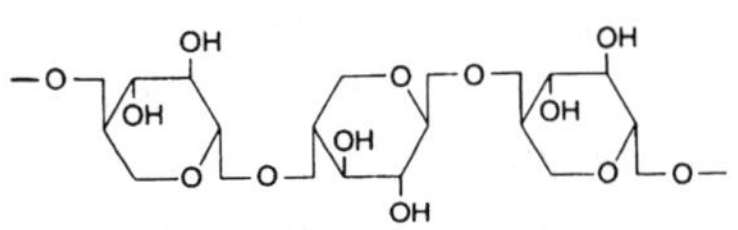

FIGURE 2.8 Molecular structure of a typical hemicellulose, xylan.

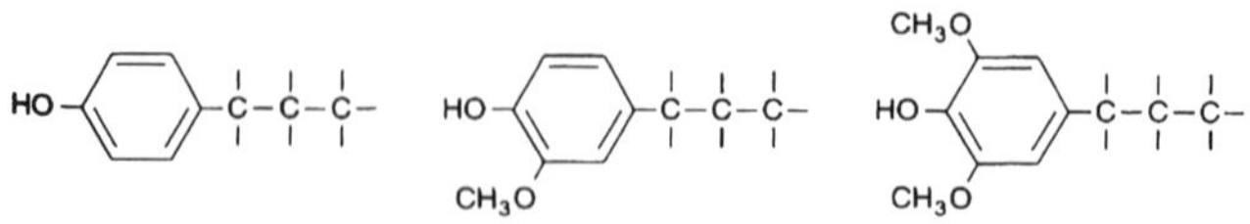

FIGURE 2.9 Some structural units of lignin.

Hemicellulose tends to yield more gases and less tar than cellulose (Milne, 2002). It is soluble in weak alkaline solutions and is easily hydrolyzed by dilute acid or base. It constitutes about 20 to 30% of the dry weight of most wood.

Lignin

Lignin is a complex highly branched polymer of phenylpropane and is an integral part of the secondary cell walls of plants. It is primarily a three-dimensional polymer of 4-propenyl phenol, 4-propenyl-2-methoxy phenol, and 4-propenyl-2.5-dimethoxyl phenol (Diebold and Bridgwater, 1997) (Figure 2.9). It is one of the most abundant organic polymers on Earth (exceeded only by cellulose). It is the third important constituent of the cell walls of woody biomass.

Lignin is the cementing agent for cellulose fibers holding adjacent cells together. The dominant monomeric units in the polymers are benzene rings. It is similar to the glue in a cardboard box, which is made by gluing together papers in special fashion. The middle lamella (Figure 2.5), which is composed primarily of lignin, glues together adjacent cells or tracheids.

Lignin is highly insoluble, even in sulphuric acid (Klass, 1998, p. 84). A typical hardwood contains about 18 to 25%, while a softwood contains 25 to 35% by dry weight.

2.4 GENERAL CLASSIFICATION OF FUELS

Classification is an important means of assessing the properties of a fuel. Fuels belonging to a particular group have similar behavior irrespective of their type or origin. Thus, when a new biomass is considered for gasification or other thermochemical conversion, we can check its classification, and then from the known properties of a biomass of that group, we can infer its conversion potential.

There are three methods of classifying and ranking fuels using their chemical constituents: atomic ratios, the ratio of ligno-cellulose constituents, and the ternary diagram. All hydrocarbon fuels may be classified or ranked according to their atomic ratios, but the second classification is limited to ligno-cellulose biomass.

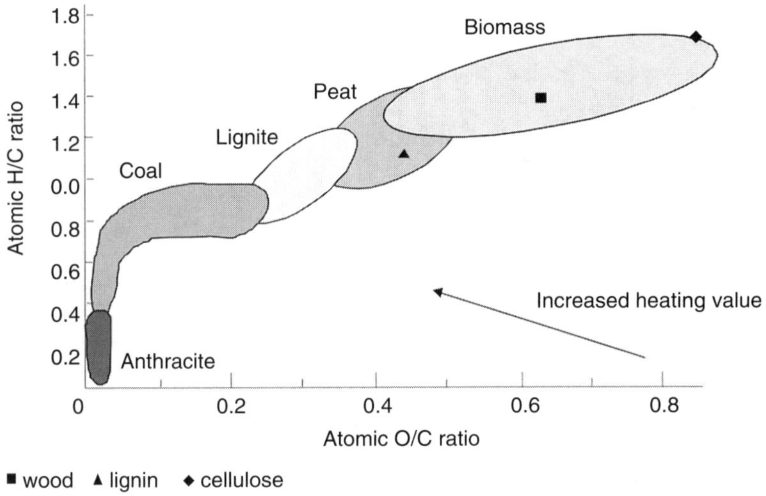

FIGURE 2.10 Classification of solid fuels by their hydrogen/carbon and oxygen/carbon ratios. *(Source: Adapted from Jones et al., 2006, p. 332.)*

2.4.1 Atomic Ratio

Classification based on the atomic ratio helps us to understand the heating value of a fuel, among other things. For example, the higher heating value (HHV) of a biomass correlates well with the oxygen-to-carbon (O/C) ratio, reducing from 38 to about 15 MJ/kg while the O/C ratio increases from 0.1 to 0.7. When the hydrogen-to-carbon (H/C) ratio increases, the effective heating value of the fuel reduces.

The atomic ratio is based on the hydrogen, oxygen, and carbon content of the fuel. Figure 2.10 plots the atomic ratios (H/C) against (O/C) on a dry ash-free basis for all fuels, from carbon-rich anthracite to carbon-deficient woody biomass. This plot, known as *van Krevelen diagram*, shows that biomass has much higher ratios of H/C and O/C than fossil fuel. For a large range of biomass, the H/C ratio might be expressed as a linear function of the (O/C) ratio (Jones et al., 2006).

$$(H/C) = 1.4125\,(O/C) + 0.5004 \tag{2.3}$$

Fresh plant biomass like leaves has very low heating values because of its high H/C and O/C ratios. The atomic ratio of a fuel decreases as its geological age increases, which means that the older the fuel, the higher its energy content. Anthracite, for example, a fossil fuel geologically formed over many thousands of years, has a very high heating value. Its lower H/C ratio gives higher heat, but the carbon intensity or the CO_2 emission from its combustion is high.

Among all hydrocarbon fuels biomass is highest in oxygen content. Oxygen, unfortunately, does not make any useful contribution to heating value and

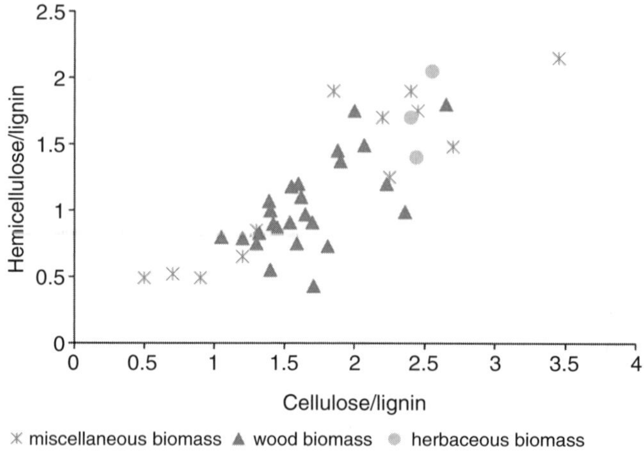

FIGURE 2.11 Classification of biomass by constituent ratios. *(Source: From Jones et al., 2006.)*

makes it difficult to transform the biomass into liquid fuels. The high oxygen and hydrogen content of biomass results in high volatile and liquid yields, respectively. High oxygen consumes a part of the hydrogen in the biomass, producing less beneficial water, and thus the high H/C content does not translate into high gas yield.

2.4.2 Relative Proportions of Ligno-Cellulosic Components

A biomass can also be classified on the basis of its relative proportion of cellulose, hemicellulose, and lignin. For example, we can predict the behavior of a biomass during pyrolysis from knowledge of these components (Jones et al., 2006). Figure 2.11 plots the ratio of hemicellulose to lignin against the ratio of cellulose to lignin. In spite of some scatter, a certain proportionality can be detected between the two. Biomass falling within these clusters behaves similarly irrespective of its type. For a typical biomass, the cellulose–lignin ratio increases from ~0.5 to ~2.7, while the hemicellulose–lignin ratio increases from 0.5 to 2.0.

2.4.3 Ternary Diagram

The ternary diagram (Figure 2.12) is not a tool for biomass classification, but it is useful for representing biomass conversion processes. The three corners of the triangle represent pure carbon, oxygen, and hydrogen—that is, 100% concentration. Points within the triangle represent ternary mixtures of these three substances. The side opposite to a corner with a pure component (C, O, or H) represents zero concentration of that component. For example, the horizontal

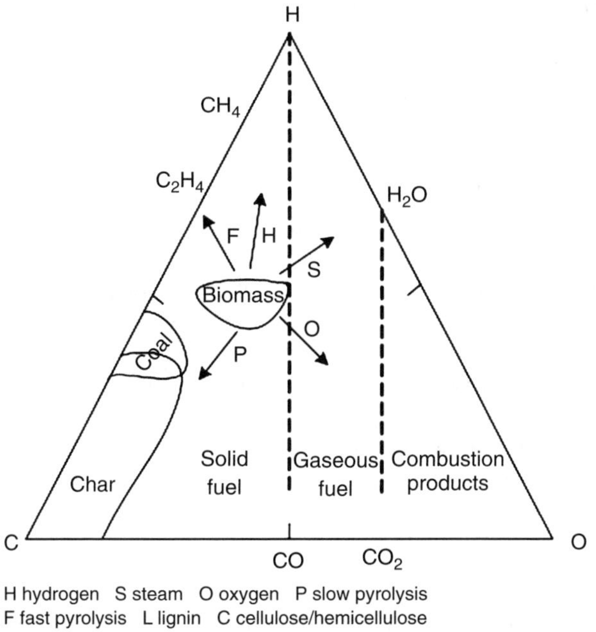

FIGURE 2.12 C-H-O ternary diagram of biomass showing the gasification process.

base in Figure 2.12 opposite to the hydrogen corner represents zero hydrogen—that is, binary mixtures of C and O.

A biomass fuel is closer to the hydrogen and oxygen corners compared to coal. This means that biomass contains more hydrogen and more oxygen than coal contains. Lignin would generally have lower oxygen and higher carbon compared to cellulose or hemicellulose. Peat is in the biomass region but toward the carbon corner, implying that it is like a high-carbon biomass. Peat, incidentally, is the youngest fossil fuel formed from biomass.

Coal resides further toward the carbon corner and lies close to the oxygen base in the ternary diagram, suggesting that it is very low in oxygen and much richer in carbon. Anthracite lies furthest toward the carbon corner because it has the highest carbon content. The diagram can also show the geological evolution of fossil fuels. With age the fuel moves further away from the hydrogen and oxygen corners and closer to the carbon corner.

As mentioned earlier, the ternary diagram can depict the conversion process. For example, carbonization or slow pyrolysis moves the product toward carbon through the formation of solid char; fast pyrolysis moves it toward hydrogen and away from oxygen, which implies higher liquid product. Oxygen gasification moves the gas product toward the oxygen corner, while steam gasification takes the process away from the carbon corner. The hydrogenation process increases the hydrogen and thus moves the product toward hydrogen.

2.5 PROPERTIES OF BIOMASS

The following sections describe some important thermophysical properties of biomass that are relevant to gasification.

2.5.1 Physical Properties

Some of the physical properties of biomass affect its pyrolysis and gasification behavior. For example, permeability is an important factor in pyrolysis. High permeability allows pyrolysis gases to be trapped in the pores, increasing their residence time in the reaction zone. Thus, it increases the potential for secondary cracking to produce char. The pores in a wood are generally oriented longitudinally. As a result, the thermal conductivity and diffusivity in the longitudinal direction are different from those in the lateral direction. This anisotropic behavior of wood can affect its thermochemical conversion. A densification process such as torrefaction (Chapter 3) can reduce the anisotropic behavior and therefore change the permeability of a biomass.

Densities

For a granular biomass, we can define four characteristic densities: true, apparent, bulk, and biomass (growth).

True Density

True density is the weight per unit volume occupied by the solid constituent of biomass. Total weight is divided by actual volume of the solid content to give its true density.

$$\rho_{true} = \frac{\text{Total mass of biomass}}{\text{Solid volume in biomass}} \tag{2.4}$$

The cell walls constitute the major solid content of a biomass. For common wood, the density of the cell wall is typically 1530 kg/m^3, and it is constant for most wood cells (Desch and Dinwoodie, 1981). The measurement of true density of a biomass is as difficult as the measurement of true solid volume. It can be measured with a pycnometer, or it may be estimated using ultimate analysis and the true density of its constituent elements. True densities of some elements are given in Table 2.4.

Apparent Density

Apparent density is based on the apparent or external volume of the biomass. This includes its pore volume (or that of its cell cavities). For a regularly shaped biomass, mechanical means such as micrometers can be used to measure different sides of a particle to obtain its apparent volume. An alternative is the use of volume displacement in water. The apparent density considers the internal

TABLE 2.4 True Density of Some Elements

Elements	C (amorphous)	Ca	Fe	K	Mg	Na	S	Si	Zn
True density (kg/m^3)	1800–2100	1540	7860	860	1740	970	2070	2320	7140

Source: Adapted from Jenkins, 1989, p. 856.

pores of a biomass particle but not the interstitial volume between biomass particles packed together.

$$\rho_{apparent} = \frac{\text{Total mass of biomass}}{\text{Apparent volume of biomass including solids and internal pores}} \quad (2.5)$$

The pore volume of a biomass expressed as a fraction of its total volume is known as its porosity, ε_p. This is an important characteristic of the biomass.

Apparent density is most commonly used for design calculations because it is the easiest to measure and it gives the actual volume occupied by a particle in a system.

Bulk Density

Bulk density is based on the overall space occupied by an amount or a group of biomass particles.

$$\rho_{bulk} = \frac{\text{Total mass of biomass particles or stack}}{\text{Bulk volume occupied by biomass particles or stack}} \quad (2.6)$$

Bulk volume includes interstitial volume between the particles, and as such it depends on how the biomass is packed. For example, after pouring the biomass particles into a vessel, if the vessel is tapped, the volume occupied by the particles settles to a lower value. The interstitial volume expressed as function of the total packed volume is known as bulk porosity, ε_{b}.

To determine the biomass bulk density, we can use standards like the American Society for Testing of Materials (ASTM) E-873-06. This process involves pouring the biomass into a standard-size box (305 mm × 305 mm × 305 mm) from a height of 610 mm. The box is then dropped from a height of 150 mm three times for settlement and refilling. The final weight of the biomass in the box divided by the box volume gives its bulk density.

The total mass of the biomass may contain the green moisture of a living plant, external moisture collected in storage, and moisture inherent in the biomass. Once the biomass is dried in a standard oven, its mass reduces. Thus, the density can be either green or oven-dry depending on if its weight includes surface moisture. The external moisture depends on the degree of wetness of the received biomass. To avoid this issue, we can completely saturate the biomass in deionized water, measure its maximum moisture density, and specify its bulk density accordingly.

Three of the preceding densities of biomass are related as follows:

$$\rho_{apparent} = \rho_{true}(1-\varepsilon_p) \tag{2.7}$$

$$\rho_{bulk} = \rho_{apparent}(1-\varepsilon_b) \tag{2.8}$$

where ε_p is the void fraction (voidage) in a biomass particle, and ε_b is the voidage of particle packing.

Biomass (Growth) Density

The term *biomass (growth) density* is used in bioresource industries to express how much biomass is available per unit area of land. It is defined as the total amount of above-ground living organic matter in trees expressed as oven-dry tons per unit area (e.g., tonnes per hectare) and includes all organic materials: leaves, twigs, branches, main bole, bark, and trees.

2.5.2 Thermodynamic Properties

Gasification is a thermochemical conversion process, so the thermodynamic properties of a biomass heavily influence its gasification. This section describes three important thermodynamic properties: thermal conductivity, specific heat, and heat of formation of biomass.

Thermal Conductivity

Biomass particles, however small they may be, are subject to heat conduction along and across their fiber, which in turn influences their pyrolysis behavior. Thus, the thermal conductivity of the biomass is an important parameter in this context. It changes with density and moisture. Based on a large number of samples, MacLean (1941) developed the following correlations (from Kitani and Hall, 1989, p. 877).

$$\begin{aligned} K_{eff}(\text{w/m.K}) &= sp.gr(0.2+0.004m_d)+0.0238 \quad \text{for } m_d > 40\% \\ &= sp.gr(0.2+0.0055m_d)+0.0238 \quad \text{for } m_d < 40\% \end{aligned} \tag{2.9}$$

where $sp.gr$ is the specific gravity of the fuel and m_d is the moisture percentage of the biomass on a dry basis.

Unlike metal and other solids, biomass is highly anisotropic. Its thermal conductivity along its fibers is different from that across them. Conductivity also depends on the biomass' moisture content, porosity, and temperature. Some of these depend on the degree of conversion as the biomass undergoes combustion or gasification. A typical wood, for example, is made of fibers, the walls of which have channels carrying gas and moisture. Thunman and Leckner (2002) write the effective thermal conductivity parallel to the direction of wood fiber as a sum of contributions from these three:

$$K_{eff} = G(x)K_s + F(x)K_w + H(x)[K_g + K_{rad}] \quad \text{W/m.K for parallel to fiber} \tag{2.10}$$

where $G(x)$, $F(x)$, and $H(x)$ are functions of the cell structure and its dimensionless length; K_s, K_w, and K_g are thermal conductivities of the dry solid (fiber wall), moisture, and gas, respectively; and K_{rad} represents the contribution of radiation to conductivity.

These components are given by the following empirical relations, which are used to calculate the directional values of thermal conductivities:

$$\begin{aligned} K_w &= -0.487 + 5.887\times10^{-3}T - 7.39\times10^{-6}T^2 \text{ W/m.K} \\ K_g &= -7.494\times10^{-3} + 1.709\times10^{-4}T - 2.377\times10^{-7}T^2 \\ &\quad + 2.202\times10^{-10}T^3 - 9.463\times10^{-14}T^4 + 1.581\times10^{-17}T^5 \end{aligned} \tag{2.11}$$

$$\begin{aligned} K_s &= 0.52 \text{ w/m.K in perpendicular direction} \\ &= 0.73 \text{ w/m.K in parallel direction of fiber} \\ K_{rad} &= 5.33\, e_{rad}\, \sigma\, T^3\, d_{pore} \text{ W/m.K} \end{aligned} \tag{2.12}$$

where e_{rad} is the emissivity in the pores having diameter d_{pore}, σ is the Stefan-Boltzmann constant, and T is the temperature in K. The contribution of gas radiation in the pores, K_{rad}, to conductivity is important only at high temperatures.

Figure 2.13 shows the variation in the thermal conductivity of wood against its dry density. The straight line represents the thermal conductivity parallel to the fibers. The curved line gives the thermal conductivity across the fibers. The straight line is calculated from Eq. (2.9); points are experimental values.

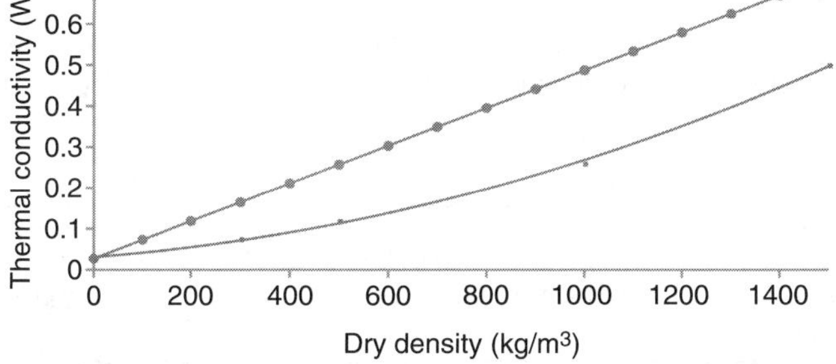

FIGURE 2.13 Thermal conductivity of biomass along the grain and across the grain increases with dry density. The plot is for dry wood. *(Source: Adapted from Thunman and Leckner, 2002.)*

Specific Heat

Specific heat is an important thermodynamic property of biomass often required for thermodynamic calculations. It is an indication of the heat capacity of a substance. Both moisture and temperature affect the specific heat of biomass. Within the temperature range of 0 to 106 °C, the specific heat of a large number of wood species (dry) can be expressed as (Jenkins, 1989, p. 876):

$$C_{p\theta} = 0.266 + 0.00116\theta \tag{2.13}$$

where temperature θ is in °C.

The effect of moisture on specific heat is expressed as

$$C_p = M_{wet} C_w + (1 - M_{wet}) C_{p\theta} \tag{2.14}$$

where M_{wet} is the moisture fraction on a wet basis, and C_w is the specific heat of water.

Heat of Formation

Heat of formation, also known as *enthalpy of formation*, is the enthalpy change when 1 mole of compound is formed at standard state (25 °C, 1 atm) from its constituting elements in their standard state. For example, hydrogen and oxygen are stable in their elemental form, so their enthalpy of formation is zero. However, an amount of energy (241.5 kJ) is released per mole when they combine to form steam.

$$H_2(\text{gas}) + 1/2\, O_2\,(\text{gas}) \rightarrow H_2O\,(\text{gas}) - 241.5\ \text{kJ/mol} \tag{2.15}$$

The heat of formation of steam is thus –241.5 kJ/mol (g). This amount of energy is taken out of the system and is therefore given a negative (–) sign in the equation to indicate an exothermic reaction.

If the compound is formed through multiple steps, the heat of formation is the sum of the enthalpy change in each process step. Gases like H_2, O_2, N_2, and Cl_2 are not compounds, and the heat of formation for them is zero. Values for the heat of formation for common compounds are shown in Table 2.5.

Heat of Combustion (Reaction)

The *heat of reaction* (HR) is the amount of heat released or absorbed in a chemical reaction with no change in temperature. In the context of combustion

TABLE 2.5 Formation Heat of Some Important Compounds

Compound	H_2O	CO_2	CO	CH_4	O_2	$CaCO_3$	NH_3
Heat of formation at 25 °C (kJ/mole)	–241.5	–393.5	–110.6	–74.8	0	–1211.8	–82.5

Source: Data collected from Perry and Green, 1997, p. 2-186.

reactions, HR is called *heat of combustion*, ΔH_{comb}, which can be calculated from the *heat of formation* (HF) as

$$HR = [\text{Sum of } HF \text{ of all products}] - [\text{Sum of } HF \text{ of all reactants}] \quad (2.16)$$

For example:

$$\begin{gathered} CH_4 + 2O_2 \rightarrow 2H_2O + CO_2 \\ \Delta H_{comb} = 2\,\Delta H_{H_2O} + \Delta H_{CO_2} - \Delta H_{CH_4} - 2\,\Delta H_{O_2} \end{gathered} \quad (2.17)$$

The ΔH_{comb} for a fuel is also defined as the enthalpy change for the combustion reaction when balanced:

$$\text{Fuel} + O_2 \rightarrow CO_2 + H_2O - HR \quad (2.18)$$

Example 2.1

Find the heat of formation of sawdust, the heating value of which is given as 476 kJ/mol. Assume its chemical formula to be $CH_{1.35}O_{0.617}$.

Solution

Using stoichiometry, the conversion reaction of sawdust (SW) can be written in the simplest terms as

$$CH_{1.35}O_{0.617} + 1.029\ O_2 \rightarrow CO_2 + 0.675\ H_2O - 476\ \text{kJ/mol SW}$$

Similar to Eq. (2.16), we can write

$$\text{HR} = [HF_{CO_2} + 0.675\ HF_{H_2O}] - [HF_{SW} + 1.029\ HF_{O_2}]$$

Taking values of *HF* (heat of formation) of CO_2, O_2 , and H_2O (g) from Table 2.5, we get

$$\begin{aligned} HR_{sw} &= [-393.5 + 0.675 \times (-241.5)] - [HF_{sw} + 1.029 \times 0] \\ &= -556.5 - HF_{sw} \end{aligned}$$

The HR for the above combustion reaction is −476 kJ/mol. So HF_{sw} = − 556.5 − (−476) = −80.5 kJ/mol

---

Ignition Temperature

Ignition temperature is an important property of any fuel because the combustion reaction of the fuel becomes self-sustaining only above it. In a typical gasifier, a certain amount of combustion is necessary to provide the energy required for drying and pyrolysis and finally for the endothermic gasification reaction. In this context, it is important to have some information on the ignition characteristics of the fuel.

Exothermic chemical reactions can take place even at room temperature, but the reaction rate, being an exponential function of temperature, is very slow

TABLE 2.6 Ignition Temperatures of Some Common Fuels

Fuel	Ignition Temperature (°C)	Volatile Matter in Fuel (dry ash-free %)	Source
Wheat straw	220	72	Grotkjær et al., 2003
Poplar wood	235	75	Grotkjær et al., 2003
Eucalyptus	285	64	Grotkjær et al., 2003
Ethanol	425		
High-volatile coal	670	34.7	Mühlen and Sowa, 1995
Medium-volatile coal	795	20.7	Mühlen and Sowa, 1995
Anthracite	930	7.3	Mühlen and Sowa, 1995

at low temperatures. The heat loss from the fuel, on the other hand, is a linear function of temperature. At low temperatures, then, any heat released through the reaction is lost to the surroundings at a rate faster than that at which it was produced. As a result, the temperature of the fuel does not increase.

When the fuel is heated by some external means, the rate of exothermic reaction increases with a corresponding increase in the heat generation rate. Above a certain temperature, the rate of heat generation matches or exceeds the rate of heat loss. When this happens, the process becomes self-sustaining and that temperature is called the *ignition temperature*.

The ignition temperature is generally lower for higher volatile matter content fuel. Because biomass particles have a higher volatile matter content than coal, they have a significantly lower ignition temperature, as Table 2.6 shows.

Ignition temperature, however, is not a unique property of a fuel, because it depends on several other factors like oxygen partial pressure, particle size, rate of heating, and a particle's thermal surroundings.

2.6 OTHER GASIFICATION-RELATED PROPERTIES OF BIOMASS

Biomass contains a large number of complex organic compounds, moisture (*M*), and a small amount of inorganic impurities known as ash (*ASH*). The organic compounds comprise four principal elements: carbon (*C*), hydrogen (*H*), oxygen (*O*), and nitrogen (*N*). Biomass (e.g., MSW and animal waste) may also have small amounts of chlorine (*Cl*) and sulfur (*S*). The latter is rarely present in biomass except for secondary sources like demolition wood, which comes from torn-down buildings and structures.

Thermal design of a biomass utilization system, whether it is a gasifier or a combustor, necessarily needs the composition of the fuel as well as its energy content. The following three primary properties describe its composition and energy content: (1) ultimate analysis, (2) proximate analysis, and (3) heating values. Experimental determination of these properties is covered by ASTM standard E-870-06.

2.6.1 Ultimate Analysis

Here, the composition of the hydrocarbon fuel is expressed in terms of its basic elements except for its moisture, *M*, and inorganic constituents. A typical ultimate analysis is

$$C + H + O + N + S + ASH + M = 100\% \tag{2.19}$$

Here, *C, H, O, N*, and *S* are the weight percentages of carbon, hydrogen, oxygen, nitrogen, and sulfur, respectively, in the fuel. Not all fuels contain all of these elements. For example, the vast majority of biomass may not contain any sulfur. The moisture or water in the fuel is expressed separately as *M*. Thus, hydrogen or oxygen in the ultimate analysis does not include the hydrogen and oxygen in the moisture, but only the hydrogen and oxygen present in the organic components of the fuel.

Recall that Figure 2.10 is a plot of the atomic ratios (H/C) and (O/C) determined from the ultimate analysis of different fuels. It shows that biomass, cellulose in particular, has very high relative amounts of oxygen and hydrogen. This results in relatively low heating values.

The sulfur content of ligno-cellulosic biomass is exceptionally low, which is a major advantage in its utilization in energy conversion when SO_2 emission is taken into account. Sulfur-bearing fuel oil, coal, and petcoke use limestone to reduce SO_2. For every mole of sulfur captured, at least 1 to 3 moles of CO_2 are released. This is because the sulfur capture reaction typically requires more than the theoretical amount of CaO, resulting in additional carbon dioxide during the production of CaO from $CaCO_3$. Thus, biomass, in addition to being CO_2 neutral, results in additional reduction in CO_2 emission for avoiding sulfur capture.

Ultimate analysis is relatively difficult and expensive compared to proximate analysis. The following ASTM standards are available for determination of the ultimate analysis of biomass components:

- *Carbon, hydrogen:* E-777 for RDF
- *Nitrogen:* E-778 for RDF
- *Sulfur:* E-775 for RDF

TABLE 2.7 Standard Methods for Biomass Compositional Analysis

Biomass Constituent	Standard Methods
Carbon	ASTM E-777 for RDF
Hydrogen	ASTM E-777 for RDF
Nitrogen	ASTM E-778 for RDF
Oxygen	By difference
Ash	ASTM D-1102 for wood; E-1755 for biomass; D-3174 for coal
Moisture	ASTM E-871 for wood; E-949 for RDF; D-3173 for coal
Hemicelluloses	TAPPI T-223 for wood pulp
Lignin	TAPPI T-222 for wood pulp; ASTM D-1106; acid insoluble in wood
Cellulose	TAPPI T-203 for wood pulp

- *Moisture:* E-871 for wood fuels
- *Ash:* D-1102 for wood fuels

Although no standard for other biomass fuels is specified, we can use the RDF standard with a reasonable degree of confidence. For determination of the carbon, hydrogen, and nitrogen component of the ultimate analysis of coal, we may use the ASTM standard D-5373-08. Table 2.7 lists standard methods of analysis for biomass materials. Table 2.8 compares the ultimate analysis of several biomass materials with that of some fossil fuels.

2.6.2 Proximate Analysis

Proximate analysis gives the composition of the biomass in terms of gross components such as moisture (M), volatile matter (VM), ash (ASH), and fixed carbon (FC). It is a relatively simple and inexpensive process. For wood fuels, we can use standard E-870-06. Separate ASTM standards are applicable for determination of the individual components of biomass:

- *Volatile matter:* E-872 for wood fuels
- *Ash:* D-1102 for wood fuels
- *Moisture:* E-871 for wood fuels
- *Fixed carbon:* determined by difference

The moisture and ash determined in proximate analysis refer to the same moisture and ash determined in ultimate analysis. However, the fixed carbon in proximate analysis is different from the carbon in ultimate analysis: In

TABLE 2.8 Comparison of Ultimate Analysis (Dry Basis) of Some Biomass and Other Fossil Fuels

Fuel	C (%)	H (%)	N (%)	S (%)	O (%)	Ash (%)	HHV (kJ/kg)	Source
Maple	50.6	6.0	0.3	0	41.7	1.4	19,958	Tillman, 1978
Douglas fir	52.3	6.3	9.1	0	40.5	0.8	21,051	Tillman, 1978
Douglas fir (bark)	56.2	5.9	0	0	36.7	1.2	22,098	Tillman, 1978
Redwood	53.5	5.9	0.1	0	40.3	0.2	21,028	Tillman, 1978
Redwood (waste)	53.4	6.0	0.1	39.9	0.1	0.6	21,314	Boley and Landers, 1969
Sewage sludge	29.2	3.8	4.1	0.7	19.9	42.1	16,000	
Rice straw	39.2	5.1	0.6	0.1	35.8	19.2	15,213	Tillman, 1978
Rice husk	38.5	5.7	0.5	0	39.8	15.5	15,376	Tillman, 1978
Sawdust	47.2	6.5	0	0	45.4	1.0	20,502	Wen et al., 1974
Paper	43.4	5.8	0.3	0.2	44.3	6.0	17,613	Bowerman, 1969
MSW	47.6	6.0	1.2	0.3	32.9	12.0	19,879	Sanner et al., 1970
Animal waste	42.7	5.5	2.4	0.3	31.3	17.8	17,167	Tillman, 1978
Peat	54.5	5.1	1.65	0.45	33.09	5.2	21,230	
Lignite	62.5	4.38	0.94	1.41	17.2	13.4	24,451	Bituminous Coal Research, 1974
PRB coal	65.8	4.88	0.86	1.0	16.2	11.2	26,436	Probstein and Hicks, 2006
Anthracite	90.7	2.1	1.0	7.6	11.4	2.5	29,963	
Petcoke	86.3	0.5	0.7	0.8	10.5	6.3	29,865	

proximate analysis it does not include the carbon in the volatile matter and is often referred to as the char yield after devolatilization.

Volatile Matter

The volatile matter of a fuel is the condensable and noncondensable vapor released when the fuel is heated. Its amount depends on the rate of heating and the temperature to which it is heated. For the determination of volatile matter, the fuel is heated to a standard temperature and at a standard rate in a controlled environment. The applicable ASTM standard for determination of volatile matter is E-872 for wood fuels and D-3175-07 for coal and coke.

Standard E-872 specifies that 50 g of test sample be taken out of no less than a 10-kg representative sample of biomass using the ASTM D-346 protocol. This sample is ground to less than 1 mm in size, and 1 g is taken from it, dried, and put in a covered crucible so as to avoid contact with air during devolatilization. The covered crucible is placed in a furnace at 950 °C and heated for seven minutes. The volatiles released are detected by luminous flame observed from the outside. After seven minutes, the crucible is taken out, cooled in a desiccator, and weighed to determine the weight loss due to devolatilization.

Standard D-3175-07, when used for nonsparking coal or coke, follows a similar process except that it requires a 1-g sample ground to 250 μm. The sample is heated in a furnace at 950 °C for seven minutes. For sparking coal or coke, the heating process deviates slightly from that specified in E-872: D-3175-07 specifies that the sample be gradually heated to 600 °C within six minutes and then put in a 950 °C furnace for six minutes. After this, the crucible containing the sample is removed and cooled for 15 minutes before it is weighed. Heating rates faster than this may yield higher volatile matter content, but that is not considered the volatile matter of the fuel's proximate analysis.

Ash

Ash is the inorganic solid residue left after the fuel is completely burned. Its primary ingredients are silica, aluminum, iron, and calcium; small amounts of magnesium, titanium, sodium, and potassium may also be present. Ash content is determined by ASTM test protocol D-1102 for wood, E-1755-01 for other biomass, and D-3174 for coal.

Standard D-1102 specifies a 2-g sample of wood (sized below 475 micron) dried in a standard condition and placed in a muffle furnace with the lid of the crucible removed. Temperature of the furnace is raised slowly to 580 to 600 °C to avoid flaming. When all the carbon is burnt, the sample is cooled and weighed. Standard E-1755-01 specifies 1 g of biomass dried, initially heated to 250 °C at 10 °C/min, and held there for 30 minutes. Following this, the temperature is increased to 575 °C and kept there until all the carbon is burnt. After that the sample is cooled and weighed.

For coal or coke, standard D-3174-04 may be used. Here a 1-g sample (pulverized below 250 micron) is dried under standard conditions and heated to 450 to 500 °C for the first one hour and then to 700 to 750 °C (950 °C for coke) for the second hour. The sample is heated for two hours or longer at that temperature to ensure that the carbon is completely burnt. It is then removed from the furnace, cooled, and weighed.

Strictly speaking, this ash does not represent the original inorganic mineral matter in the fuel, as some of the ash constituents can undergo oxidation during burning. For exact analysis, correction may be needed. The ash content of biomass is generally very small, but may play a significant role in biomass utilization especially if it contains alkali metals such as potassium or halides such as chlorine. Straw, grasses, and demolition wood are particularly susceptible to this problem. These components can lead to serious agglomeration, fouling, and corrosion in boilers or gasifiers (Mettanant et al., 2009).

The ash obtained from biomass conversion does not necessarily come entirely from the biomass itself. During collection, biomass is often scraped off the forest floor and then undergoes multiple handlings, during which it can pick up a considerable amount of dirt, rock, and other impurities. In many plants, these impurities constitute the major inorganic component of the biomass feedstock.

Moisture

High moisture is a major characteristic of biomass. The root of a plant biomass absorbs moisture from the ground and pushes it into the sapwood. The moisture travels to the leaves through the capillary passages. Photosynthesis reactions in the leaves use some of it, and the rest is released to the atmosphere through transpiration. For this reason there is more moisture in the leaves than in the tree trunk.

The total moisture content of some biomass can be as high as 90% (dry basis), as seen in Table 2.9. Moisture drains much of the deliverable energy from a gasification plant, as the energy used in evaporation is not recovered. This important input design parameter must be known for assessment of the cost of or energy penalty in drying the biomass. The moisture in biomass can remain in two forms: (1) free, or external; and (2) inherent, or equilibrium.

Free moisture is that above the equilibrium moisture content. It generally resides outside the cell walls. Inherent moisture, on the other hand, is absorbed within the cell walls. When the walls are completely saturated the biomass is said to have reached the fiber saturation point, or equilibrium moisture. Equilibrium moisture is a strong function of the relative humidity and weak function of air temperature. For example, the equilibrium moisture of wood increases

TABLE 2.9 Moisture Content of Some Biomass

Biomass	Corn Stalks	Wheat Straw	Rice Straw	Rice Husk	Dairy Cattle Manure	Wood Bark	Sawdust	Food Waste	RDF Pellets	Water Hyacinth
Moisture (wet basis)	40–60	8–20	50–80	7–10	88	30–60	25–55	70	25–35	95.3

Source: Compiled from Kitani and Hall, 1989, p. 863.

from 3 to 27% when the relative humidity increases from 10 to 80% (Jenkins, 1996, p. 864).

Moisture content (*M*) is determined by the test protocol given in ASTM standards D-871-82 for wood, D-1348-94 for cellulose, D-1762-84 for wood charcoal, and E-949-88 for RDF (total moisture). For equilibrium moisture in coal one could used D-1412-07. In these protocols, a weighed sample of the fuel is heated in an air oven at 103 °C and weighed after cooling. To ensure complete drying of the sample, the process is repeated until its weight remains unchanged. The difference in weight between a dry and a fresh sample gives the moisture content in the fuel.

Standard E-871-82, for example, specifies that a 50-g wood sample be dried at 103 °C for 30 minutes. It is left in the oven at that temperature for 16 hours before it is removed and weighed. The weight loss gives the moisture (*M*) of the proximate analysis.

Standard E-1358-06 provides an alternative means of measurement using microwave. However, this alternative represents only the physically bound moisture; moisture released through chemical reactions during pyrolysis constitutes volatile matter. The moisture content of some biomass fuels is given in Table 2.10.

Moisture Basis

Biomass moisture is often expressed on a dry basis. For example, if W_{wet} kg of wet biomass becomes W_{dry} after drying, its dry basis (M_{dry}) is expressed as

$$M_{dry} = \frac{W_{wet} - W_{dry}}{W_{dry}} \tag{2.20}$$

This can give a moisture percentage greater than 100% for very wet biomass, which might be confusing. For that reason, the basis of moisture should always be specified.

TABLE 2.10 Comparison of Proximate Analysis of Biomass Measured by Two Methods

Fuel	FC (% dry)	VM (% dry)	ASH (% dry)	Technique
Corn cob	18.5	80.1	1.4	ASTM
	16.2	80.2	30.6	TG
Rice husk	16.7	65.5	17.9	ASTM
	19.9	60.6	19.5	TG

Source: From Klass, 1998, p. 239.

The wet-basis moisture is

$$M_{wet} = \frac{W_{wet} - W_{dry}}{W_{wet}} \tag{2.21}$$

The wet basis (M_{wet}) and the dry basis (M_{dry}) are related as

$$M_{dry} = \frac{M_{wet}}{1 - M_{wet}} \tag{2.22}$$

Fixed Carbon

Fixed carbon (*FC*) in a fuel is determined from the following equation, where *M*, *VM*, and *ASH* stand for moisture, volatile matter, and ash, respectively.

$$FC = 1 - M - VM - ASH \tag{2.23}$$

This represents the solid carbon in the biomass that remains in the char in the pyrolysis process after devolatilization. With coal, *FC* includes elemental carbon in the original fuel, plus any carbonaceous residue formed while heating, in the determination of *VM* (standard D-3175).

Biomass carbon comes from photosynthetic fixation of CO_2 and thus all of it is organic. During the determination of *VM*, a part of the organic carbon is transformed into a carbonaceous material called pyrolytic carbon. Since *FC* depends on the amount of *VM*, it is not determined directly. *VM* also varies with the rate of heating. In a real sense, then, fixed carbon is not a fixed quantity, but its value, measured under standard conditions, gives a useful evaluation parameter of the fuel. For gasification analysis, *FC* is an important parameter because in most gasifiers the conversion of fixed carbon into gases determines the rate of gasification and its yield. This conversion reaction, being the slowest, is used to determine the size of the gasifier.

Char

Char, though a carbon residue of pyrolysis or devolatilization, is not pure carbon; it is not the fixed carbon of the biomass. Known as *pyrolytic char*, it contains some volatiles and ash in addition to fixed carbon. Biomass char is very reactive. It is highly porous and does not cake. This noncaking property makes it easy to handle.

2.6.3 Thermogravimetric Analysis

Because of the time and expense involved in proximate analysis by ASTM D-3172, Klass (1998) proposed an alternative using thermogravimetry (TG) or differential thermogravimetry (DTG). In these techniques, a small sample of the fuel is heated in a specified atmosphere at the desired rate in an electronic microbalance. This gives a continuous record of the weight change of the fuel sample in a TG apparatus. The DTG apparatus gives the rate of change in the weight of the fuel sample continuously. Thus, from the measured weight loss-versus-time graphs, we can determine the fuel's moisture, volatile mater, and ash content. The fixed carbon can be found from Eq. (2.23). This method, though not an industry standard, can quickly provide information regarding the thermochemical conversion of a fuel. Table 2.10 compares results of proximate analysis (dry basis) of some biomass from the ASTM and TG methods.

TG analysis provides additional information on reaction mechanisms, kinetic parameters, thermal stability, and heat of reaction. A detailed database of thermal analysis is given in Gaur and Reed (1995).

2.6.4 Bases of Expressing Biomass Composition

The composition of a fuel is often expressed on different bases depending on the situation. The following four bases of analysis are commonly used:

- As received
- Air dry
- Total dry
- Dry and ash-free

A comparison of these is shown in Figure 2.14.

As-Received Basis

When using the as-received basis, the results of ultimate and proximate analyses may be written as follows:

$$\text{Ultimate: } C + H + O + N + S + ASH + M = 100\% \quad (2.24)$$

$$\text{Proximate: } VM + FC + M + ASH = 100\% \quad (2.25)$$

where *VM, FC, M,* and *ASH* represent the weight percentages of volatile matter, fixed carbon, moisture, and ash, respectively, measured by proximate analysis; and *C, H, O, N,* and *S* represent the weight percentages of carbon, hydrogen,

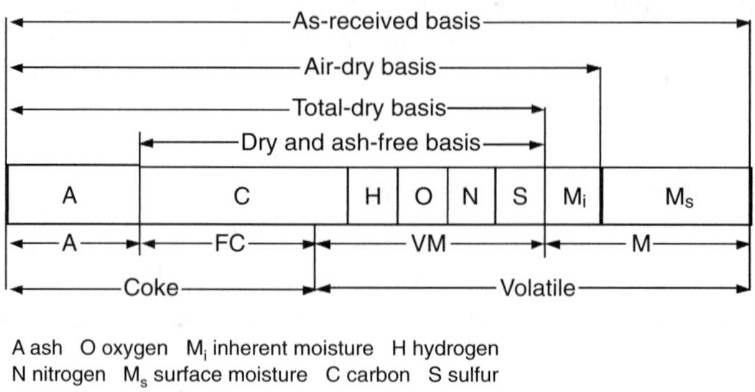

FIGURE 2.14 Bases for expressing fuel composition.

oxygen, nitrogen, and sulfur, respectively, as measured by ultimate analysis. The ash and moisture content of the fuel is the same in both analyses. As received can be converted into other bases.

Air-Dry Basis

When the fuel is dried in air its surface moisture is removed while its inherent moisture is retained. So, to express the constituent on an air-dry basis, the amount is divided by the total mass less the surface moisture. For example, the carbon percentage on the air-dry basis is calculated as

$$C_{ad} = \frac{100C}{100 - M_a}\% \tag{2.26}$$

where M_a is the mass of surface moisture removed from 100 kg of moist fuel after drying in air. Other constituents of the fuel can be expressed similarly.

Total-Dry Basis

Fuel composition on the air-dry basis is a practical parameter and is easy to measure, but to express it on a totally moisture-free basis we must make allowance for surface as well as inherent moisture. This gives the carbon on a total-dry basis, C_{td}:

$$C_{td} = \frac{100C}{100 - M}\% \tag{2.27}$$

where M is the total moisture (surface + inherent) in the fuel: $M = M_a + M_i$.

Dry Ash–Free Basis

Ash is another component that at times is eliminated along with moisture. This gives the fuel composition on a dry ash–free (DAF) basis. Following the aforementioned examples, the carbon percentage on a dry ash–free basis, C_{daf}, can be found.

$$C_{daf} = \frac{100C}{100 - M - ASH}\% \tag{2.28}$$

where $(100 - M - ASH)$ is the mass of biomass without moisture and ash.

The percentages of all constituents on any basis totals 100. For example:

$$C_{daf} + H_{daf} + O_{daf} + N_{daf} + S_{daf} = 100\% \tag{2.29}$$

2.6.5 Heating Value of Fuel

The heating value of biomass is relatively low, especially on a volume basis, because its density is very low.

Higher Heating Value

Higher heating value (HHV) is defined as the amount of heat released by the unit mass or volume of fuel (initially at 25 °C) once it is combusted and the products have returned to a temperature of 25 °C. It includes the latent heat of vaporization of water. HHV can be measured in a bomb calorimeter using ASTM standard D-2015 (withdrawn by ASTM 2000, and not replaced). It is also called gross calorific value (GCV). In North America the thermal efficiency of a system is usually expressed in terms of HHV, so it is important to know the HHV of the design fuel.

Lower Heating Value

The temperature of the exhaust flue gas of a boiler is generally in the range 120 to 180 °C. The products of combustion are rarely cooled to the initial temperature of the fuel, which is generally below the condensation temperature of steam. So the water vapor in the flue gas does not condense, and therefore the latent heat of vaporization of this component is not recovered. The effective heat available for use in the boiler is a lower amount, which is less than the chemical energy stored in the fuel.

The lower heating value (LHV), also known as the net calorific value (NCV), is defined as the amount of heat released by fully combusting a specified quantity less the heat of vaporization of the water in the combustion product.

The relationship between HHV and LHV is given by

$$LHV = HHV - h_g\left(\frac{9H}{100} + \frac{M}{100}\right) \tag{2.30}$$

where *LHV, HHV, H*, and *M* are lower heating value, higher heating value, hydrogen percentage, and moisture percentage, respectively, on an as-received basis. Here, h_g is the latent heat of steam in the same units as *HHV* (i.e., 970 BTU/lb, 2260 kJ/kg, or 540 kCal/kg).

Many European countries define the efficiency of a thermal system in terms of LHV. Thus, an efficiency expressed in this way appears higher than that expressed in HHV (as is the norm in many countries, including the United States and Canada), unless the basis is specified.

Bases for Expressing Heating Values

Similar to fuel composition, heating value (HHV or LHV) may be also expressed on any of the following bases:

- As-received basis (ar)
- Dry basis (db), also known as moisture-free basis (mf)
- Dry ash–free basis (daf), also known as moisture ash–free basis (maf)

If M_f kg of fuel contains Q kJ of heat, M_w kg of moisture, and M_{ash} kg of ash, HHV can be written on different bases as follows:

$$HHV_{ar} = \frac{Q}{M_f}\ \text{kJ/kg}$$
$$HHV_{db} = \frac{Q}{(M_f - M_w)}\ \text{kJ/kg} \tag{2.31}$$
$$HHV_{daf} = \frac{Q}{(M_f - M_w - M_{ash})}\ \text{kJ/kg}$$

Estimation of Biomass Heating Values

Experimental methods are the most reliable means of determining the heating value of biomass. If these are not possible, empirical correlations like the Dulong-Berthelot equation, originally developed for coal with modified coefficients for biomass, may be used. Channiwala and Parikh (2002) developed the following unified correlation for HHV based on 15 existing correlations and 50 fuels, including biomass, liquid, gas, and coal.

$$HHV = 349.1C + 1178.3H + 100.5S - 103.4O - 15.1N - 21.1ASH\ \text{kJ/kg} \tag{2.32}$$

where *C, H, S, O, N*, and *ASH* are percentages of carbon, hydrogen, sulfur, oxygen, nitrogen, and ash as determined by ultimate analysis on a dry basis. This correlation is valid within the range:

- $0 < C < 92\%$; $0.43 < H < 25\%$
- $0 < O < 50$; $0 < N < 5.6\%$
- $0 < ASH < 71\%$; $4745 < HHV < 55{,}345$ kJ/kg

Ultimate analysis is necessary with this correlation, but it is expensive and time consuming. Zhu and Venderbosch (2005) developed an empirical method to estimate HHV without ultimate analysis. This empirical relationship between the stoichiometric ratio (SR) and the HHV is based on data for 28 fuels that include biomass, coal, liquid, and gases. The relation is useful for preliminary design:

$$HHV = 3220 \times \text{Stoichiometric ratio}, \quad \text{kJ/kg} \tag{2.33}$$

where the stoichiometric ratio is the theoretical mass of the air required to burn 1 kg fuel.

Stoichiometric Amount of Air for Complete Combustion

Noting that dry air contains 23.16% oxygen, 76.8% nitrogen, and 0.04% inert gases by weight, the dry air required for complete combustion of a unit weight of dry hydrocarbon, M_{da}, is given by

$$M_{da} = [0.1153\,C + 0.3434\,(H - O/8) + 0.0434\,S]\ \text{kg/kg dry fuel} \tag{2.34}$$

where *C, H, O,* and *S* are the percentages of carbon, hydrogen, oxygen, and sulfur, respectively, on a dry basis.

2.6.6 Composition of the Product Gas of Gasification

The product gas of gasification is generally a mixture of several gases, including moisture or steam. Its composition may be expressed in any of the following ways:

- Mass fraction, m_i
- Mole fraction, n_i
- Volume fraction, V_i
- Partial pressure, P_i

It may also be expressed on a dry or a wet basis. The wet basis is the composition gas expressed on the basis of total mass of the gas mixture including any moisture in it. The dry basis is the composition with the moisture entirely removed.

The following example illustrates the relationship between different ways of expressing the product gas composition.

Example 2.2

The gasification of a biomass yields *M* kg/s product gas, with the production of its individual constituents as follows:

Hydrogen—M_H, kg/s
Carbon monoxide—M_{CO}, kg/s
Carbon dioxide—M_{CO_2}, kg/s
Methane—M_{CH_4}, kg/s
Other hydrocarbon (e.g., C_3H_8)—M_{HC}, kg/s
Nitrogen—M_N, kg/s
Moisture—M_{H_2O}, kg/s

Find the composition of the product gas in mass fraction, mole fraction, and other fractions.

Solution

Since the total gas production rate, *M*, is

$$M = M_H + M_{CO} + M_{CO_2} + M_{CH_4} + M_H + M_N + M_{H_2O} \quad \text{kg/s} \tag{i}$$

the mass fraction of each species is found by dividing the individual production rate by the total. For example, the mass fraction of hydrogen is $m_H = M_H/M$.

The mole of an individual species is found by dividing its mass by its molecular weight:

$$\text{Moles of hydrogen, } n_H = \text{mass/molecular weight of } H_2 = m_H/2 \tag{ii}$$

The total number of moles of all gases is found by adding the moles of *i* species of gases, $n = \Sigma\,(n_{i)})$ moles. So the mole fraction of hydrogen is $x_H = n_H/n$. Similarly for any gas, the mole fraction is

$$x_i = n_i/n \quad \text{(iii)}$$

where the subscript refers to the *i*th species.

The volume fraction of a gas can be found by noting that the volume that 1 kmol of any gas occupies at NTP (at 0 °C and 1 atm) is 22.4 m^3. So, taking the example of hydrogen, the volume of 1 kmol of hydrogen in the gas mixture is 22.4 nm^3 at NTP.

The total volume of the gas mixture is V = summation of volumes of all constituting gases in the mixture = Σ[number of moles (n_i) × 22.4] nm^3 = 22.4 n. The volume fraction of hydrogen in the mixture is volume of hydrogen/total volume of the mixture:

$$V_H = 22.4n_H/(22.4\ \Sigma n_i) = n_H/n = x_H \quad \text{(iv)}$$

Thus, we note that

Volume fraction = mole fraction

The partial pressure of a gas is the pressure it exerts if it occupies the entire mixture volume, V. Ideal gas law gives the partial pressure of a gas component, i, as

$$P_i = n_i RT/V \quad \text{Pa}$$

The total pressure, P, of the gas mixture containing total moles, n, is

$$P = n\,RT/V \quad \text{Pa}$$

So we can write

$$x_i = \frac{n_i}{n} = \frac{P_i}{P} = \frac{v_i}{V} \quad \text{(v)}$$

Partial pressure as fraction of total pressure = mole fraction = volume fraction

The partial pressure of hydrogen is $P_H = x_H\,P$.

The molecular weight of the mixture gas, MW_m, is known from the mass fraction and the molecular weight of individual gas species.

$$MW_m = \Sigma[x_i MW_i] \quad \text{(vi)}$$

where MW_i is the molecular weight of gas component i with mole fraction x_i.

Symbols and Nomenclature

ASH = weight percentage of ash (%)
C = weight percentage of carbon (%)
C_p = specific heat of biomass (J/g.K)
$C_{p\theta}$ = specific heat of biomass at temperature θ °C (J/g.C)
C_w = specific heat of water (J/g.K)
d_{pore} = pore diameter (m)
e_{rad} = emissivity in the pores (–)

FC = weight percentage of fixed carbon (%)
$G(x)$, $F(x)$, $H(x)$ = functions of the cell structure and its dimensionless length of the biomass in Eq. (2.10) (–)
HR = heat of combustion or heat of reaction (kJ/mol)
HF = heat of formation (kJ/mol)
H = weight percentage of hydrogen (%)
HHV = high heating value of fuel (kJ/kg)
h_g = latent heat of vaporization (kJ/kg)
K_{eff} = effective thermal conductivity of biomass (W/m.K)
K_s = thermal conductivity of the solid in dry wood (W/m.K)
K_w = thermal conductivity of the moisture in dry wood (W/m.K)
K_g = thermal conductivity of the gas in dry wood (W/m.K)
K_{rad} = radiative contribution to the conductivity of wood (W/m.K)
LHV = low heating value of fuel (kJ/kg)
M_{wet} = biomass moisture expressed in wet basis (–)
M_{dry} = biomass moisture expressed in dry basis (–)
m_d = moisture percentage (by weight, %)
M = weight percentage of moisture (%)
M_a = mass of surface moisture in biomass (kg)
M_i = mass of inherent moisture in biomass (kg)
M_f = mass of fuel (kg)
M_w = mass of moisture in the fuel (kg)
M_{ash} = mass of ash in the fuel (kg)
m_i = mass fraction of the ith gas (–)
MW = molecular weight of gas mixture (–)
n = number of moles (–)
n_i = mole fraction of the ith gas (–)
N = weight percentage of nitrogen (%)
O = weight percentage of oxygen (%)
P_i = partial pressure of the ith gas (Pa)
P = total pressure of the gas (Pa)
Q = heat content of fuel (kJ)
R = universal gas constant (8.314 J/mol.K)
$sp.gr$ = specific gravity (–)
S = weight percentage of sulfur (%)
T = temperature (K)
W_{wet} = weight of wet biomass (kg)
W_{dry} = weight of dry biomass (kg)
VM = weight percentage of volatile matter (%)
V = volume of gas (m^3)
V_i = volume fraction of the ith gas (–)
ΔH_{comb} = heat of combustion or reaction, kJ/mol
ρ_{true} = true density of biomass (kg/m^3)
$\rho_{apparent}$ = apparent density of biomass (kg/m^3)
ρ_{bulk} = bulk density of biomass (kg/m^3)
ε_b = bulk porosity of biomass (–)
ε_p = porosity of biomass (–)

σ = Steven-Boltzmann's constant (5.67×10^{-8} W/ m^2K^4)
θ = temperature (°C)

Subscripts

ad, ar, db = subscripts representing air dry, as-received basis, and dry basis
daf = dry ash–free basis
td = total-dry basis
i = *i*th component
m = mixture

Chapter 3

Pyrolysis and Torrefaction

3.1 INTRODUCTION

Pyrolysis is a thermochemical decomposition of biomass into a range of useful products, either in the total absence of oxidizing agents or with a limited supply that does not permit gasification to an appreciable extent. It is one of several reaction steps or zones observed in a gasifier. During pyrolysis, large complex hydrocarbon molecules of biomass break down into relatively smaller and simpler molecules of gas, liquid, and char (Figure 3.1).

Pyrolysis has similarity to and some overlap with processes like cracking, devolatilization, carbonization, dry distillation, destructive distillation, and thermolysis, but it has no similarity with the gasification process, which involves chemical reactions with an external agent known as *gasification medium*. Pyrolysis of biomass is typically carried out in a relatively low temperature range of 300 to 650 °C compared to 800 to 1000 °C for gasification.

Torrefaction is a relatively new process that heats the biomass in the absence of air to improve its usefulness as a fuel. Interest in torrefaction is rising on account of its several advantages.

This chapter explains the basics of pyrolysis and torrefaction. A brief discussion of the design implications of the two is also presented.

3.1.1 Historical Background

Charcoal from wood via pyrolysis was essential for extraction of iron from iron-ore in the pre-industrial era. Figure 3.2 shows a typical beehive oven used in early times to produce charcoal from biomass using a slow pyrolysis process. This practice continued until wood supplies nearly ran out and coal, produced inexpensively from underground mines, replaced charcoal for iron production.

The modern petrochemical industry owes a great deal to the invention of a process of kerosene production using pyrolysis. In the mid-1840s, Abraham Gesner, a physician practicing in Halifax, Canada (Figure 3.3), began searching

Biomass Gasification and Pyrolysis. DOI: 10.1016/B978-0-12-374988-8.00003-9

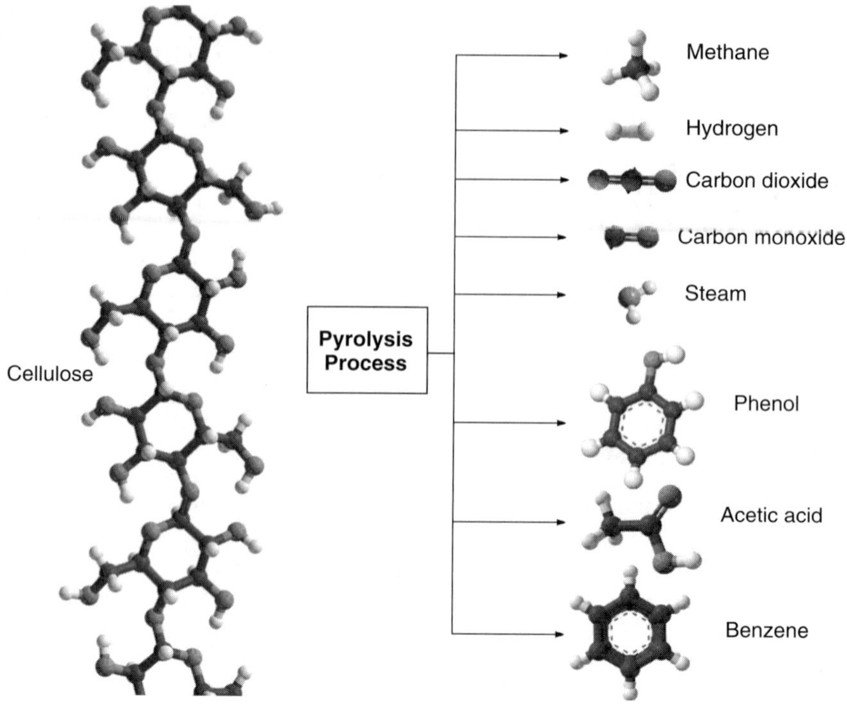

FIGURE 3.1 Process of decomposition of large hydrocarbon molecules into smaller ones during pyrolysis.

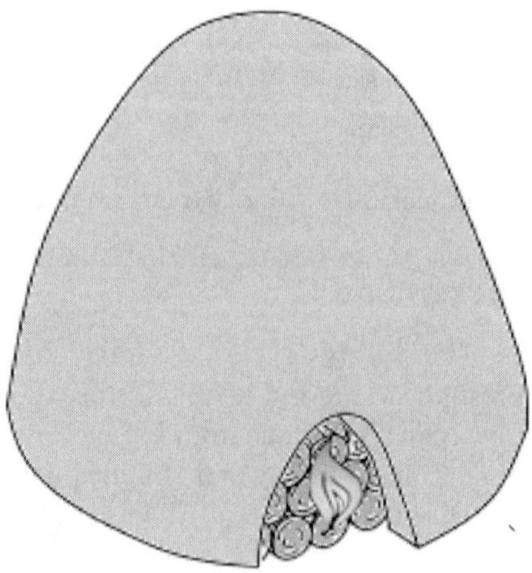

FIGURE 3.2 Beehive oven for charcoal production through slow pyrolysis of wood.

FIGURE 3.3 Abraham Gesner, inventor of kerosene and his kerosene lamp.

for a cleaner-burning mineral oil to replace the sooty whale oil used for illumination at the time. By carefully distilling a few lumps of coal at 427 °C, purifying the product by treating it with sulfuric acid and lime, and then redistilling it, he obtained several ounces of a clear liquid (Gesner, 1861). When this liquid was burned in an oil lamp similar to the one shown in Figure 3.3, it produced a clear bright light that was much superior to the smoky light produced by the burning of whale oil, the primary fuel used during those times on the eastern seaboard of the United States and in Atlantic Canada. Dr. Gesner called his fuel *kerosene*—from the Greek words for wax and oil. Later, in the 1850s, when crude oil began to flow in Pennsylvania and Ontario, Gesner extracted kerosene from that as well.

The invention of kerosene, the first transportable liquid fuel, brought about a revolution in lighting that touched even the remotest parts of the world. It also had a major positive impact on the ecology. For example, in 1846 more than 730 ships hunted whales to meet the huge demand for whale oil. In just a few years after the invention of kerosene, the hunt was reduced to only a few ships, saving whales from possible extinction.

3.2 PYROLYSIS

Pyrolysis involves heating biomass or other feed in the absence of air or oxygen at a specified rate to a maximum temperature, known as the *pyrolysis temperature*, and holding it there for a specified time. The nature of its product depends on several factors, including pyrolysis temperature and heating rate.

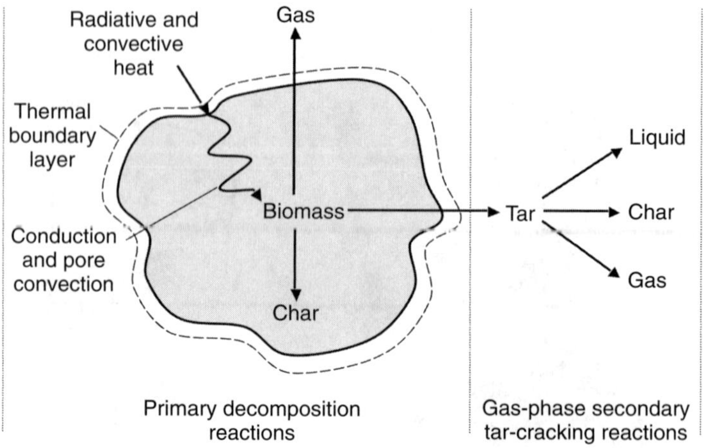

FIGURE 3.4 Pyrolysis in a biomass particle.

The initial product of pyrolysis is made of condensable gases and solid char. The condensable gas may break down further into noncondensable gases (CO, CO_2, H_2, and CH_4), liquid, and char (Figure 3.4). This decomposition occurs partly through gas-phase homogeneous reactions and partly through gas-solid–phase heterogeneous thermal reactions. In gas-phase reactions, the condensable vapor is cracked into smaller molecules of noncondensable permanent gases such as CO and CO_2.

The pyrolysis process may be represented by a generic reaction such as

$$C_nH_mO_p\,(\text{Biomass}) \xrightarrow{\text{Heat}} \Sigma_{liquid} C_xH_yO_z + \Sigma_{gas} C_aH_bO_c + \text{H}_2\text{O} + C\,(\text{char}) \tag{3.1}$$

Pyrolysis is an essential prestep in a gasifier. This step is relatively fast, especially in reactors with rapid mixing.

Figure 3.5 illustrates the process by means of a schematic of a typical pyrolysis plant. Biomass is fed into a pyrolysis chamber containing hot solids (fluidized bed) that heat the biomass to the pyrolysis temperature, at which decomposition starts. The condensable and noncondensable vapors released from the biomass leave the chamber, while the solid char produced remains partly in the chamber and partly in the gas. The gas is separated from the char and cooled downstream of the reactor. The condensable vapor condenses as bio-oil or pyrolysis oil; the noncondensable gases leave the chamber as product gas. These gases may be fired in a burner to heat the pyrolysis chamber, as shown in Figure 3.5, or released for other purposes. Similarly, the solid char may be collected as a commercial product or burned in a separate chamber to produce heat that is necessary for pyrolysis. As this gas is free from oxygen,

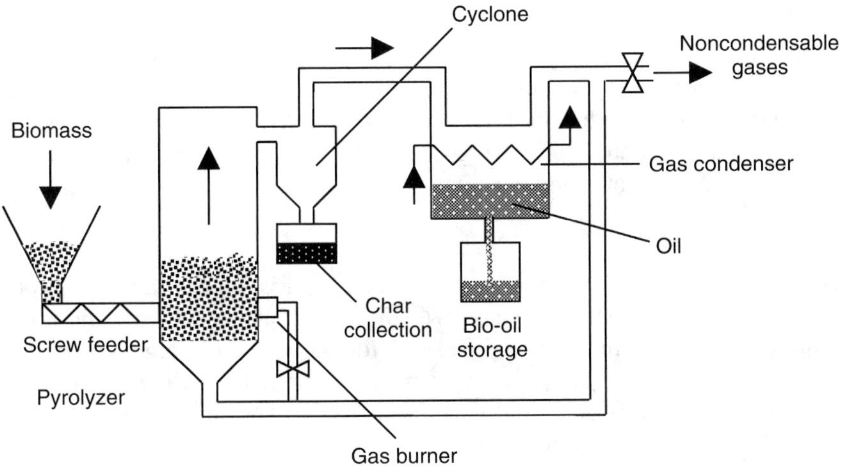

FIGURE 3.5 Simplified layout of a pyrolysis plant.

part of it may be recycled into the pyrolysis chamber as a heat carrier or fluidizing medium. There are, of course, variations of the process, which will be discussed later.

3.2.1 Pyrolysis Products

As mentioned earlier, pyrolysis involves a breakdown of large complex molecules into several smaller molecules. Its product is classified into three principal types:

- Solid (mostly char or carbon)
- Liquid (tars, heavier hydrocarbons, and water)
- Gas (CO_2, H_2O, CO, C_2H_2, C_2H_4, C_2H_6, C_6H_6, etc.)

The relative amounts of these products depend on several factors including the heating rate and the final temperature reached by the biomass.

The pyrolysis product should not be confused with the "volatile matter" of a fuel as determined by its *proximate analysis*. In proximate analysis, the liquid and gas yields are often lumped together as "volatile matter," and the char yield as "fixed carbon." Since the relative fraction of the pyrolysis yields depends on many operating factors, determination of the volatile matter of a fuel requires the use of standard conditions as specified in test codes such as ASTM D-3172 and D-3175. The procedure laid out in D-3175, for example, involves heating a specified sample of the fuel in a furnace at 950 °C for seven minutes to measure its volatile matter.

Solid

Char is the solid yield of pyrolysis. It is primarily carbon (~85%), but it can also contain some oxygen and hydrogen. Unlike fossil fuels, biomass contains very little inorganic ash. The lower heating value (LHV) of biomass char is about 32 MJ/kg (Diebold and Bridgwater, 1997), which is substantially higher than that of the parent biomass or its liquid product.

Liquid

The liquid yield, known as tar, bio-oil, or biocrude, is a black tarry fluid containing up to 20% water. It consists mainly of homologous phenolic compounds. Bio-oil is a mixture of complex hydrocarbons with large amounts of oxygen and water. While the parent biomass has an LHV in the range of 19.5 to 21 MJ/kg dry basis, its liquid yield has a lower LHV, in the range of 13 to 18 MJ/kg wet basis (Diebold et al., 1997).

Bio-oil is produced by rapidly and simultaneously depolymerizing and fragmenting the cellulose, hemicellulose, and lignin components of biomass. In a typical operation, the biomass is subjected to a rapid increase in temperature followed by an immediate quenching to "freeze" the intermediate pyrolysis products. Rapid quenching is important, as it prevents further degradation, cleavage, or reaction with other molecules (see Section 3.4.2 for more details).

Bio-oil is a microemulsion, in which the continuous phase is an aqueous solution of the products of cellulose and hemicellose decomposition, and small molecules from lignin decomposition. The discontinuous phase is largely composed of pyrolytic lignin macromolecules (Piskorz et al., 1988). Bio-oil typically contains molecular fragments of cellulose, hemicellulose, and lignin polymers that escaped the pyrolysis environment (Diebold and Bridgwater, 1997). The molecular weight of the condensed bio-oil may exceed 500 Daltons (Diebold and Bridgwater, 1997). Compounds found in bio-oil fall into the following five broad categories (Piskorz et al., 1988):

- Hydroxyaldehydes
- Hydroxyketones
- Sugars and dehydrosugars
- Carboxylic acids
- Phenolic compounds

Gas

Primary decomposition of biomass produces both condensable gases (vapor) and noncondensable gases (primary gas). The vapors, which are made of heavier molecules, condense upon cooling, adding to the liquid yield of pyrolysis. The noncondensable gas mixture contains lower-molecular-weight gases like carbon dioxide, carbon monoxide, methane, ethane, and ethylene. These do not condense on cooling. Additional noncondensable gases produced through secondary cracking of the vapor (see Section 3.4.2) are called *secondary gases*. The final noncondensable gas product is thus a mixture of both primary and

Table 3.1 Comparison of Heating Values of Five Fuels

Fuel	Petcoke	Bituminous Coal	Sawdust	Bio-Oil	Pyrolysis Gas
Units	MJ/kg	MJ/kg	MJ/kg dry	MJ/kg	MJ/Nm^3
Heating value	~29.8	~26.4	~20.5	13–18	11–20

secondary gases. The LHV of primary gases is typically 11 MJ/Nm^3, but that of pyrolysis gases formed after severe secondary cracking of the vapor is much higher: 20 MJ/Nm^3 (Diebold and Bridgwater, 1997). Table 3.1 compares the heating values of pyrolysis gas with those of bio-oil, raw biomass, and two fossil fuels.

3.2.2 Types of Pyrolysis

Based on heating rate, pyrolysis may be broadly classified as slow and fast. It is considered slow if the time, $t_{heating}$, required to heat the fuel to the pyrolysis temperature is much longer than the characteristic pyrolysis reaction time, t_r, and vice versa. That is:

- Slow pyrolysis: $t_{heating} \gg t_r$
- Fast pyrolysis: $t_{heating} \ll t_r$

These criteria may be expressed in terms of heating rate as well, assuming a simple linear heating rate ($T_{pyr}/t_{heating}$, K/s). The characteristic reaction time, t_r, for a single reaction is taken as the reciprocal of the rate constant, k, evaluated at the pyrolysis temperature (Probstein and Hicks, 2006, p. 63).

There are a few other variants depending on the medium in and pressure at which the pyrolysis is carried out. Given specific operating conditions, each process has its characteristic products and applications. In the following list, the first two types are based on the heating rate while the third is based on the environment or medium in which the pyrolysis is carried out: (1) slow pyrolysis, (2) fast pyrolysis, and (3) hydropyrolysis.

Slow and fast pyrolysis are carried out generally in the absence of a medium. Two other types are conducted in a specific medium: (1) hydrous pyrolysis (in H_2O) and (2) hydropyrolysis (in H_2). These types are used mainly for the production of chemicals.

In slow pyrolysis, the residence time of vapor in the pyrolysis zone (vapor residence time) is on the order of minutes or longer. This process is used primarily for char production and is broken down into two types: (1) carbonization and (2) conventional.

In fast pyrolysis, the vapor residence time is on the order of seconds or milliseconds. This type of pyrolysis, used primarily for the production of bio-oil and gas, is of two main types: (1) flash and (2) ultra-rapid.

Table 3.2 Characteristics of Some Pyrolysis Processes

Pyrolysis Process	Residence Time	Heating Rate	Final Temperature (°C)	Products
Carbonization	Days	Very low	400	Charcoal
Conventional	5–30 min	Low	600	Char, bio-oil, gas
Fast	<2 s	Very high	~500	Bio-oil
Flash	<1 s	High	<650	Bio-oil, chemicals, gas
Ultra-rapid	<0.5 s	Very high	~1000	Chemicals, gas
Vacuum	2–30 s	Medium	400	Bio-oil
Hydropyrolysis	<10 s	High	<500	Bio-oil
Methano-pyrolysis	<10 s	High	>700	Chemicals

Table 3.2 compares the characteristics of different pyrolysis processes, and shows carbonization as the slowest and ultra-rapid as the fastest. Carbonization produces mainly charcoal; fast pyrolysis processes target production of liquid or gas.

Slow Pyrolysis

Carbonization is a slow pyrolysis process, in which the production of charcoal or char is the primary goal. It is the oldest form of pyrolysis, in use for thousands of years. The biomass is heated slowly in the absence of oxygen to a relatively low temperature (~400 °C) over an extended period of time, which in ancient times ran for several days to maximize the char formation. Figure 3.2 is a sketch of a typical beehive oven in which large logs were stacked and covered by a clay wall. A small fire at the bottom provided the required heat, which essentially stayed in the well-insulated closed chamber. Carbonization allows adequate time for the condensable vapor to be converted into char and noncondensable gases.

Conventional pyrolysis involves all three types of pyrolysis product (gas, liquid, and char). As such, it heats the biomass at a moderate rate to a moderate temperature (~600 °C). The product residence time is on the order of minutes.

Fast Pyrolysis

The primary goal of fast pyrolysis is to maximize the production of liquid or bio-oil. The biomass is heated so rapidly that it reaches the peak (pyrolysis) temperature before it decomposes. The heating rate can be as high as 1000 to 10,000 °C/s, but the peak temperature should be below 650 °C if bio-oil is the product of interest. However, the peak temperature can be up to 1000 °C if the production of gas is of primary interest. Fluidized beds similar to the one shown in Figures 3.5 and 3.9(a) and (b) (see p. 82), may be used for fast pyrolysis.

Four important features of the fast pyrolysis process that help increase the liquid yield are: (1) very high heating rate, (2) reaction temperature within the range of 425 to 600 °C, (3) short residence time (< 3 s) of vapor in the reactor, and (4) rapid quenching of the product gas.

Flash Pyrolysis

In flash pyrolysis biomass is heated rapidly in the absence of oxygen to a relatively modest temperature range of 450 to 600 °C. The product, containing condensable and noncondensable gas, leaves the pyrolyzer within a short residence time of 30 to 1500 ms (Bridgwater, 1999). Upon cooling, the condensable vapor is then condensed into a liquid fuel known as bio-oil. Such an operation increases the liquid yield while reducing the char production. A typical yield of bio-oil in flash pyrolysis is 70 to 75% of the total pyrolysis product.

Ultra-Rapid Pyrolysis

Ultra-rapid pyrolysis involves extremely fast mixing of biomass with a heat-carrier solid, resulting in a very high heat-transfer and hence heating rate. A rapid quenching of the primary product follows the pyrolysis, occurring in its reactor. A gas–solid separator separates the hot heat-carrier solids from the noncondensable gases and primary product vapors, and returns them to the mixer. They are then heated in a separate combustor. Then a nonoxidizing gas transports the hot solids to the mixer, as illustrated in Figure 3.9(d) (see p. 82). A precisely controlled short uniform residence time is an important feature of ultra-rapid pyrolysis. To maximize the product yield of gas, the pyrolysis temperature is around 1000 °C for gas and around 650 °C for liquid.

Pyrolysis in the Presence of a Medium

Normal pyrolysis is carried out in the absence of a medium such as air, but a special type is conducted in a medium such as water or hydrogen.

In hydropyrolysis, thermal decomposition of biomass takes place in an atmosphere of high-pressure hydrogen. Hydropyrolysis can increase the volatile yield and the proportion of lower-molar-mass hydrocarbons (Rocha et al., 1997). This process is different from the hydrogasification of char. Its higher volatile yield is attributed to hydrogenation of free-radical fragments sufficient to stabilize them before they repolymerize and form char (Probstein and Hicks, 2006, p. 99).

Hydrous pyrolysis is the thermal cracking of the biomass in high-temperature water. It is used by a commercial company, Changing World Technology, to convert turkey offal into light hydrocarbon that can be used for production of fuel, fertilizer, or chemicals. In a two-stage process, the first stage takes place in water at 200 to 300 °C under pressure; in the second stage the produced hydrocarbon is cracked into lighter hydrocarbon at a temperature of around 500 °C (Appel et al., 2004). High oxygen content is an important shortcoming of bio-oil. Hydropyrolysis can produce bio-oil with reduced oxygen.

3.3 PYROLYSIS PRODUCT YIELD

The product of pyrolysis depends on the design of the pyrolyzer, the physical and chemical characteristics of the biomass, and important operating parameters such as

- Heating rate
- Final temperature (pyrolysis temperature)
- Residence time in the reaction zone

Besides these, the tar and the yields of other products depend on (1) pressure, (2) ambient gas composition, and (3) presence of mineral catalysts (Shafizadeh, 1984).

By changing the final temperature and the heating rate, it is possible to change the relative yields of the solid, liquid, and gaseous products of pyrolysis. Rapid heating yields higher volatiles and more reactive char than produced by a slower heating process; slower heating rate and longer residence time result in secondary char produced from a reaction between the primary char and the volatiles.

3.3.1 Effect of Biomass Composition

The composition of the biomass, especially its hydrogen-to-carbon (H/C) ratio, has an important bearing on the pyrolysis yield. Each of the three major constituents of a ligno-cellulosic biomass has its preferred temperature range of decomposition. Analysis of data from thermogravimetric apparatus (TGA) differential thermogravimetry (DTG) on some selected biomass suggests the following temperature ranges for initiation of pyrolysis (Kumar and Pratt, 1996):

- Hemicellulose: 150–350 °C
- Cellulose: 275–350 °C
- Lignin: 250–500 °C

The individual constituents undergo pyrolysis differently, making varying contributions to yields. For example, cellulose and hemicellulose are the main sources of volatiles in ligno-cellulose biomass. Of these, cellulose is a primary source of condensable vapor. Hemicellulose, on the other hand, yields more noncondensable gases and less tar than is released by cellulose (Reed, 2002, p. II-109). Owing to its aromatic content, lignin degrades slowly, making a major contribution to the char yield.

Cellulose decomposes over a narrow temperature range of 300 to 400 °C. In the absence of any catalyst, pure cellulose pyrolyzes predominantly to a monomer, levoglucosan (Diebold and Bridgwater, 1997). Above 500 °C, the levoglucosan vaporizes, with negligible char formation, thus contributing mainly to gas and oil yields. Hemicelluloses are the least-stable components of wood, perhaps because of their lack of crystallinity (Reed, 2002, p. II-102).

Unlike cellulose, lignin decomposes over a broader temperature range of 280 to 500 °C, with the maximum release rate occurring at 350 to 450 °C (Kudo and Yoshida, 1957). Lignin pyrolysis produces more aromatics and char than produced by cellulose (Soltes and Elder, 1981). It yields about 40% of its weight as char under a slow heating rate at 400 °C (Klass, 1998). Lignin makes some contribution to the liquid yield (~35%), which contains aqueous components and tar. It yields phenols via cleavage of ether and carbon–carbon linkages (Mohan et al., 2006). The gaseous product of lignin pyrolysis is only about 10% of its original weight.

Particle Size

The composition, size, shape, and physical structure of the biomass exert some influence on the pyrolysis product through their effect on heating rate. Finer biomass particles offer less resistance to the escape of condensable gases, which therefore escape relatively easily to the surroundings before undergoing secondary cracking. This results in a higher liquid yield. Larger particles, on the other hand, facilitate secondary cracking due to the higher resistance they offer to the escape of the primary pyrolysis product. For this reason, older methods of charcoal production used stacks of large-size wood pieces in a sealed chamber (Figure 3.2).

3.3.2 Effect of Pyrolysis Temperature

During pyrolysis, a fuel particle is heated at a defined rate from the ambient to a maximum temperature, known as the *pyrolysis temperature*. The fuel is held there until completion of the process. The pyrolysis temperature affects both composition and yield of the product. Figure 3.6 is an example of how, during the pyrolysis of a biomass, the release of various product gases changes with different temperatures. We can see that the release rates vary widely for different gaseous constituents.

The amount of char produced also depends on the pyrolysis temperature. Low temperatures result in more char; high temperatures result in less. Figure 3.7 shows how the amount of char produced from the pyrolysis of a birch wood particle decreases with increasing temperature.

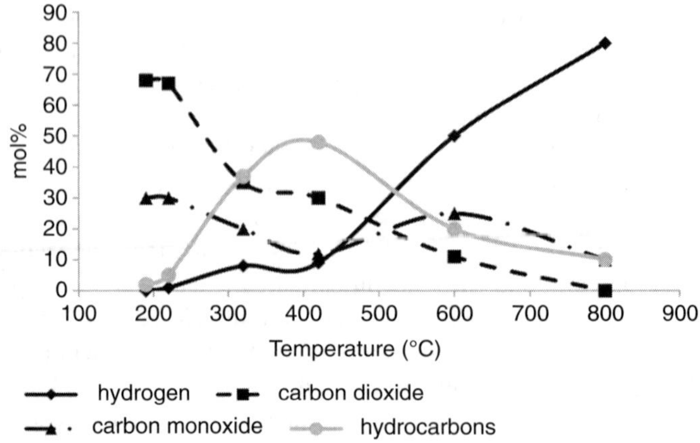

FIGURE 3.6 Release of gases during dry distillation of wood. *(Source: Based on data from Nikitin et al., 1962.)*

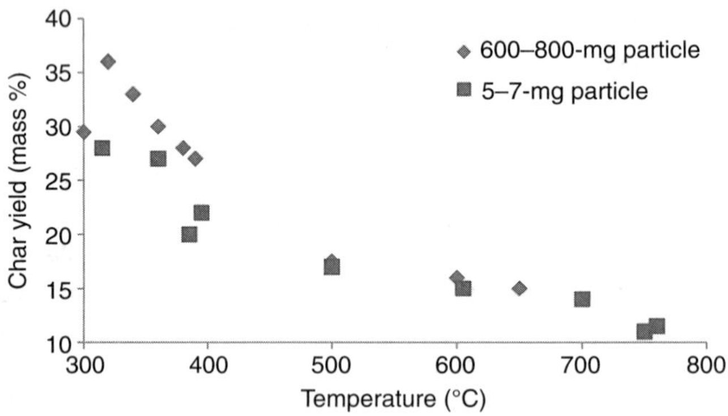

FIGURE 3.7 Char yield from pyrolysis decreases with increasing temperature. Data are for two sizes (or mass) of birch wood particle. *(Source: Adapted from Davidson, 2001.)*

3.3.3 Effect of Heating Rate

The rate of heating of the biomass particles has an important influence on the yield and composition of the product. Rapid heating to a moderate temperature (400–600 °C) yields higher volatiles and hence more liquid, while slower heating to that temperature produces more char. For example, Debdoubi et al. (2006) observed that, when the heating rate increased from 5 to 250 °C/min to 400 to 500 °C/min, the liquid yield from Esparto increased from 45 to 68.5%.

The heating rate alone, however, does not define the product. The residence time of the product in the reactor is also important. During slow heating, a slow or gradual removal of volatiles from the reactor permits a secondary reaction to occur between char particles and volatiles, leading to a secondary char formation.

The operating parameters of a pyrolyzer are adjusted to meet the requirement of the final product of interest. Tentative design norms for heating in a pyrolyzer include the following:

- To maximize char production, use a slow heating rate (<0.01–2.0 °C/s), a low final temperature, and a long gas residence time.
- To maximize liquid yield, use a high heating rate, a moderate final temperature (450–600 °C), and a short gas residence time.
- To maximize gas production, use a slow heating rate, a high final temperature (700–900 °C), and a long gas residence time.

Production of charcoal through carbonization uses the first norm. Fast pyrolysis uses the second to maximize liquid yield. The third norm is used when gas production is to be maximized.

3.4 PYROLYSIS KINETICS

A study of pyrolysis kinetics provides important information for the engineering design of a pyrolyzer or a gasifier. It also helps explain how the different processes in a pyrolyzer affect product yields and composition. Three major processes that influence the pyrolysis rate are chemical kinetics, heat transfer, and mass transfer. This section describes the physical and chemical aspects that govern the process.

3.4.1 Physical Aspects

From a thermal standpoint, we may divide the pyrolysis process into four stages. Although divided by temperature, the boundaries between them are not sharp; there is always some overlap:

Drying (~100 °C). During the initial phase of biomass heating at low temperature, the free moisture and some loosely bound water is released. The free moisture evaporates, and the heat is then conducted into the biomass interior (Figure 3.4). If the humidity is high, the bound water aids the melting of the lignitic fraction, which solidifies on subsequent cooling. This phenomenon is used in *steam bending* of wood, which is a popular practice for shaping it for furniture (Diebold and Bridgwater, 1997).

Initial Stage (100–300 °C). In this stage, exothermic dehydration of the biomass takes place with the release of water and low-molecular-weight gases like CO and CO_2.

Intermediate Stage (>200 °C). This is *primary pyrolysis*, and it takes place in the temperature range of 200 to 600 °C. Most of the vapor or precursor to bio-oil is produced at this stage. Large molecules of biomass particles decompose into char (*primary char*), condensable gases (vapors and precursors of the liquid yield), and noncondensable gases.

Final stage (~300–900 °C). The final stage of pyrolysis involves secondary cracking of volatiles into char and noncondensable gases. If they reside in the biomass long enough, relatively large-molecular-weight condensable gases can crack, yielding additional char (called *secondary char*) and gases. This stage typically occurs above 300 °C (Reed, 2002, p. III-6). The condensable gases, if removed quickly from the reaction site, condense outside in the downstream reactor as tar or bio-oil. It is apparent from Figure 3.6 that a higher pyrolysis temperature favors production of hydrogen, which increases quickly above 600 °C. An additional contribution of the shift reaction (Eq. 1.8) further increases the hydrogen yield above 900 °C.

Temperature has a major influence on the product of pyrolysis. The carbon dioxide yield is high at lower temperatures and decreases at higher temperatures. The release of hydrocarbon gases peaks at around 450 °C and then starts decreasing above 500 °C, boosting the generation of hydrogen.

Hot char particles can catalyze the primary cracking of the vapor released within the biomass particle and the secondary cracking occurring outside the particle but inside the reactor. To avoid cracking of condensable gases and thereby increase the liquid yield, rapid removal of the condensable vapor is very important. The shorter the residence time of the condensable gas in the reactor, the less the secondary cracking and hence the higher the liquid yield.

Some overlap of the stages in the pyrolysis process is natural. For example, owing to its low thermal conductivity (0.1–0.05 W/m.K), a large log of wood may be burning outside while the interior may still be in the drying phase, and water may be squeezed out from the ends. During a forest fire this phenomenon is often observed. The observed intense flame comes primarily from the combustion of the pyrolysis products released from the wood interior rather than from the burning of the exterior surface.

3.4.2 Chemical Aspects

As mentioned earlier, a typical biomass has three main components: (1) cellulose, (2) hemicellulose, and (3) lignin. These constituents have different rates of degradation and preferred temperature ranges of decomposition.

Cellulose

Decomposition of cellulose is a complex multistage process. A large number of models have been proposed to explain it. The Broido-Shafizadeh model (Bradbury et al., 1979) is the best-known and can be applied, at least qualitatively, to most biomass (Bridgwater et al., 2001).

Figure 3.8 is a schematic of the Broido-Shafizadeh model, according to which the pyrolysis process involves an intermediate prereaction (I) followed by two competing first-order reactions:

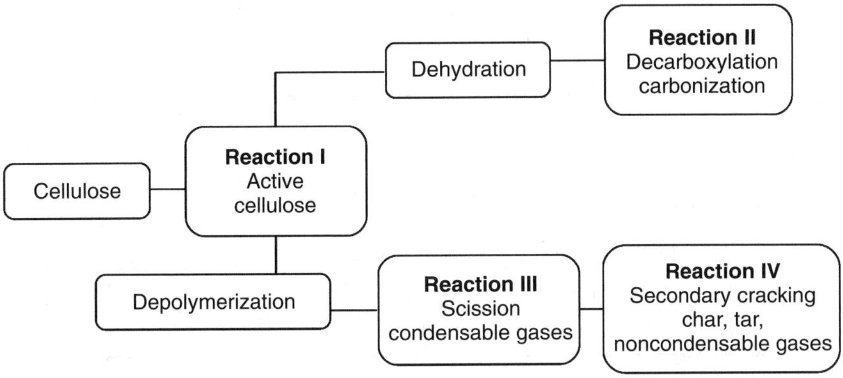

FIGURE 3.8 Modified Broido-Shafizadeh model of cellulose, which can be reasonably applied to the whole biomass.

- *Reaction II:* dehydration (dominates at low temperature and slow heating rates)
- *Reaction III:* depolymerization (dominates at fast heating rates)

Reaction II involves dehydration, decarboxylation, and carbonization through a sequence of steps to produce char and noncondensable gases like water vapor, carbon dioxide, and carbon monoxide. It is favored at low temperatures, of less than 300 °C (Soltes and Elder, 1981, p. 82), and slow heating rates (Reed, 2002, p. II-113).

Reaction III involves depolymerization and scission, forming vapors including tar and condensable gases. Levoglucosan is an important intermediate product in this path (Klass, 1998, p. 228), which is favored under faster heating rates (Reed, 2002, p. II-113) and higher temperatures of over 300 °C (Soltes and Elder, 1981, p. 82).

The condensable vapor, if permitted to escape the reactor quickly, can condense as bio-oil or tar. On the other hand, if it is held in contact with biomass within the reactor, it can undergo secondary reactions (reaction IV), cracking the vapor into secondary char, tar, and gases (Figure 3.8). Reactions II and III are preceded by reaction I, which forms a very short-lived intermediate product called *active cellulose* that is liquid at the reaction temperature but solid at room temperature (Boutin and Lédé, 2001; Bradbury et al., 1979; Bridgwater et al., 2001).

There are diverse views on the existence of reaction I, as this unstable species is not seen in the final product in most pyrolysis processes. It is, however, apparent in *ablative pyrolysis*, where wood is dragged over a hot metal surface to produce the feeling of smooth lubrication due to the presence of the intermediate liquid product, "active cellulose."

Depolymerization (reaction III) (Figure 3.8) has activation energies higher than those of dehydration (reaction II) (Bridgwater et al, 2001). Thus, a lower

Table 3.3 Rate Constants for Pyrolysis of Cellulose According to the Broido-Shafizadeh Model

Reaction: $\frac{dm}{dt} = A_i(V_i^3 - V_i)e^{-E_i/RT}$	A_i (s^{-1})	E_i (kJ/mol)
I—First degradation (active cellulose)[1]	2.8×10^{19}	243
II—Dehydration (char and gas)[1]	1.31×10^{10}	153
III—Depolymerization (tars)[1]	3.16×10^{14}	198
IV—Secondary cracking (Gas, Char)[2]	4.28×10^{6}	107.5

[1]Data excerpted from Bradbury et al., 1979.
[2]Data excerpted from Liden et al., 1988.

temperature and a longer residence time favor this reaction, producing primarily char, water, and carbon dioxide. On the other hand, owing to its higher activation energy, reaction III is favored at higher temperatures, fast heating rate, and longer residence times, yielding mainly gas. Moderate temperature and short vapor residence time avoid secondary cracking, producing mainly condensable vapor—the precursor of bio-oil, which is of great commercial importance. For cellulose pyrolysis, Table 3.3 gives some suggested reaction rate constants for reactions I, II, III, and IV.

The Broido-Shafizadeh model, though developed for one biomass component (cellulose), can be applied to the pyrolysis of an entire biomass such as wood. If a log of wood is heated very slowly, it shows glowing ignition, because reaction II predominates under this condition, producing mostly char, which ignites in contact with air without a yellow flame. If the wood is heated faster, it burns with a yellow flame, because at a higher heating rate, reaction III predominates, producing more vapors or tar, both of which burn in air with a bright flame.

Hemicellulose

Hemicellulose produces more gas and less tar but also produces less char in comparison to cellulose. However, it produces the same amount of aqueous product of pyroligneous acid (Soltes and Elder, 1981, p. 84). Hemicellulose undergoes rapid thermal decomposition (Demirbas, 2000), which starts at a temperature lower than that for cellulose or lignin. It contains more combined moisture than lignin has, and its softening point is lower as well. The exothermic peak of hemicellulose appears at a temperature lower than that for lignin (Demirbas, 2000). In slow pyrolysis of wood, hemicellulose pyrolysis begins at 130 to 194 °C, with most of the decomposition occurring above 180 °C (Mohan et al., 2006, p. 126).

Lignin

Pyrolysis of lignin typically produces about 55% char (Soltes and Elder, 1981), 15% tar, 20% aqueous components (pyroligneous acid), and about 12% gases. It is more difficult to dehydrate lignin than cellulose or hemicellulose (Mohan, 2006, p. 127). The tar produced from it contains a mixture of phenolic compounds, one of which, phenol, is an important raw material of green resin (a resin produced from biomass). The aqueous portion comprises methanol, acetic acid, acetone, and water. The decomposition of lignin in wood can begin at 280 °C, continuing to 450 to 500 °C, and can reach a peak rate at 350 to 450 °C (Kudo and Yoshida, 1957).

3.4.3 Kinetic Models of Pyrolysis

To optimize the process parameters and maximize desired yields, knowledge of the kinetics of pyrolysis is important. However, it is very difficult to obtain reliable data of kinetic rate constants that can be used for a wide range of biomass and for different heating rates. This is even more difficult for fast pyrolysis as it is a nonequilibrium and nonsteady state process. For engineering design purposes, a "black-box" approach can be useful, at least for the first approximation. The following discussion presents a qualitative understanding of the process based on data from relatively slow heating rates.

Kinetic models of the pyrolysis of ligno-cellulosic fuels like biomass may be broadly classified into three types (Blasi, 1993):

One-stage global single reaction. The pyrolysis is modeled by a one-step reaction using experimentally measured weight-loss rates.

One-stage, multiple reactions. Several parallel reactions are used to describe the degradation of biomass into char and several gases. A one-stage simplified kinetic model is used for these parallel reactions. It is useful for determination of product distribution.

Two-stage semiglobal. This model includes both primary and secondary reactions, occurring in series.

One-Stage Global Single-Reaction Model

This reaction model is based on a single overall reaction:

$$\text{Biomass} \rightarrow \text{Volatile} + \text{Char}$$

The rate of pyrolysis depends on the unpyrolyzed mass of the biomass. Thus, the decomposition rate of mass, m_b, in the primary pyrolysis process may be written as

$$\frac{dm_b}{dt} = -k\,(m_b - m_c) \tag{3.2}$$

Here, m_c is the mass of char remaining after complete conversion (kg), k is the first-order reaction rate constant (s^{-1}), and t is the time (s).

The fractional change, X, in the mass of the biomass may be written in nondimensional form as

$$X = \frac{(m_0 - m_b)}{(m_0 - m_c)} \tag{3.3}$$

where m_0 is the initial mass of the biomass (kg).

Substituting fractional conversion for the mass of biomass in Eq. (3.2),

$$\frac{dX}{dt} = k(1 - X) \tag{3.4}$$

Solving this equation we get

$$X = 1 - A\exp(-kt) \tag{3.5}$$

where A is the pre-exponential coefficient, E is the activation energy (J/mol), R is the gas constant (J/mol.K), and T is the temperature (K).

Owing to the difficulties in extracting data from dynamic thermogravimetric analysis, reliable data on the pre-exponential factor A and the activation energy E are not easily available for fast pyrolysis (Reed, 2002, p. II-103). However, for slow heating we can obtain some reasonable values. If the effect of secondary cracking and the heat-transfer limitation can be restricted, the weight-loss rate of pure cellulose during pyrolysis can be represented by an irreversible, one-stage global first-order equation.

Table 3.4 lists values of the activation energy E and the pre-exponential factor A, according to the one-step global reaction model for the pyrolysis of various biomass types at a relatively slow heating rate.

Table 3.4 Kinetic Rate Constants for the One-Step Single-Reaction Global Model

Fuel	Temperature (K)	E (kJ/mol)	A (s^{-1})	Reference
Cellulose	520–1270	166.4	3.9×10^{11}	Lewellen et al., 1977
Hemicellulose	520–1270	123.7	1.45×10^{9}	Min, 1977
Lignin	520–1270	141.3	1.2×10^{8}	Min, 1977
Wood	321–720	125.4	1.0×10^{8}	Nolan et al., 1972
Almond shell	730–880	95–121	1.8×10^{6}	Font et al., 1990
Beech sawdust	450–700	84 ($T > 600$K)	2.3×10^{4}	Barooah and Long, 1976

Other models are not discussed here, but details are available in several publications, including Blasi (1993).

3.5 HEAT TRANSFER IN A PYROLYZER

The preceding discussions assume that the heat or mass transport rate is too high to offer any resistance to the overall rate of pyrolysis. This is true in the temperature range of 300 to 400 °C (Thurner and Mann, 1981), but at higher temperatures heat and mass transport influence the overall rate and so cannot be neglected. This section deals with heat transport during pyrolysis.

During pyrolysis, heat is transported to the particle's outer surface by radiation and convection. Thereafter, it is transferred to the interior of the particle by conduction and pore convection (Figure 3.4). The following modes of heat transfer are involved in this process (Babu and Chaurasia, 2004b).

- Conduction inside the particle
- Convection inside the particle pores
- Convection and radiation from the particle surface

In a commercial pyrolyzer or gasifier, the system heats up a heat-transfer medium first; that, in turn, transfers the heat to the biomass. The heat-transfer medium can be one or a combination of the following:

- Reactor wall (for vacuum reactor)
- Gas (for entrained-bed or entrained-flow reactor)
- Heat-carrier solids (for fluidized bed)

Bubbling fluidized beds use mostly solid–solid heat transfer. Circulating fluidized beds and transport reactors make use of gas-solid heat transfer in addition to solid–solid heat transfer.

Since heat transfer to the interior of the biomass particle is mostly by thermal conduction, the low thermal conductivity of biomass (~0.1 W/m.K) is a major deterrent to the rapid heating of its interior. For this reason, even when the heating rate of the particle's exterior is as fast as 10,000 °C/s, the interior can be heated at a considerably slower rate for a coarse particle. Because of the associated slow heating of the interior, the secondary reactions within the particles become increasingly important as the particle size increases, and as a result the liquid yield reduces (Scott and Piskorz, 1984). For example, Shen et al. (2009) noted that oil yield decreased with particle size within the range of 0.3 to 1.5 mm, but no effect was noted when the size was increased to 3.5 mm. Experimental results (Seebauer et al., 1997), however, do not show much effect of particle size on the biomass.

3.5.1 Mass Transfer Effect

Mass transfer can influence the pyrolysis product. For example, a sweep of gas over the fuel quickly removes the products from the pyrolysis environment.

Thus, secondary reactions such as thermal cracking, repolymerization, and recondensation are minimized (Sensoz and Angin, 2008).

3.5.2 Is Pyrolysis Autothermal?

An important question for designers is whether a pyrolyzer can meet its own energy needs or is dependent on external energy. The short and tentative answer is that a pyrolyzer as a whole is not energy self-sufficient. The reaction heat is inadequate to meet all energy demands, which include heat required to raise the feed and any inert heat-transfer media to the reaction temperature, heat consumed by endothermic reactions, and heat losses from the reactor. In most cases it is necessary to burn the noncondensable gases and the char produced to provide the heat required. If that is not adequate, other heat sources are necessary to supply the energy required for pyrolysis. The following section discusses the heat requirement of reactions taking place in a pyrolyzer.

The dehydration (reaction II) process is exothermic, while depolymerization (reaction III) and secondary cracking (reaction IV) are endothermic (Bridgwater et al., 2001). Among reactions between intermediate products of pyrolysis, some are exothermic and some are endothermic. In general, pyrolysis of hemicellulose and lignin is exothermic. Cellulose pyrolysis is endothermic at lower temperatures (<400–450 °C), and it becomes exothermic at higher temperatures owing to the following exothermic reactions (Klass, 1998).

$$CO + 3H_2 \rightarrow CH_4 + H_2O - 226\ \text{kJ/gmol} \tag{3.6}$$

$$CO + 2H_2 \rightarrow CH_3OH - 105\ \text{kJ/gmol} \tag{3.7}$$

$$0.17\ C_6H_{10}O_5 \rightarrow C + 0.85\ H_2O - 80\ \text{kJ/gmol} \tag{3.8}$$

$$CO + H_2O \rightarrow CO_2 + H_2 - 42\ \text{kJ/gmol} \tag{3.9}$$

(All equations refer to a temperature of 1000 K, and $C_6H_{10}O_5$ represents the cellulose monomer.)

For this reason a properly designed system initially requires external heat only until the required temperature is reached.

Char production from cellulose (Eq. 3.8) is slightly exothermic. However, at a higher temperature, when sufficient hydrogen is produced by reaction (Eq. 3.9), other exothermic reactions (Eqs. 3.6 and 3.7) can proceed. At low temperatures and short residence times of volatiles, only endothermic primary reactions are active (heat of reaction –225 kJ/kg), while at high temperatures exothermic secondary reactions (heat of reaction 20 kJ/kg) are active (Blasi, 1993).

In conclusion, for design purposes one may neglect the heat of reaction for the pyrolysis process, but it is necessary to calculate the energy required for vaporization of products and for heating feedstock gases to the pyrolysis temperature (Boukis et al., 2007).

3.6 PYROLYZER TYPES

Pyrolyzers have been used since ancient times to produce charcoal (Figure 3.2). Early pyrolyzers operated in batch mode using a very slow rate of heating and for long periods of reaction to maximize the production of char. If the objective of pyrolysis was to produce the maximum amount of liquid or gas, then the rate of heating, the peak pyrolysis temperature, and the duration of pyrolysis had to be chosen accordingly. These choices also decided what kind of reactor was to be used. Table 3.5 shows how the yield is maximized for different choices of heating rate, temperature, and gas residence time.

Modern pyrolyzers are more concerned with gas and liquid products, and require a continuous process. A number of different types of pyrolysis reactor have been developed. Based on the gas–solid contacting mode, they can be broadly classified as fixed bed, fluidized bed, and entrained bed, and then further subdivided depending on design configuration. The following are some of the major pyrolyzer designs in use:

- Fixed or moving bed
- Bubbling fluidized bed
- Circulating fluidized bed
- Ultra-rapid
- Rotating cone
- Ablative
- Vacuum

Except for the moving bed other gasifier types are illustrated in Figure 3.9.

3.6.1 Fixed-Bed Pyrolyzer

Fixed-bed pyrolysis, operating in batch mode, is the oldest pyrolyzer type. Heat for the thermal decomposition of biomass is supplied either from an external source or by allowing limited combustion as in a beehive oven (Figure 3.2).

Table 3.5 Effect of Operating Variables on Pyrolysis Yield

Maximium Yield	Maximum Temperature	Heating Rate	Gas Residence Time
Char	Low	Slow	Long
Liquid	Low (~500 °C[1])	High	Short
Gas	High	Low	Long

[1]Data excerpted from Bridgwater, 1999.
Source: Compiled from data in Demirbas, 2001.

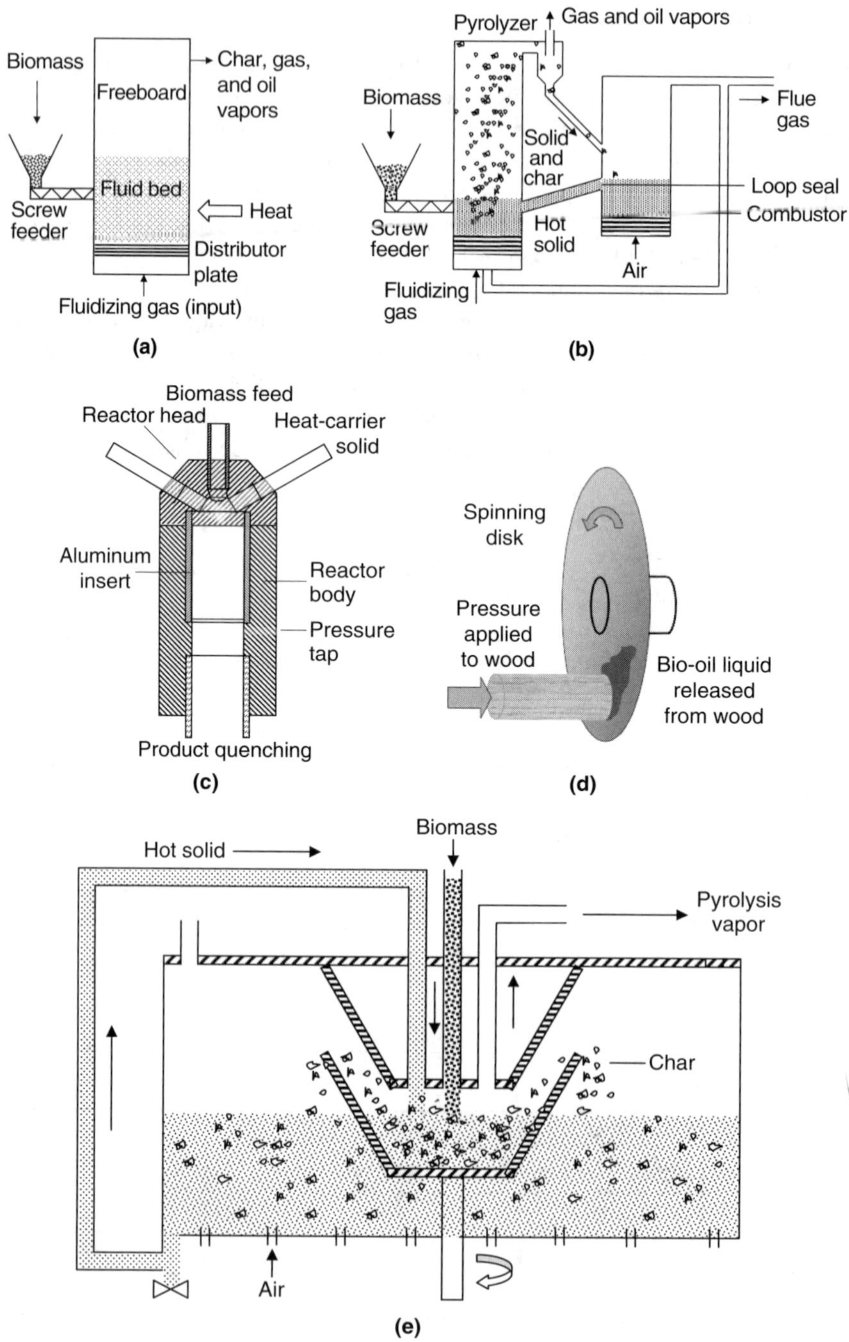

FIGURE 3.9 A variety of pyrolyzer designs: (a) bubbling fluidized bed, (b) circulating fluidized bed, (c) ultra-rapid, (d) ablative, (e) rotating cone, and

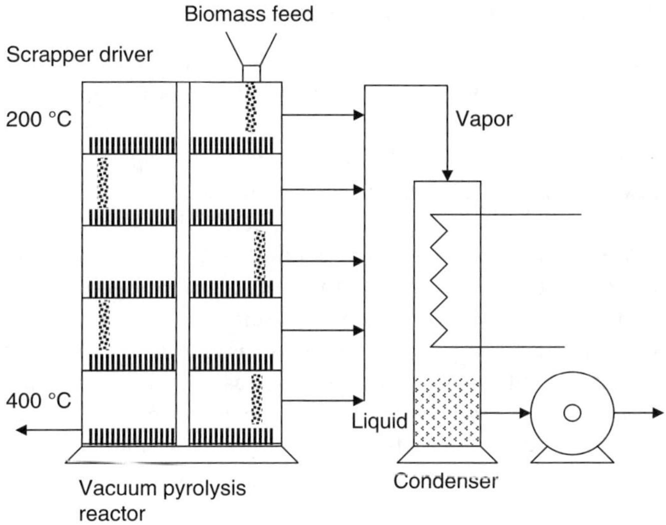

(f)

FIGURE 3.9 *Continued* (f) vacuum.

The product may flow out of the pyrolyzer because of volume expansion while the char remains in the reactor. In some designs a sweep gas is used for effective removal of the product gas from the reactor. This gas is necessarily inert without oxygen. The main product of this type is char owing to the relatively slow heating rate and the long residence time of the product in the pyrolysis zone.

3.6.2 Bubbling-Bed Pyrolyzer

Figure 3.9(a) shows a bubbling fluidized-bed pyrolyzer. Crushed biomass (2–6 mm) is fed into a bubbling bed of hot sand. The bed is fluidized by an inert gas such as recycled flue gas. Intense mixing of inert bed solids (sand is commonly used) offers good and uniform temperature control. It also provides high heat transfer to biomass solids. The residence time of the solids is considerably higher than that of the gas in the pyrolyzer.

The required heat for pyrolysis may be provided either by burning a part of the product gas in the bed, as shown in Figure 3.5, or by burning the solid char in a separate chamber and transferring that heat to the bed solids (Figure 3.9b). The pyrolysis product would typically contain about 70 to 75% liquid on dry wood feed. As shown in the figure, the char in the bed solids acts as a vapor-cracking catalyst, so its separation through elutriation or otherwise is important if the secondary cracking is to be avoided to maximize the liquid product. The entrained char particles are separated from the product gas using single- or multistage cyclones. A positive feature of a bubbling fluidized-bed pyrolyzer is that it is relatively easy to scale up.

3.6.3 Circulating Fluidized-Bed Pyrolyzer

A circulating fluidized-bed (CFB) pyrolyzer, shown in Figure 3.9(b), works on the same principle as the bubbling fluidized bed except that the bed is highly expanded and solids continuously recycle around an external loop comprising a cyclone and loop seal. The bed operates in a special hydrodynamic regime known as *fast bed.* It provides good temperature control and is uniform around the entire height of the unit. The superficial gas velocity in a CFB is considerably higher than that in a bubbling bed. High velocity combined with excellent mixing allows a CFB to have large throughputs of biomass. Here, gas and solids move up the reactor with some degree of internal refluxing. As a result, the residence time of average biomass particles is longer than that of the gas, but the difference is not as high as it is in a bubbling bed. A major advantage of this system is that char entrained from the reactor is easily separated and burnt in an external fluidized bed. The combustion heat is transferred to the inert bed solids that are recycled to the reactor by means of a loop seal.

Rapid thermal pyrolysis (RTP), a commercial process developed by Ensyn, probably originated from the ultra-rapid fluidized-bed pyrolyzer developed at the University of Western Ontario in Canada. RTP uses a riser reactor. Here, biomass is introduced into a vessel and rapidly heated to 500 °C by a tornado of upflowing hot sand; it is then cooled within seconds. The heating rate is on the order of 1000 °C/s, and the reactor residence time is from a few hundredths of a millisecond to a maximum of 5 seconds, which gives a liquid yield as high as 83% for wood (Hulet et al., 2005).

3.6.4 Ultra-Rapid Pyrolyzer

High heating rate and short residence time in the pyrolysis zone are two key requirements of high liquid yield. The ultra-rapid pyrolyzer, shown in Figure 3.9(c), developed by the University of Western Ontario provides extremely short mixing (10–20 ms), reactor residence (70–200 ms), and quench (~20 ms) times. Because the reactor temperature is also low (~650 °C), one can achieve a liquid yield as high as 90% (Hulet et al., 2005). The inert gas nitrogen is heated 100 °C above the reactor temperature and injected at very high velocity into the reactor to bombard a stream of biomass injected in the reactor. The reactor can also use a heat-carrier solid like sand that is heated externally and bombarded on a biomass stream through multiple jets. Such a high-velocity impact in the reactor results in an exceptionally high heating rate. The biomass is thus heated to the pyrolysis temperature in a few milliseconds. The pyrolysis product leaves the reactor from the bottom and is immediately cooled to suppress a secondary reaction or cracking of the oil vapor. This process is therefore able to maximize the liquid yield during pyrolysis.

3.6.5 Ablative Pyrolyzer

This process, shown in Figure 3.9(d), involves creation of high pressure between a biomass particle and a hot reactor wall. This allows uninhibited heat transfer from the wall to the biomass, causing the liquid product to melt out of the biomass the way frozen butter melts when pressed against a hot pan. The biomass sliding against the wall leaves behind a liquid film that evaporates and leaves the pyrolysis zone, which is the interface between biomass and wall. As a result of high heat transfer and short gas residence time, a liquid yield as high as 80% is reported (Diebold and Power, 1988). The pressure between biomass and wall is created either by mechanical means or by centrifugal force. In a mechanical system a large piece of biomass is pressed against a rotating hot plate.

3.6.6 Rotating-Cone Pyrolyzer

In this process, biomass particles are fed into the bottom of a rotating cone (360–960 rev/min) together with an excess of heat-carrier solid particles (see Figure 3.9(e)). Centrifugal force pushes the particles against the hot wall; the particles are transported spirally upward along the wall. Owing to its excellent mixing, the biomass undergoes rapid heating (5000 K/s) and is pyrolyzed within the small annular volume. The product gas containing bio-oil vapor leaves through another tube, while the solid char and sand spill over the upper rim of the rotating cone into a fluidized bed surrounding it, as shown in Figure 3.9(e). The char burns in the fluidized bed, and this combustion helps heat the cone as well as the solids that are recycled to it to supply heat for pyrolysis. Special features of this reactor include very short solids residence time (0.5 seconds) and a small gas-phase residence time (0.3 seconds). These typically provide a liquid yield of 60 to 70% on dry feed (Hulet et al., 2005). The absence of a carrier gas is another advantage of this process. The complex geometry of the system may raise some scale-up issues.

3.6.7 Vacuum Pyrolyzer

A vacuum pyrolyzer, as shown in Figure 3.9(f), comprises a number of stacked heated circular plates. The top plate is at about 200 °C while the bottom one is at about 400 °C. Biomass fed to the top plate drops into successive lower plates by means of scrapers. The biomass undergoes drying and pyrolysis while moving over the plates. No carrier gas is required in this pyrolyzer. Only char is left when the biomass reaches the lowest plate. Though the heating rate of the biomass is relatively slow, the residence time of the vapor in the pyrolysis zone is short. As a result, the liquid yield in this process is relatively modest, about 35 to 50% on dry feed, with a high char yield. This pyrolyzer design is complex, especially given the fouling potential of the vacuum pump.

3.7 PYROLYZER DESIGN CONSIDERATIONS

This section discusses pyrolyzer design considerations in the production of liquid fuel and charcoal through pyrolysis.

3.7.1 Production of Liquid through Pyrolysis

Pyrolysis is one of several means of production of liquid fuel from biomass. The maximum yield of organic liquid (pyrolytic oil or bio-oil) from thermal decomposition may be increased to as high as 70% (dry weight) if the biomass is rapidly heated to an intermediate temperature and if a short residence time in the pyrolysis zone is allowed to reduce secondary reactions. Table 3.2 earlier in the chapter shows how heating rate, pyrolysis temperature, and residence time affect the nature of the pyrolysis product. These findings may be summarized as follows:

- A slower heating rate, a lower temperature, and a longer residence time maximize the yield of solid char.
- A higher heating rate, a higher temperature, and a shorter residence time maximize the gas yield.
- A higher heating rate, an intermediate temperature, and a shorter residence time maximize the liquid yield.

There is an optimum pyrolysis temperature for maximum liquid yield. The yield is highest at 500 °C and drops sharply above and below this temperature (Boukis et al., 2007). The residence time is generally in the range of 0.1 to 2.0 seconds. These values depend on several factors, including the type of biomass (Klass, 1998). We can use a kinetic model for a reasonable yield assessment. The one proposed by Liden et al. (1988) is successful in predicting pyrolysis liquid yields over a wide range of conditions.

Heat transfer is a major consideration in the design of a pyrolyzer. The heat balance for a typical pyrolyzer may be written as

$$\begin{aligned}&[\text{Heat released by char combustion}]+[\text{Heat in incoming stream}]\\&\quad=[\text{Heat required for pyrolysis}]+[\text{Heat loss}]\end{aligned} \tag{3.10}$$

Assessing heat loss accurately is difficult before the unit is designed. So, for preliminary assessment, we can take this to be 10% of the heat in the incoming stream (Boukis et al., 2007, p. 1377).

Fast, or flash, pyrolysis is especially suitable for pyrolytic liquefaction of biomass. The product is a mixture of several hydrocarbons, which allows production of fuel and chemicals through appropriate refining methods. The heating value of the liquid produced is in the same range (15–19 MJ/kg) as that of the parent biomass. The pyrolytic liquid contains several water-soluble sugars and polysaccharide-derivative compounds and water-insoluble pyrolytic lignin.

Pyrolytic liquid contains a much higher amount of oxygen (~50%) than does most fuel oil. It is also heavier (specific gravity ~1.3) and more viscous. Unlike fuel oil, pyrolytic oil increases in viscosity with time because of polymerization. This oil is not self-igniting like fuel oil, and as such it cannot be blended with diesel for operating a diesel engine.

Pyrolytic oil is, however, a good source of some useful chemicals, like natural food flavoring, that can be extracted, leaving the remaining product for burning. Alternately, we can subject the pyrolytic oil to hydrocracking to produce gasoline and diesel.

3.7.2 Production of Charcoal through Pyrolysis

Carbon is a preferred product of biomass pyrolysis at a moderate temperature. Thermodynamic equilibrium calculation shows that the char yield of most biomass may not exceed 35%. Table 3.6 gives the theoretical equilibrium yield of biomass at different temperatures. Assuming that cellulose represents biomass, the stoichiometric equation for production of charcoal (Antal, 2003) may be written as

$$C_6H_{10}O_5 \rightarrow 3.74C + 2.65H_2O + 1.17CO_2 + 1.08CH_4 \qquad (3.11)$$

Charcoal production from biomass requires slow heating for a long duration but at a relatively low temperature of around 400 °C. An extreme example of a pyrolysis or carbonization is in the coke oven in an iron and steel plant, which pyrolyzes (carbonizes) coking coal to produce hard coke used for iron extraction. This is an indirectly heated fixed-bed pyrolyzer that operates at a temperature exceeding 1000 °C and for a long period of time to maximize gas and solid coke production.

Table 3.6 Thermodynamic Equilibrium Concentration of Cellulose Pyrolysis at Different Temperatures

Product	Temperature (°C)				
(%)	200	300	400	500	600
C	32	28	27	27	25.2
H_2O	36.5	32.5	9.5	27	22.5
CH_4	8.5	10	10.5	10	9
CO_2	23.9	28	32	35	36
CO	0	0	0.1	1.2	4.5

Source: Derived from data in Antal, 2003.

FIGURE 3.10 Comparison of several biomass species before and after torrefaction. Top row: raw biomass, rice husk, sawdust, peanut husk, bagasse, and water-hyacinth; bottom row: the same species after torrefaction. *(Source: Adapted from previous work of the author.)*

3.8 TORREFACTION

Torrefaction, a process different from carbonization, is a mild pyrolysis process carried out in a temperature range of 230 to 300 °C in the absence of oxygen. This thermal pretreatment of biomass improves its energy density, reduces its oxygen-to-carbon (O/C) ratio, and reduces its hygroscopic nature. During this process the biomass dries and partially devolatilizes, decreasing its mass while largely preserving its energy content. The torrefaction process removes H_2O and CO_2 from the biomass. As a result, both the O/C and the H/C ratios of the biomass decrease. In raw biomass, high oxygen content prompts its over-oxidation during gasification, increasing the thermodynamic losses of the process. Torrefaction could reduce this loss by reducing the oxygen in the biomass. Torrefaction also increases the relative carbon content of the biomass. The properties of a torrefied wood depends on torrefaction temperature, time, and on the type of wood feed.

A popular example of torrefaction is the process of roasting coffee beans. As the green beans are heated to 200 to 300 °C, their surface darkens (*www.coffeeresearch.org/coffee*). Figure 3.10 contains photographs of rice husk, peanut husk, bagasse, and water hyacinth before and after torrefaction. The color change is present in all biomass but to different degrees.

Torrefaction also modifies the structure of the biomass, making it more friable or brittle. This is caused by the depolymerization of hemicellulose. As a result, the process of size reduction becomes easier, lowering its energy consumption and the cost of handling. This makes it easier to co-fire biomass in a pulverized-coal fired boiler or gasify it in an entrained-flow reactor.

Torrefaction causes some reduction in the energy content of the biomass because of partial devolatilization, but given the much higher reduction in mass, the energy density of the biomass increases. Table 3.7 shows an example of torrefaction. Here, we note that by losing only 11 to 17% energy, the biomass

Table 3.7 Changes in the Bagasse Properties after Torrefaction at 250 °C

	Torrefaction Time (min)		
Property	**15**	**30**	**45**
Mass yield (%)	69	68.33	62
Energy yield (%)	88.86	91.06	83.23
Energy density (% energy yield/% mass yield)	1.29	1.33	1.34
Energy required (MJ/kg product)	2.34	2.58	2.99
Higher heating value (HHV) (MJ/kg product)	19.88	20.57	20.72
Rise in HHV (%)	22.35	24.96	25.51
HHV (MJ/kg raw material)	15.44	15.44	15.44
Net energy (MJ/kg product)	17.54	17.99	17.73

Note: Moisture absorption after 2 hr in water: Raw bagasse, 186%; torrified bagasse, 7.63%.
Source: Adapted from Pimchua et al., 2009.

(bagasse) lost 31 to 38% of its original mass. Thus, there is a 29 to 33% increase in energy density (energy per unit mass) of the biomass. This increases its higher heating value (HHV) to about 20 MJ/kg. Even if we take into account the energy used in the torrefaction process, we can see from Table 3.7 that there is a net rise in the energy density of the fuel.

Another special feature of torrefaction is that it reduces the hygroscopic property of biomass; therefore, when torrefied biomass is stored, it absorbs less moisture than that absorbed by fresh biomass. For example, while raw bagasse absorbed 186% moisture when immersed in water for two hours, it absorbed only 7.6% moisture under this condition after torrefying the bagasse for 60 minutes at 250 °C (Pimchua et al., 2009). The reduced hygroscopic (or enhanced hydrophobic) nature of torrefied biomass mitigates one of the major shortcomings for energy use of biomass.

3.8.1 Advantages of Torrefaction

Torrefied wood performs better than original wood (or another biomass) in both gasification and combustion. Major features and advantages of torrefaction are as follows:

- It increases the O/C ratio of the wood, which improves its gasification efficiency.
- It reduces power requirements for size reduction, and improves handling.
- It offers cleaner-burning fuel with little acid in the smoke.

- A fuel gas that has an enhanced heating value may be obtained through gasification.
- Torrefied wood absorbs less moisture when stored.
- One can produce superior-quality biomass pellets with higher volumetric energy density.

Based on these features, we can easily speculate that torrefaction will allow biomass to be used in entrained-flow gasification, direct combustion in a PC-fired boiler, and production of biopellet.

3.8.2 Mechanism of Torrefaction

In biomass, hemicellulose is like the cement in reinforced concrete, and cellulose is like the steel rods. The strands of microfibrils (cellulose) are supported by the hemicellulose. Decomposition of hemicellulose during torrefaction is like the melting away of the cement from the reinforced concrete. Thus, the size reduction of biomass consumes less energy after torrefaction.

During torrefaction the weight loss of biomass comes primarily from the decomposition of its hemicellulose constituents. Hemicellulose decomposes mostly within the temperature range 150 to 280 °C, which is the temperature window of torrefaction. As we can see from Figure 3.11, the hemicellulose component undergoes the greatest amount of degradation within the 200 to 300 °C temperature window. Lignin, the binder component of biomass, starts softening above its glass-softening temperature (~130 °C), which helps densification (pelletization) of torrefied biomass. Unlike hemicellulose, cellulose

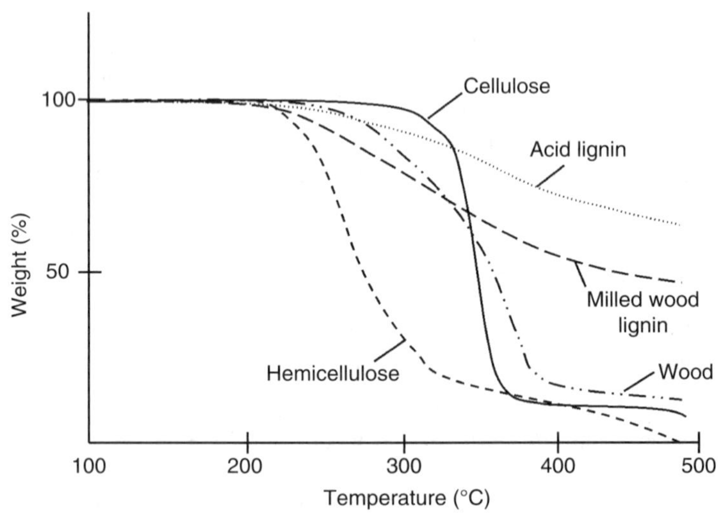

FIGURE 3.11 Weight loss in wood cellulose, hemicellulose, and lignin during torrefaction.

shows limited devolatilzation and carbonization and that too does not start below 250 °C.

Thus, hemicellulose decomposition is the primary mechanism of torrefaction. At lower temperatures (< 160 °C), as biomass dries it releases H_2O and CO_2. Water and carbon dioxide, which make no contribution to the energy in the product gas, constitute a dominant portion of the weight loss during torrefaction. Above 180 °C, the reaction becomes exothermic, releasing gas with small heating values. The initial stage (< 250 °C) involves hemicellulose depolymerization, leading to an altered and rearranged polysugar structures (Bergman et al., 2005a). At higher temperatures (250–300 °C) these form chars, CO, CO_2, and H_2O. The hygroscopic property of biomass is partly lost in torrefaction because of the destruction of OH groups through dehydration, which prevents the formation of hydrogen bonds.

3.8.3 Design Considerations for Torrefaction

In a typical torrefaction process the biomass is heated gently to the desired torrefaction temperature (θ_{tor}), held there for a specified reaction time, and then cooled down. The torrefaction temperature and the reaction time are two of the most important parameters in this process. The torrefaction temperature θ_{tor} generally reduces with reaction time $t_{heating}$.

The design norm for torrefaction is

$$\begin{gathered} 200\,°\mathrm{C} < \theta_{tor} < 300\,°\mathrm{C} \\ \frac{\theta_{tor} - 200}{t_{heating}} < 1\,°\mathrm{C/s} \end{gathered} \tag{3.12}$$

where θ_{tor} is the torrefaction temperature in °C, and $t_{heating}$ is the heating time above 200 °C. A typical reaction time is about 30 minutes. The properties of torrefied wood depend on (1) the type of wood, (2) the reaction temperature, and (3) the reaction time.

Torrefaction loses more oxygen and hydrogen than carbon. Hence, the H/C and O/C ratios decrease. However, it should not be confused with carbonization, which takes place at a much higher temperature and produces charcoal with even lower H/C and O/C ratios.

3.8.4 Torrefied Pellet

The pelletizing process resolves some typical problems of biomass fuels: transport and storing costs are minimized, handling is improved, and the volumetric calorific value is increased. Pelletization may not increase the energy density on a mass basis, but it can increase the energy content of the fuel on a volume basis. For example, while the energy density on a mass basis for raw wood, torrefied wood, wood pellet, and torrefied pellet was 10.5, 19.9, 16.2, and

21.6 kJ/kg (LHV as received basis), respectively, it was 5.0, 4.6, 10.5, and 18.4 GJ/m^3, respectively, on a volume basis (Bergman, 2005c). Thus, pelletization of torrefied wood greatly increases the transportation and handling cost of biomass. Pelletization of torrefied biomass is better than torrefaction of pelletized wood from the standpoint of process energy consumption and product stability.

Symbols and Nomenclature

A = pre-exponential factor (s^{-1})
E = activation energy (J/mol)
k = reaction rate (s^{-1})
m_b = mass of biomass at time t (kg)
m_c = mass of char residue (kg)
m_o = initial mass of biomass (kg)
R = universal gas constant (J/mol.K)
T = temperature (K)
T_{pyr} = pyrolysis temperature (K)
$t_{heating}$ = heating time (s)
t_r = reaction time (s)
X = fractional change in mass of biomass
θ_{tor} = torrefaction temperature (°C)

Chapter 4

Tar Production and Destruction

4.1 INTRODUCTION

Tar is a major nuisance in both gasification and pyrolysis. It is a thick, black, highly viscous liquid that condenses in the low-temperature zones of a gasifier, clogging the gas passage and leading to system disruptions. Tar is highly undesirable, as it can create the following problems:

- Condensation and subsequent plugging of downstream equipment
- Formation of tar aerosols
- Polymerization into more complex structures

Nevertheless, tar is an unavoidable by-product of the thermal conversion process. This chapter discusses what tar is, how it is formed, and how to influence its formation such that plants and equipment can live with this "necessary evil" while minimizing its detrimental effects.

4.2 BASICS OF TAR

Tar is a complex mixture of condensable hydrocarbons, including, among others, oxygen-containing, 1- to 5-ring aromatic, and complex polyaromatic hydrocarbons (Devi et al., 2003). Neeft et al. (1999) defined tar as "all organic contaminants with a molecular weight larger than 78, which is the molecular weight of benzene." The International Energy Agency (IEA) Bioenergy Agreement, the U.S. Department of Energy (DOE), and the DGXVII of the European Commission agreed to identify all components of product gas having a molecular weight higher than benzene as tar (Knoef, 2005, p. 278).

A common perception about tar is that it is a product of gasification and pyrolysis that can potentially condense in colder downstream sections of the unit. While this is a fairly good description, a more specific and scientific definition may be needed for technical, scientific, and legal work. Presently, there is no universally accepted definition of tar. As many as 30 definitions are available in the literature (Knoef, 2005, p. 279). Of these, that of the IEA's gasification task force, as follows, appears most appropriate (Milne et al., 1998):

Biomass Gasification and Pyrolysis. DOI: 10.1016/B978-0-12-374988-8.00004-0

The organics, produced under thermal or partial-oxidation regimes (gasification) of any organic material, are called "tar" and are generally assumed to be largely aromatic.

4.2.1 Acceptable Limits for Tar

Tar remains vaporized until the gas carrying it is cooled, when it either condenses on cool surfaces or remains in fine aerosol drops (<1 micron). This makes the product gas unsuitable for use in gas engines, which have a low tolerance for tar. Thus, there is a need for tar reduction in product gas when the gas is to be used in an engine. This can be done through appropriate design of the gasifier and the right choice of operating conditions, including reactor temperature and heating rate. Even these adjustments may not reduce tars in the gas to the required level, necessitating further downstream cleanup.

Standard gas cleaning involves filtration and/or scrubbing, which not only removes tar but also strips the gas of particulate matters and cools it to room temperature. These practices clean the gas adequately, making it acceptable to most gas engines. However, they result in a great reduction in overall efficiency in the production of electricity or mechanical power using a gas engine. Furthermore, gas cleaning greatly adds to the capital investment of the plant.

Biomass gasification is at times used for distributed power generation in remote locations in small- to medium-capacity plants. For such plants, the addition of a scrubber or a filtration system significantly increases the overall plant costs. This limitation makes biomass-based distributed power-generation projects highly sensitive to the cost of tar cleanup.

The presence of tar in the product gas from gasification can potentially decide the usefulness of the gas. The following are the major applications of the product gas:

- Direct-combustion systems
- Internal-combustion engines
- Syngas production
- Pipeline transport over long distances

In direct-combustion systems, the gas produced is burnt directly in a nearby unit. Co-firing of gasified biomass in fossil-fuel-fired boilers is an example. Industrial units like ovens, furnaces, and kilns are also good examples of direct firing. In such applications, it is not necessary to cool the gas after production. The gas is fired directly in a burner while it is still hot, in the temperature range of 600 to 900 °C. Thus, there is little chance of tar condensation. However, the pipeline between the gasifier exit and the burner inlet should be such that the gas does not cool down below the dewpoint of tar. If that happens, tar deposition might clog the pipes, leading to hazardous conditions.

In applications where the raw gas is burnt directly without cooling, there is no need for cleaning. Such systems have no restrictions on the amount of tar

TABLE 4.1 Upper Limits of Biomass Gas Tar and Particulates

Application	Particulate (g/Nm^3)	Tar (g/Nm^3)
Direct combustion	No limit specified	No limit specified
Syngas production	0.02	0.1
Gas turbine	0.1–120	0.05–5
IC engine	30	50–100
Pipeline transport		50–500 for compressor
Fuel cells		<1.0

Source: Data compiled from Milne et al., 1998.

and particulates as long as the gas travels freely to the burner, and as long as the burner design does not impose any restrictions of its own. However, the flue gas produced after combustion must meet local emission requirements.

Internal-combustion engines, such as diesel or Otto engines, are favorite applications of gasified biomass, especially for distributed power generation. In such applications the gas must be cooled, but there is a good chance of condensation of the tar in the engine or in fuel-injection systems. Furthermore, the piston-cylinder system of an internal-combustion engine is not designed to handle solids, which imposes tighter limits on the tar as well as on the particulate level in the gas. Particulate and tar concentrations in the product gas should therefore be below the tolerable limits, which are 30 mg/Nm^3 for particulates and 100 mg/Nm^3 for tar (Milne et al., 1998, p. 41). The gas turbine, another user of biomass gas, imposes even more stringent restrictions on the cleanliness of the gas because its blades are more sensitive to deposits from the hot gas passing through them after combustion. Here, the particulate concentration should be between 0.1 and 120 mg/Nm^3 (Milne et al., 1998).

The limits for particulates and tar in syngas applications are even more stringent, as tar poisons the catalyst. For these applications, Graham and Bain (1993) suggest an upper limit as low as 0.02 mg/Nm^3 for particulates and 0.1 mg/Nm^3 for tar. Interest in fuel cells is rising, especially for the direct production of electricity from hydrogen through gasification. The limiting level of tar in the gas fed into a fuel cell is specific to the organic constituents of the gas. Table 4.1 presents data on the tolerance levels of tar and particulate contents for several applications of gas.

The amount of tar in product gas depends on the gasification temperature as well as on the gasifier design. Typical tar levels in gases from downdraft and updraft biomass gasifiers are 1 g/Nm^3 and 50 g/Nm^3, respectively (Table 4.2). Tar levels in product gas from bubbling and circulating fluidized-bed gasifiers are about 10 g/Nm^3. Table 4.2 also shows that the amount of tar

TABLE 4.2 Typical Levels of Tar in Biomass Gasifier by Type

Gasifier Type	Average Tar Concentration in Product Gas (g/Nm^3)	Tar as % of Biomass Feed
Downdraft	<1.0	<2.0
Fluidized bed	10	1–5
Updraft	50	10–20
Entrained flow	negligible	

Source: Data compiled from Milne and Evans, 1998, p. 15.

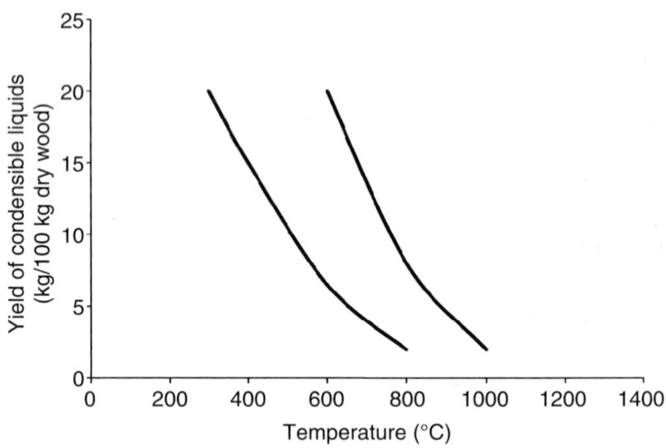

FIGURE 4.1 Effect of maximum gasification temperature on tar yield.

produced varies from 1 to 20% of the feed of the biomass. For a given gasifier, the amount of tar reduces with temperature, as shown in Figure 4.1.

4.2.2 Tar Formation

Tar is produced primarily through depolymerization during the pyrolysis stage of gasification. Biomass (or other feed), when fed into a gasifier, first undergoes pyrolysis that can begin at a relatively low temperature of 200 °C and complete at 500 °C. In this temperature range the cellulose, hemicellulose, and lignin components of biomass break down into *primary tar*, which is also known as *wood oil* or *wood syrup*. This contains oxygenates and primary organic condensable molecules called *primary tar* (Milne et al, 1998, p. 13). Char is also produced at this stage. Above 500 °C the primary tar components start reforming into smaller, lighter noncondensable gases and a series of heavier molecules called *secondary tar*. The noncondensable gases include CO_2, CO, and H_2O. At still higher temperatures, primary tar products are destroyed and tertiary products are produced.

4.2.3 Tar Composition

As we can see in Table 4.3, tar is a mixture of various hydrocarbons. It may also contain oxygen-containing compounds, derivatives of phenol, guaiacol, veratrol, syringol, free fatty acids, and esters of fatty acids (Razvigorova et al., 1994). The yield and composition of tar depends on the reaction temperature, the type of reactor, and the feedstock. Table 4.3 shows that benzene is the largest component of a typical tar.

Tar may be classified into four major product groups: primary, secondary, alkyl tertiary, and condensed tertiary (Evans and Milne, 1997). Short descriptions of these follow.

Primary Tar

Primary tar is produced during primary pyrolysis. It comprises oxygenated, primary organic, condensable molecules. Primary products come directly from the breakdown of the cellulose, hemicellulose, and lignin components of biomass. Milne et al. (1998) listed a large number of compounds of acids, sugars, alcohols, ketones, aldehydes, phenols, guaiacols, syringols, furans, and mixed oxygenates in this group.

TABLE 4.3 Typical Composition of Tar

Component	Weight (%)
Benzene	37.9
Toluene	14.3
Other 1-ring aromatic hydrocarbons	13.9
Naphthalene	9.6
Other 2-ring aromatic hydrocarbons	7.8
3-ring aromatic hydrocarbons	3.6
4-ring aromatic hydrocarbons	0.8
Phenolic compounds	4.6
Heterocyclic compounds	6.5
Others	1.0

Source: Adapted from Milne et al., 1998.

Secondary Tar

As the gasifier's temperature rises above 500 °C, primary tar begins to rearrange, forming more noncondensable gases and heavier molecules called *secondary tar*, of which phenols and olefins are important constituents.

Tertiary Tar Products

The *alkyl tertiary product* includes methyl derivatives of aromatics, such as methyl acenaphthylene, methylnaphthalene, toluene, and indene (Evans and Milne, 1997).

Condensed tertiary aromatics make up a polynuclear aromatic hydrocarbon (PAH) series without substituents (atoms or a group of atoms substituted for hydrogen in the parent chain of hydrocarbon). This series contains benzene, naphthalene, acenaphthylene, anthracene/phenanthrene, and pyrene.

The secondary and tertiary tar products come from the primary tar. The primary products are destroyed before the tertiary products appear (Milne et al., 1998).

Figure 4.2 shows that with increasing temperature the primary tar product decreases but the tertiary product increases. Above 500 °C the secondary tar increases at the expense of the primary tar. Once the primary tar is nearly destroyed, tertiary tar starts appearing with increasing temperature. At this stage the secondary tar begins to decrease. Thus, high temperatures destroy the primary tar but not the tertiary tar products.

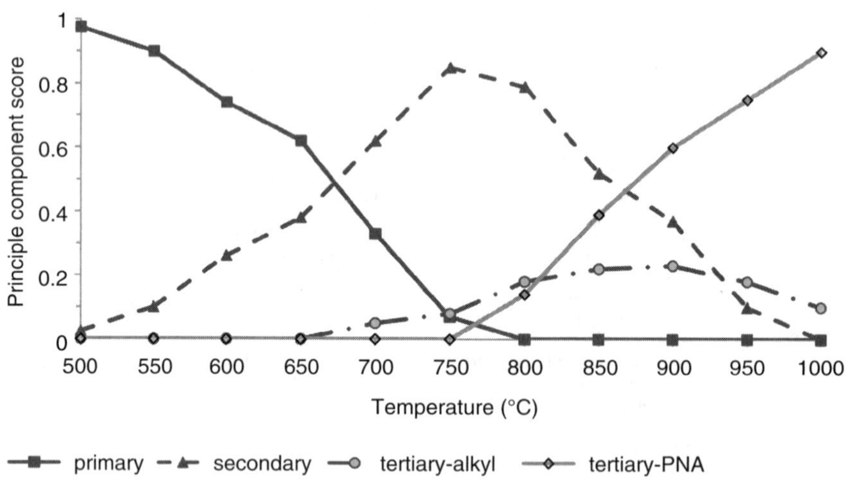

FIGURE 4.2 Variation in primary, secondary, and tertiary tar products with temperature measured at 0.3 seconds residence time. *(Source: Adapted from Evans and Milne, 1997, p. 804.)*

4.3 TAR REDUCTION

The tar in coal gasification comprise benzene, toluene, xylene, and coal tar, all of which have good commercial value and can be put to good use. Tar from biomass, on the other hand, is mostly oxygenated and has little commercial use. Thus, it is a major headache in gasifiers, and a major roadblock in the commercialization of biomass gasification. Research over the years has improved the situation greatly, but the problem has not completely disappeared. Tar removal remains an important part of the development and design of biomass gasifiers.

Several options are available for tar reduction. These may be divided into two broad groups: (1) in-situ (or primary) tar reduction, which avoids tar formation; and (2) post-gasification (or secondary) reduction, which strips the product gas of the tar already produced. They are shown in Figure 4.3.

In-situ reduction is carried out by various means so that the generation of tar inside the gasifier is lessened, thereby eliminating the need for any removal to occur downstream. As this process is carried out in the gasifier, it influences the product gas quality. Post-gasification reduction, on the other hand, does not interfere with the process in the reactor, and therefore the quality of the product gas is unaffected.

At times it may not be possible to remove the tar to the desired degree while retaining the quality of the product gas. In such cases a combination of in-situ and post-gasification reduction can prove very effective. The tar reduced is removed after the product gas leaves the gasifier. Details of these two approaches are given in the following sections.

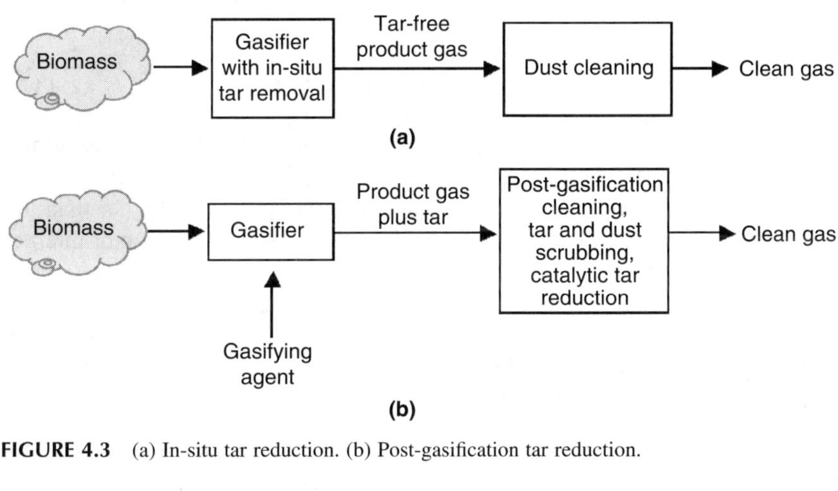

FIGURE 4.3 (a) In-situ tar reduction. (b) Post-gasification tar reduction.

4.3.1 In-Situ Tar Reduction

In this approach the operating conditions in the gasifier are adjusted such that tar formation is reduced. Alternately, the tar produced is converted into other products before it leaves the gasifier. Reduction is achieved by

- Modification of the operating conditions of the gasifier
- Addition of catalysts or alternative bed materials in the fluidized bed
- Modification of the gasifier design

Biomass type also influences the tar product. The appropriate choice of one or a combination of these factors can greatly reduce the amount of tar in the product gas leaving the gasifier. Reforming, thermal cracking, and steam cracking are three major reactions responsible for tar destruction (Delgado et al., 1996). They convert tar into an array of smaller and lighter hydrocarbons as shown here:

$$\text{Tar} \Rightarrow \left| \begin{array}{l} \text{Reforming} \\ \text{Thermal cracking} \\ \text{Steam cracking} \end{array} \right| \Rightarrow [CO_2 + CO + H_2 + CH_4 + \ldots + \text{Coke}]$$

Tar reforming. We can write the reforming reaction as in Eq. (4.1) by representing tar as C_nH_x. The cracking reaction takes place in steam gasification, where steam cracks the tar, producing simpler and lighter molecules like H_2 and CO.

$$C_nH_x + nH_2O \rightarrow (n + x/2)H_2 + nCO \quad (4.1)$$

Dry tar reforming. The dry reforming reaction takes place when CO_2 is the gasifying medium. Here tar is broken down into H_2 and CO (Eq. 4.2). Dry reforming is more effective than steam reforming when dolomite is used as the catalyst (Sutton et al., 2001).

$$C_nH_x + nCO_2 \rightarrow (x/2)H_2 + 2nCO \quad (4.2)$$

Thermal cracking. Thermal cracking can reduce tar, but it is not as attractive as reforming because it requires high (>1100 °C) temperature and produces soot (Dayton, 2002). Because this temperature is higher than the gas exit temperature for most biomass gasifiers, external heating or internal heat generation with the addition of oxygen may be necessary. Both options have major energy penalties.

Steam cracking. In steam cracking, the tar is diluted with steam and is briefly heated in a furnace in the absence of oxygen. The saturated hydrocarbons are broken down into smaller hydrocarbons.

The following sections elaborate the operating conditions used in in-situ reduction of tar.

Operating Conditions

Operating parameters influencing tar formation and conversion include reactor temperature, reactor pressure, gasification medium, equivalence ratio, and residence time.

Temperature

Reactor operating temperature influences both the quantity and composition of tar. The quantity in general decreases with an increase in reaction temperature, as does the amount of unconverted char. Thus, high-temperature operation is desirable on both counts. The production of oxygen-containing compounds like phenol, cresol, and benzofuran reduces with temperature, especially below 800 °C. With increasing temperature the amount of 1- and 2-ring aromatics with substituents decreases, but that of 3- and 4-ring aromatics increases. Aromatic compounds without substituents (naphthalene, benzene, etc.) are favored at high temperatures. The naphthalene and benzene content of the gas increases with temperature (Devi et al., 2003). High temperature also reduces the ammonia content of the gas and improves the char conversion, but has a negative effect of reducing the product gas' useful heating value.

An increase in the freeboard temperature in a fluidized-bed gasifier can also reduce the tar in the product gas. A reduction in tar was obtained by Narváez et al. (1996) by injecting secondary air into the freeboard. This may be due to increased combustion in the freeboard. Raising the temperature through secondary air injection in the freeboard may have a negative impact on heating value.

Reactor Pressure

With increasing pressure, the amount of tar decreases, but the fraction of PAH increases (Knight, 2000).

Gasification Medium

Four mediums—air, steam, carbon dioxide, and a steam–oxygen mixture—can be used for gasification; they may have different effects on tar formation and conversion. The ratio of fuel to medium is an important parameter that influences the product of gasification, including tar. This parameter is expressed differently for different mediums. For example, for air gasification the parameter is the equivalence ratio (ER); for steam gasification it is the steam-to-biomass ratio (S/B); and for steam–oxygen gasification it is the gasifying ratio (Table 4.4). Table 4.5 shows the range of tar production for three gasification mediums for typical values of their characteristic parameters.

In general, the yield of tar in steam gasification is greater than that in steam–oxygen gasification. Of these, air gasification is the lowest tar producer (Gil et al., 1999). The tar yield in a system depends on the amount of gasifying medium per unit biomass gasified.

TABLE 4.4 Gasification Mediums and Characteristic Parameters

Medium	Parameter
Air	ER = ratio of air used to stoichiometric air
Steam	Steam-to-biomass (S/B) ratio
Carbon dioxide	CO_2-to-biomass ratio
Steam and oxygen	Gasifying ratio (GR): (steam + O_2)-to-biomass ratio

Gasification in air: Both the yield and the concentration of tar in the product gas decreases with an increase in the ER. Higher ER (see Section 6.6.2 for a definition) allows greater amounts of oxygen to react with the volatiles in the *flaming pyrolysis* zone (see Figure 4.5, page 111). Above an equivalence ratio of 0.27 phenols are nearly all converted and less tar is formed (Kinoshita et al. 1994). This decrease is greater at higher temperatures. At a higher ER, the fraction of PAH, benzene, naphthalene, and other 3- and 4-ring aromatics increases. While higher ER reduces the tar, it reduces the quality of the gas as well. The heating value of the gas is reduced because of nitrogen dilution from air.

Gasification in steam: When steam reacts with biomass to produce H_2 (Eq. 4.3), the tar-reforming reaction reduces the tar.

$$C_nH_x + nH_2O \rightarrow (n + x/2)H_2 + nCO \tag{4.3}$$

A large reduction in tar yield was seen over an S/B ratio range of 0.5 to 2.5 (Herguido et al., 1992). Further reduction is possible in the presence of catalyst, which encourages the tar-reforming reaction (García et al., 1999). Three main types of catalyst are dolomite, alkali metal, and nickel.

Gasification in a steam–oxygen mixture: The addition of oxygen with steam further improves tar reduction. Additionally, it provides the heat

TABLE 4.5 Comparison of Tar Production in Three Gasification Mediums

Medium	Operating Condition	Tar Yield (g/Nm^3)	LHV (MJ/Nm^3 dry)	Tar Yield ($g/kg\ BM_{daf}$)
Steam	S/B = 0.9	30–80	12.7–13.3	70
Steam and oxygen	GR = 0.9, H_2O/O_2 = 3	4–30	12.5–13.0	8–40
Air	ER = 0.3; H/C = 2.2	2–20	4.5–6.5	6–30

Source: Data compiled from Gil et al., 1999.

needed to make the gasification reaction autothermal. Instead of the S/B ratio, the ratio of steam–oxygen to biomass, known as the *gasification ratio* (GR), is used. The tar yield reduces with an increase in the gasifying ratio. For example, an 85% reduction in tar is obtained when the GR is increased from 0.7 to 1.2 (Aznar et al., 1997). Light tars are produced at a low GR.

Gasification in carbon dioxide: The tar may be reformed on the catalyst surface in a carbon dioxide medium. Such a reaction is called *dry reforming* and is shown here (Sutton et al., 2001).

$$C_nH_x + nCO_2 \rightarrow 2nCO + (x/2)H_2 \tag{4.4}$$

The effect of gasifying agents on tar reduction or tar yield are compared in Table 4.5 (Gil et al., 1999).

Residence Time

Residence time has a nominal effect on tar yield in a fluidized-bed gasifier. Kinoshita et al. (1994) noted that with increasing residence time (bed height/superficial gas velocity), the yield of oxygenated compounds and 1- and 2-ring compounds (benzene and naphthalene excepted) decreased, but the yield of 3- and 4-ring compounds increased.

Tar Reduction by Additives in Fluidized-Bed Gasifiers

Catalysts accelerate the two main chemical reactions of tar reduction. In a steam-reforming reaction, we have

$$C_nH_x + nH_2O \xrightarrow{\text{Catalyst}} (n + x/2)H_2 + nCO \tag{4.5}$$

In a dry-reforming reaction, we have

$$C_nH_x + nCO_2 \xrightarrow{\text{Catalyst}} (x/2)H_2 + 2nCO \tag{4.6}$$

Catalysts can facilitate tar reduction reactions either in the primary reactor (gasifier) or downstream in a secondary reactor. In either case, dolomite, olivine, alkali, nickel, and char have found successful use in catalysts for tar reduction.

Dolomite

Dolomite ($MgCO_3$, $CaCO_3$) is relatively inexpensive and is readily available. It is more active if calcined and used downstream in the secondary reactor at above 800 °C (Sutton et al., 2001). The reforming reaction of tar on a dolomite surface occurs at a higher rate with CO_2 than with steam. Under proper conditions it can entirely convert the tar, but cannot convert methane if that is to be avoided for syngas production. Carbon deposition deactivates dolomite, which, being less expensive, may be discarded.

Olivine

Olivine is a magnesium-iron silicate mineral (Mg, Fe_2)SiO_4 that comes in sizes (100–400 micron) and density ranges (2500–2900 kg/m^3) similar to those of sand. Thus, it is conveniently used with sand in a fluidized-bed gasifier. The catalytic activity of olivine is comparable to that of calcined dolomite. When using olivine, Mastellone and Arena (2008) noted a complete destruction of tar from a fluidized-bed gasifier for plastic wastes, while Rapagnà et al. (2000) obtained a 90% reduction in a biomass-fed unit.

Alkali

Alkali metal catalysts are premixed with biomass before they are fed into the gasifier. Some of them are more effective than others. For example, the order of effectiveness of some alkali catalysts can be shown as follows:

$$K_2CO_3 > Na_2CO_3 > (Na_3H(CO_3)_2 \times 2H_2O) > Na_2B_4O_7 \times 10H_2O \qquad (4.7)$$

Unlike dolomite, alkali catalysts can reduce methane in the product gas, but it is difficult to recover them after use. Furthermore, alkali cannot be used as a secondary catalyst. Its use in a fluidized bed makes the unit prone to agglomeration (Mettanant et al., 2009).

Nickel

Many commercial nickel catalysts are available in the market for reduction of tar as well as methane in the product gas. They contain various amounts of nickel. For example, catalyst R-67-7H of Haldor Topsøe has 12 to 14% Ni on an Mg/Al_2O_3 support (Sutton et al., 2001). Nickel catalysts are highly effective and work best in the secondary reactor. Use of dolomite or alkali as the primary catalyst and nickel as the secondary catalyst has been successfully demonstrated for tar and methane reduction. Catalyst activity is influenced by temperature, space–time, particle size, and composition of the gas atmosphere. The optimum operating temperature for a nickel catalyst in a downstream fluidized bed is 780 °C (Sutton et al., 2001). Steam-reforming nickel catalysts for heavy hydrocarbons are effective for reduction of tar while nickel catalysts for light hydrocarbons are effective for methane reduction. Deactivation due to carbon deposition and particle growth is a problem for nickel-reforming catalysts.

Char

Char, a carbonaceous product of pyrolysis, also catalyses tar reforming when used in the secondary reactor. Chembukulam et al. (1981) obtained a nearly total reduction in tar with this. As a major gasification element, char is not easily available in a gasifier's downstream. Design modification is needed to incorporate char as a catalyst.

Gasifier Design

The design of the gasifier can be a major influence on the amount of tar in the product gas. For example, a counter current moving-bed gasifier with an internal recycle and a separate combustion zone can reduce the tar content to as low as 0.1 g/Nm^3 (Susanto and Beenackers, 1996), while in an updraft gasifier the tar can well exceed 100 g/Nm^3. To understand how gasifier design might influence tar production, we will examine the tar production process.

As we saw in Figure 4.2, primary tar is produced at fairly low temperatures (200–500 °C). It is a mixture of condensable hydrocarbons that undergoes molecular rearrangement (reforming) at higher temperatures (700–900 °C), producing some noncondensable gases and secondary tar. Tar is produced at an early stage when biomass (or another fuel) undergoes pyrolysis following drying. Char is produced further downstream in the process and is often the final solid residue left over from gasification. The gasifier design determines where pyrolysis takes place, how the tar reacts with oxidants, and the temperature of the reactions. This in turn determines the net tar production in the gasifier.

Updraft, downdraft, fluidized bed, and entrained bed are the four major types of gasifier with their distinct tar formation. Table 4.2 earlier in the chapter compares their tar production, and a brief discussion of formation of tar in these reactors follows here.

Updraft Gasifier

Biomass is fed from the top and a gasifying medium (air) is fed from the bottom. The product gas leaves from the top while solids leave from the bottom. Figure 4.4 illustrates the motion of biomass, gas, and tar. The temperature is highest close to the grate, where oxygen meets with char and burns the char. The hot gas travels up, providing heat to the endothermic gasification reactions, and meets pyrolyzing biomass at a low temperature (200–500 °C). Primary tar is produced in this temperature range (Figure 4.4). This tar travels upward through cooler regions and therefore has no opportunity for conversion into gases and secondary tar. For this reason, updraft gasifiers generate the highest amount of tar—typically 10 to 20% by weight of the feed.

Downdraft Gasifier

Figure 4.5 shows the tar production in a downdraft gasifier. Here, both gas and feed travel downward. The temperature is highest in the downstream combustion zone. The tar is produced after drying at lower temperatures (200–500 °C) close to the feed point. The oxygen in the air, along with the tar, travels downward to the hotter zone. Owing to the availability of oxygen and high temperature, the tar readily burns in a flame, raising the gas temperature to 1000 to 1400 °C. The flame occurs in the interstices between feed particles, which remain at 500 to 700 °C (Milne et al., 1998, p. 14). This phenomenon is called

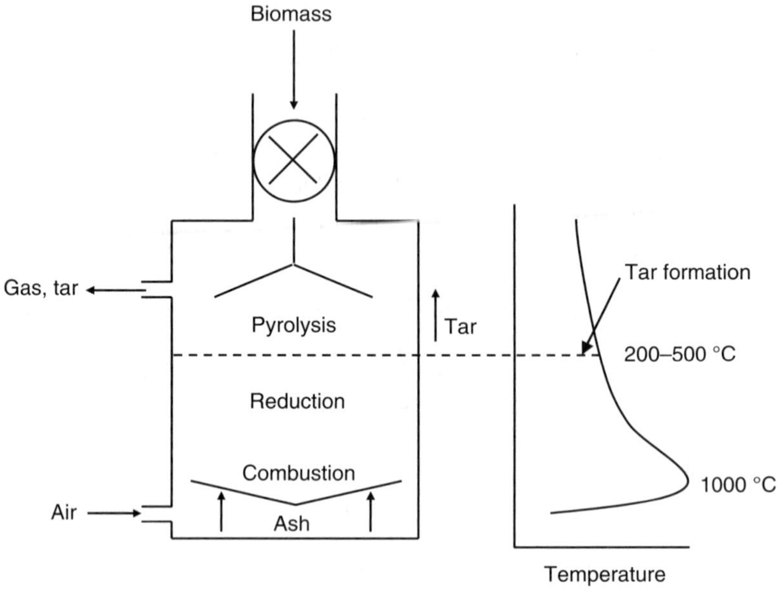

FIGURE 4.4 Tar production in an updraft gasifier. Tar passes through only the low-temperature (200–500 °C) zone, so it has no opportunity to crack.

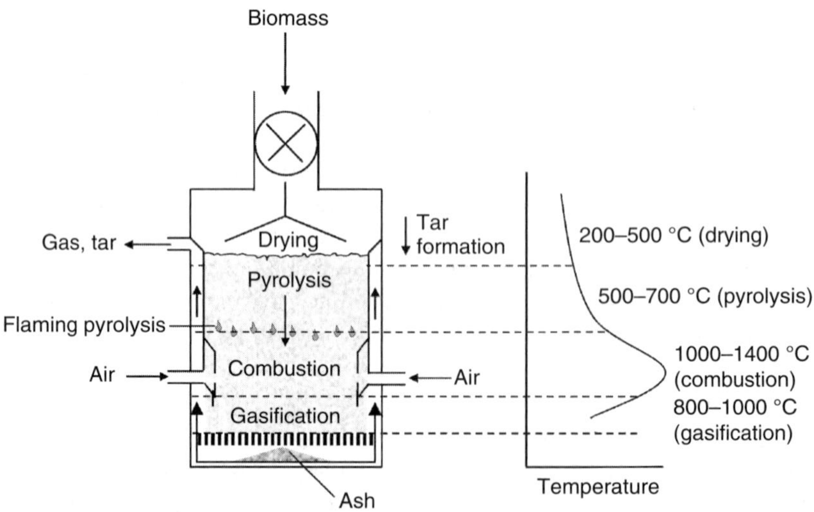

FIGURE 4.5 Tar generation in a downdraft gasifier. The tar produced passes through the highest-temperature zone and so is easily cracked.

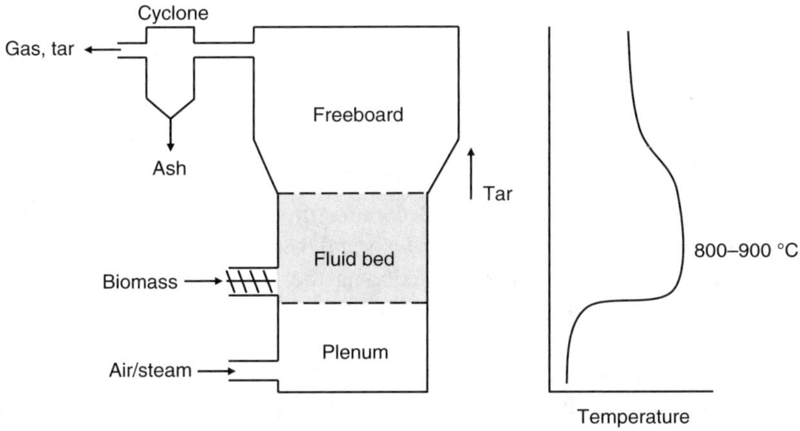

FIGURE 4.6 Bubbling fluidized-bed gasifier. The tar is not produced at any specific location and so it passes through the average-temperature zone.

flaming pyrolysis. Since the pyrolysis product, tar, contacts oxygen while passing through the highest-temperature zone, it has the greatest opportunity to be converted into noncondensable gases. For this reason, a downdraft gasifier has the lowest tar production (<1 g/Nm^3).

Fluidized-Bed Gasifier

In a typical fluidized bed (bubbling or circulating) air enters from the bottom, but is fuel fed from the side or top. In either case, the fuel is immediately mixed throughout the bed owing to its exceptionally high degree of mixing (Figure 4.6). Thus, the fresh oxygen (in air) entering the grid comes into immediate contact with fresh biomass particles undergoing pyrolysis as well as with spent char particles from the biomass, which has been in the bed for some time. Oxygen's contact with the fresh biomass burns the tar released, while its contact with the spent char particles causes the char to burn.

Though the solids are back-mixed, the gases flow upward in plug-flow mode. This means that further up in the bed neither older char particles nor fresh pyrolyzing biomass particles come in contact with the oxygen. Any tar released moves up in the bed and leaves along with the product gas. For this reason, tar generation in a fluidized-bed gasifier is between the two extremes represented by updraft and downdraft gasifiers, averaging about 10 mg/Nm^3.

Entrained-Flow Gasifier

Tar production is negligible, as whatever is released passes through a very-high-temperature (>1000 °C) zone and is therefore nearly all converted into gases.

Design Modifications for Tar Removal

Modification of a reactor design for tar removal involves the following:

- Secondary air injection
- Separation of the pyrolysis zone from the char gasification zone
- Passage of pyrolysis products through the char

Char is an effective means of tar decomposition. A moving-bed two-stage gasifier that uses the first stage for pyrolysis and the second stage for conversion of tar in a bed of char succeeds in reducing the tar by 40 times (Bui et al., 1994). Air addition in the second stage increases the temperature and thereby reduces the tar (Knoef, 2005, p. 170).

A large commercial unit (70-MW fuel power) uses this concept, where biomass dries and pyrolyzes in a horizontal moving bed, heated by waste heat from a diesel engine. The tar concentration of the product gas is about 50 g/Nm^3. This gas passes through the neck of a vertical chamber, where injection of preheated gas raises the temperature above 1100 °C, reducing the tar amounts to 0.5 g/Nm^3. It then passes through a fixed bed of char or carbon being gasified. Tar in the gas leaving the gasifier is very low (<0.025 g/Nm^3). It is further cleaned to 0.005 g/Nm^3 in a bag filter (Knoef, 2005, p. 159).

Another design involves twin fluidized beds. Biomass fed into the first bed is pyrolyzed. The char then travels to a parallel fast fluidized-bed combustor that burns part of it. A commercial unit (8-MW fuel power) operates on this principle, where gas leaving the gasifier contains 1.5 to 4.5 g/Nm^3 tar. A fabric filter that separates dust and some tar reduces its concentration to 0.75 g/Nm^3, which is finally reduced to 0.010 to 0.04 g/Nm^3 in a scrubber.

4.3.2 Post-Gasification—Secondary Reduction of Tar

As indicated earlier, the level of cleaning needed for the product gas depends greatly on its end use. For example, combustion in an engine or a gas turbine needs substantially cleaner product gas than that required by a boiler. Most commercial plants use particulate filters or scrubbers to attain the required level of cleanliness. A substantial amount of tar can be removed from the gas in a post-gasification cleanup section. It can be either catalytically converted into useful gases like hydrogen or simply captured and scrubbed away. The two basic post-gasification methods are physical removal and cracking (catalytic or thermal).

Physical Tar Removal

Physical cleaning is similar to the removal of dust particles from a gas. It requires the tar to be condensed before separation. The energy content of the tar is often lost in this process such that it remains as mist or drops on suspended particles in gas. Physical tar removal can be accomplished by cyclones, barrier filters, wet electrostatic precipitators, wet scrubbers, or alkali salts. The choice depends on the following:

- Inlet concentration of particulate and tar
- Inlet particle size distribution (PSD)
- Particulate tolerance of the downstream application of the gas

The size distribution of the inlet particulates is difficult to measure, especially for finer particulates, but its measurement is important in choosing the right collection devices. For example, submicron (<1 micron) particulates need a wet electrostatic precipitator, but this device is significantly more expensive than others. A fabric filter may work for fines, but it may fail if there is any chance of condensation.

Cyclones

Cyclones are not very practical for tar removal because of the tar's stickiness and because cyclones cannot remove small (<1 micron) tar droplets (Knoef, 2005, p. 196). A fabric filter has been used with the help of a precoat, which is removed along with the dust cake formed on the filter.

Barrier Filters

Barrier filters present a physical barrier in the path of tar and particulates while allowing the clean gas to pass through. One of their special features is that they allow coating of their surface with appropriate catalytic agents to facilitate tar cracking. These filters are of two types: candle and fabric.

Candle filters are porous, ceramic, or metallic. The porosity of the material is chosen such that the finest particles do not pass through. Particles failing to pass through the filter barrier deposit on the wall (Figure 4.7), forming a layer

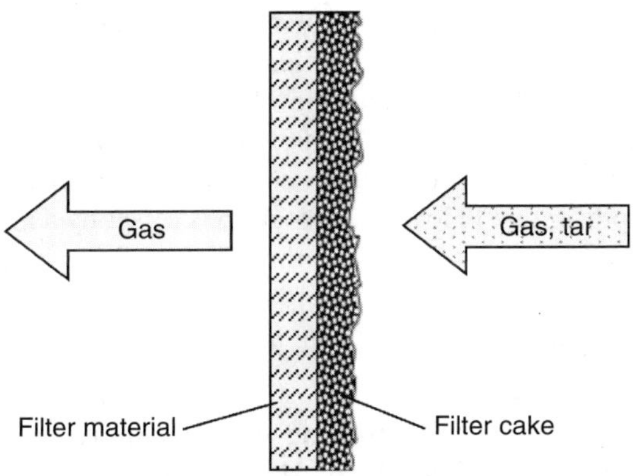

FIGURE 4.7 Operation of a barrier filter.

of solids called a "filter cake." Gas passes through the porous layer as well as through the filter. One major problem with the filter cake is that as it grows in thickness, the pressure drop across the filter increases. Thus, provision is made for its occasional removal. A popular means of removal is pressure pulse in opposite directions.

Besides their high-pressure drop, barrier filters also suffer from the problem that if a filter is broken or cracked, dust and tar-laden gas preferentially flow through that passage, adversely affecting downstream equipment. The condensation of tar on the filter elements can block the filter, and this is a major concern. Ceramic filters can be designed to operate in temperatures as high as 800 to 900 °C.

Fabric filters are made of woven fabric as opposed to porous materials as in candle filters. Unlike candle filters, they can operate only in lower temperatures (<350 °C). Here the filter cake is removed by either back-flushing as with a candle filter, or shaking. Condensation of tar on the fabric is a problem here if the gas is cooled excessively.

Wet Electrostatic Precipitators

Wet electrostatic precipitators (ESPs) are used in some gasification plants. The gas is passed through a strong electric field with electrodes. High voltage charges the solid and liquid particles. As the flue gas passes through a chamber containing anode plates or rods with a potential of 30 to 75 kV, the particles in the flue gas pick up the charge and are collected downstream by positively charged cathode collector plates. Grounded plates or walls also attract the charged particles and are often used for design simplicity. Although collection efficiency does not decrease as particles build up on the plates, periodic mechanical wrapping is required to clean the plates to prevent the impediment of the gas flow or the short-circuiting of the electrodes through the built-up ash.

The collected solid particles are cleaned by mechanical means, but a liquid like tar needs cleaning by a thin film of water. Wet electrostatic precipitators have very high (> 90%) collection efficiency over the entire range of particle size down to about 0.5 micron, and they have very low pressure drop (few inches water gauge). Sparking due to high voltage is a concern with an ESP, especially when it is used to clean highly combustible syngas. Thus, the savings from lower fan power due to low pressure drop is offset by a higher safety cost. Additionally, the capital cost for ESP is three to four times higher than that for a wet scrubber.

Wet Scrubbers

Here, water or an appropriate scrubbing liquid is sprayed on the gas. Solid particles and tar droplets collide with the drops, forming larger droplets because of coalescence. These larger droplets are easily separated from the gas by a demisterlike cyclone. The gas needs to be cooled until it is below 100 °C before cleaning. The tar-laden scrubbing liquid may be fed back into the gasifier or its combustion section. Alternatively, it may be regenerated by stripping the tar away.

Some commercial methods, such as the OLGA and TARWTC technologies, use proprietary oil as the scrubbing liquid. The tar liquid is then reinjected into the gasifier for further conversion (Knoef, 2005, p. 196). Scrubbers have a high (>90%) collection efficiency, but the efficiency drops sharply below 1-micron-sized particles. They consume a large amount of fan power owing to the large (~50-inch water gauge) pressure drop across the scrubber. While their operating cost is high, their capital cost is much less than that for ESPs.

A system with a tar removal scrubber produces cleaned gas with a lower outlet temperature and a higher energy content, but it contains tars that are more difficult to remove. The main challenge of tar removal is the formation of "tar balls," which are long-chained hydrocarbons that have a tendency to agglomerate and stick together, fouling equipment in the initial stages of tar condensing and collecting.

The tar-laden stripper gas, if fed into the gasifier, lowers its dewpoint well below that of water. This allows condensation of the tar, while flue gas containing tar vapor can be recycled back to the combustion section of the gasifier for combustion.

Alkali Remover

Compared to fossil fuels, biomass is rich in alkali salts that typically vaporize at high gasifier temperatures but condense downstream below 600 °C. Because condensation of alkali salts causes serious corrosion problems, efforts are made to strip the gas of alkali. If the gas can be cooled to below 600 °C, the alkali will condense onto fine solid particles (<5 microns) that can be captured in a cyclone, ESPs, or filters. Some applications do not permit cooling of the gas. In such cases, the hot gas may be passed through a bed of active bauxite maintained at 650 to 725 °C.

Disposal of Collected Tar

Tar removal processes produce liquid wastes with higher organic compound concentrations, which increase the complexity of water treatment. Wastewater contaminants include dissolved organics, inorganic acids, NH_3, and metals. Collected tars are classified as hazardous waste, especially if they are formed at high temperatures (Stevens, 2001). Several technologies are available for treatment of these contaminants before their final disposal. Hasler et al. (1997) presented a description of the available technologies that comprise extraction with organic solvent, distillation, adsorption on activated carbon, wet oxidation, oxidation with hydrogen peroxide (H_2O_2), oxidation with ozone (O_3), incineration, and biological treatment.

Cracking

Cracking involves breaking large molecules into smaller ones. It converts tar into permanent gases such as H_2 or CO. The energy content of the tar is thus mostly recovered through the smaller molecules formed. Unlike in physical cleaning, the tar need not be condensed for cracking. This process involves heating the tar to a high temperature (~1200 °C) or exposing it to catalysts at lower temperatures (~800 °C). There are two major types of cracking: thermal and catalytic.

Thermal Cracking

Thermal cracking without a catalyst is possible at a high temperature (~1200 °C). The temperature requirement depends on the constituents of the tar. For example, oxygenated tars may crack at around 900 °C (Stevens, 2001). Oxygen or air may be added to allow partial combustion of the tar to raise its temperature, which is favorable for thermal cracking. Thermal decomposition of biomass tars in electric arc plasma is another option. This is a relatively simple process but it produces gas with a lower energy content.

Catalytic Cracking

Catalytic cracking is commercially used in many plants for the removal of tar and other undesired elements from product gas. It generally involves passing the dirty gas over catalysts. The main chemical reactions taking place in a catalytic reactor are represented by Eq. (4.5) in the presence of steam (steam reforming) and Eq. (4.6) in the presence of CO_2 (dry reforming). The main reactions for tar conversion are endothermic, so a certain amount of combustion reactions are allowed in the reactor by adding air.

Nonmetallic catalysts include less-expensive disposable catalysts: dolomite, zeolite, calcite, and so forth. They can be used as bed materials in a fluidized bed through which tar-laden gas is passed at a temperature of 750 to 900 °C. Attrition and deactivation of the catalyst are a problem (Lammars et al., 1997). A proprietary nonmetallic catalyst, D34, has been used with success in a fluidized bed at 800 °C followed by a wet scrubber (Knoef, 2005, p. 153).

Metallic catalysts include Ni, Ni/Mo, Ni/Co/Mo, NiO, Pt, and Ru on supports like silica-alumina and zeolite (Aznar et al., 1997). Some of them are used in the petrochemical industry and are readily available. A Ni/Co/Mo blend converts NH_3 along with tars. Catalysts deactivate during tar cracking and so need reactivation. Typically the catalysts are placed in a fixed or fluidized bed. Tar-laden gas is passed through at a temperature of 800 to 900 °C.

Dolomite (calcined) and olivine sand are very effective in in-situ tar reduction. This type of catalytic cracking takes place at the typical temperature of a fluidized bed. Good improvement in gas yield and tar reduction is noted when catalytic bed materials are used.

Chapter 5

Gasification Theory and Modeling of Gasifiers

5.1 INTRODUCTION

The design and operation of a gasifier require an understanding of the gasification process and how its design, feedstock, and operating parameters influence the performance of the plant. A good comprehension of the basic reactions is fundamental to the planning, design, operation, troubleshooting, and process improvement of a gasification plant, as is learning the alphabet to read a book. This chapter introduces the basics of the gasification process through a discussion of the reactions involved and the kinetics of the reactions with specific reference to biomass. It also explains how this knowledge can be used to develop a mathematical model of the gasification process.

5.2 GASIFICATION REACTIONS AND STEPS

Gasification is the conversion of solid or liquid feedstock into useful and convenient gaseous fuel or chemical feedstock that can be burned to release energy or used for production of value-added chemicals.

Gasification and combustion are two closely related thermochemical processes, but there is an important difference between them. Gasification packs energy into chemical bonds in the product gas; combustion breaks those bonds to release the energy. The gasification process adds hydrogen to and strips carbon away from the feedstock to produce gases with a higher hydrogen-to-carbon (H/C) ratio, while combustion oxidizes the hydrogen and carbon into water and carbon dioxide, respectively.

A typical biomass gasification process may include the following steps:

- Drying
- Thermal decomposition or pyrolysis
- Partial combustion of some gases, vapors, and char
- Gasification of decomposition products

Pyrolysis is a thermal decomposition process that partially removes carbon from the feed but does not add hydrogen. Gasification, on the other hand,

Biomass Gasification and Pyrolysis. DOI: 10.1016/B978-0-12-374988-8.00005-2

requires a gasifying medium like steam, air, or oxygen to rearrange the molecular structure of the feedstock in order to convert the solid feedstock into gases or liquids; it can also add hydrogen to the product. The use of a medium is essential for the gasification process.

5.2.1 Gasifying Mediums

Gasifying agents react with solid carbon and heavier hydrocarbons to convert them into low-molecular-weight gases like CO and H_2. The main gasifying agents used for gasification are

- Oxygen
- Steam
- Air

Oxygen is a popular gasifying agent, though it is primarily used for the combustion step. It may be supplied to a gasifier either in pure form or through air. The heating value and the composition of the gas produced in a gasifier are strong functions of the nature and amount of the gasifying agent used. A ternary diagram (Figure 5.1) of carbon, hydrogen, and oxygen (see Section 2.4.3) demonstrates the conversion paths of formation of different products in a gasifier.

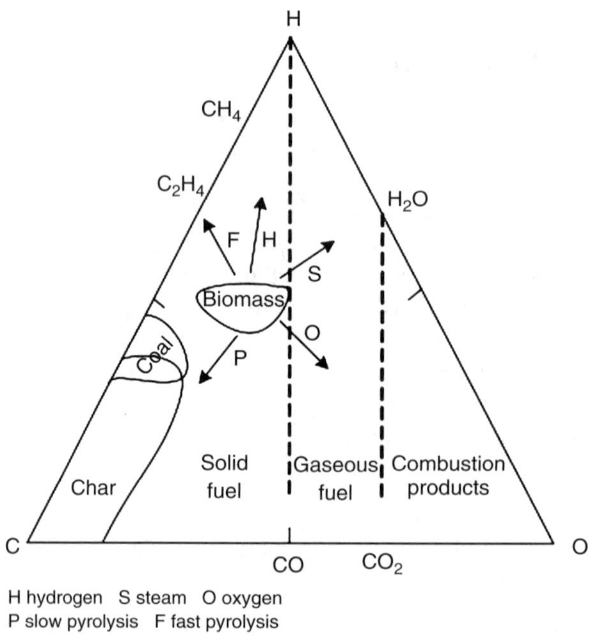

FIGURE 5.1 C-H-O diagram of the gasification process.

TABLE 5.1 Heating Values for Product Gas Based on Gasifying Medium

Medium	Heating Value (MJ/Nm^3)
Air	4–7
Steam	10–18
Oxygen	12–28

If oxygen is used as the gasifying agent, the conversion path moves toward the oxygen corner. Its products include CO for low oxygen and CO_2 for high oxygen. When the amount of oxygen exceeds a certain (stoichiometric) amount, the process moves from gasification to combustion, and the product is "flue gas" instead of "fuel gas." Neither flue gas nor the combustion product contains residual heating value when cooled. A move toward the oxygen corner (Figure 5.1) leads to a lowering of hydrogen content and an increase in carbon-based compounds such as CO and CO_2 in the product gas.

If steam is used as the gasification agent, the path is upward toward the hydrogen corner in Figure 5.1. Then the product gas contains more hydrogen per unit of carbon, resulting in a higher H/C ratio. Some of the intermediate-reaction products like CO and H_2 also help to gasify the solid carbon.

The choice of gasifying agent affects the heating value of the product gas. If air is used instead of oxygen, the nitrogen in it greatly dilutes the product. From Table 5.1, we can see that oxygen gasification has the highest heating value followed by steam and air gasification.

5.3 THE GASIFICATION PROCESS

A typical gasification process generally follows the sequence of steps listed on the next page (illustrated schematically in Figure 5.2).

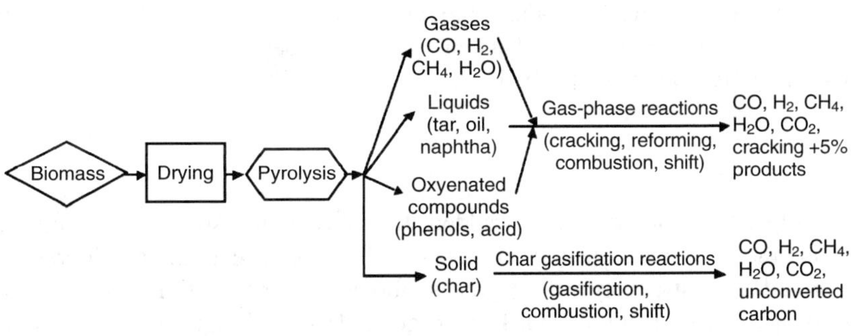

FIGURE 5.2 Potential paths for gasification.

- Preheating and drying
- Pyrolysis
- Char gasification
- Combustion

Though these steps are frequently modeled in series, there is no sharp boundary between them, and they often overlap. The following paragraphs discuss these sequential phases of biomass gasification.

In a typical process, biomass is first heated (dried) and then it undergoes thermal degradation or pyrolysis. The products of pyrolysis (i.e., gas, solid, and liquid) react among themselves as well as with the gasifying medium to form the final gasification product. In most commercial gasifiers, the thermal energy necessary for drying, pyrolysis, and endothermic reactions comes from a certain amount of exothermic combustion reactions allowed in the gasifier. Table 5.2 lists some of the important chemical reactions taking place in a gasifier.

5.3.1 Drying

The typical moisture content of freshly cut wood ranges from 30 to 60%, and for some biomass it can exceed 90% (see Table 2.9). Every kilogram of moisture in the biomass takes away a minimum of 2260 kJ of extra energy from the gasifier to vaporize water, and that energy is not recoverable. For a high level of moisture this loss is a concern, especially for energy applications. While we cannot do much about the inherent moisture residing within the cell structure, efforts may be made to drive away the external or surface moisture. A certain amount of predrying is thus necessary to remove as much moisture from the biomass as possible before it is fed into the gasifier. For the production of a fuel gas with a reasonably high heating value, most gasification systems use dry biomass with a moisture content of 10 to 20%.

The final drying takes place after the feed enters the gasifier, where it receives heat from the hot zone downstream. This heat dries the feed, which releases water. Above 100 °C, the loosely bound water that is in the biomass is irreversibly removed. As the temperature rises, the low-molecular-weight extractives start volatilizing. This process continues until a temperature of approximately 200 °C is reached.

5.3.2 Pyrolysis

In pyrolysis no external agent is added. In a slow pyrolysis process, the solid product moves toward the carbon corner of the ternary diagram, and more char is formed. In fast pyrolysis, the process moves toward the C-H axis opposite the oxygen corner (Figure 5.1). The oxygen is largely diminished, and thus we expect more liquid hydrocarbon.

TABLE 5.2 Typical Gasification Reactions at 25 °C

Reaction Type	Reaction
Carbon Reactions	
R1 (Boudouard)	$C + CO_2 \leftrightarrow 2CO + 172$ kJ/mol[1]
R2 (water-gas or steam)	$C + H_2O \leftrightarrow CO + H_2 + 131$ kJ/mol[2]
R3 (hydrogasification)	$C + 2H_2 \leftrightarrow CH_4 - 74.8$ kJ/mol[2]
R4	$C + 0.5\ O_2 \rightarrow CO - 111$ kJ/mol[1]
Oxidation Reactions	
R5	$C + O_2 \rightarrow CO_2 - 394$ kJ/mol[2]
R6	$CO + 0.5O_2 \rightarrow CO_2 - 284$ kJ/mol[4]
R7	$CH_4 + 2O_2 \leftrightarrow CO_2 + 2H_2O - 803$ kJ/mol[3]
R8	$H_2 + 0.5\ O_2 \rightarrow H_2O - 242$ kJ/mol[4]
Shift Reaction	
R9	$CO + H_2O \leftrightarrow CO_2 + H_2 - 41.2$ kJ/mol[4]
Methanation Reactions	
R10	$2CO + 2H_2 \rightarrow CH_4 + CO_2 - 247$ kJ/mol[4]
R11	$CO + 3H_2 \leftrightarrow CH_4 + H_2O - 206$ kJ/mol[4]
R14	$CO_2 + 4H_2 \rightarrow CH_4 + 2H_2O - 165$ kJ/mol[2]
Steam-Reforming Reactions	
R12	$CH_4 + H_2O \leftrightarrow CO + 3H_2 + 206$ kJ/mol[3]
R13	$CH_4 + 0.5\ O_2 \rightarrow CO + 2H_2 - 36$ kJ/mol[3]

[1]*Source:* Higman and van der Burgt, 2008, p. 12.
[2]*Source:* Klass, 1998, p. 276.
[3]*Source:* Higman and van der Burgt, 2008, p. 3.
[4]*Source:* Knoef, 2005, p. 15.

Pyrolysis, which precedes gasification, involves the thermal breakdown of larger hydrocarbon molecules of biomass into smaller gas molecules (condensable and noncondensable) with no major chemical reaction with air, gas, or any other gasifying medium. For a detailed description of this process, see Chapter 3.

One important product of pyrolysis is tar formed through condensation of the condensable vapor produced in the process. Being a sticky liquid, tar creates a great deal of difficulty in industrial use of the gasification product. A discussion of tar formation and ways of cracking or reforming it into useful noncondensable gases is presented in Chapter 4.

5.3.3 Char Gasification Reactions

The gasification step that follows pyrolysis involves chemical reactions among the hydrocarbons in fuel, steam, carbon dioxide, oxygen, and hydrogen in the reactor, as well as chemical reactions among the evolved gases. Of these, char gasification is the most important. The char produced through pyrolysis of biomass is not necessarily pure carbon. It contains a certain amount of hydrocarbon comprising hydrogen and oxygen.

Biomass char is generally more porous and reactive than coke. Its porosity is in the range of 40 to 50% while that of coal char is 2 to 18%. The pores of biomass char are much larger (20–30 micron) than those of coal char (~5 angstrom) (Encinar et al., 2001). Thus, its reaction behavior is different from that of chars derived from coal, lignite, or peat. For example, the reactivity of peat char decreases with conversion or time, while the reactivity of biomass char increases with conversion (Figure 5.3). This reverse trend can be attributed to the increasing catalytic activity of the biomass char's alkali metal constituents (Risnes et al., 2001).

Gasification of biomass char involves several reactions between the char and the gasifying mediums. Following is a description of some of those reactions with carbon, carbon dioxide, hydrogen, steam, and methane.

$$\text{Char} + O_2 \rightarrow CO_2 \text{ and } CO \tag{5.1}$$

$$\text{Char} + CO_2 \rightarrow CO \tag{5.2}$$

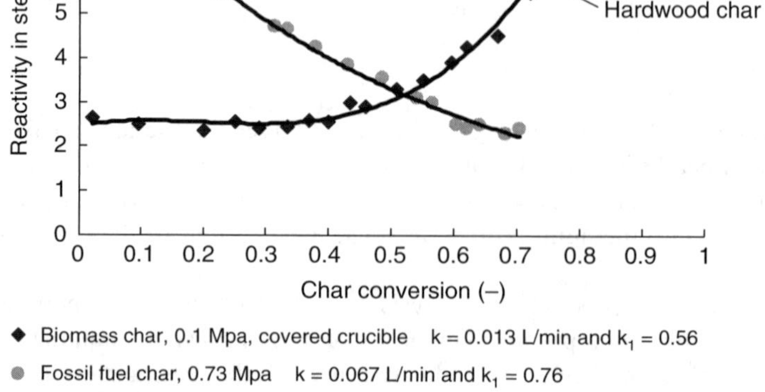

FIGURE 5.3 Reactivities of peat char for gasification in steam decrease with conversion; reactivities of hardwood char increase with conversion. *(Source: Data from Liliedahl and Sjostrom, 1997.)*

$$\text{Char} + H_2O \rightarrow CH_4 \text{ and } CO \tag{5.3}$$

$$\text{Char} + H_2 \rightarrow CH_4 \tag{5.4}$$

Equations (5.1) through (5.4) show how gasifying agents like oxygen, carbon dioxide, and steam react with solid carbon to convert it into lower-molecular-weight gases like carbon monoxide and hydrogen. Some of the reactions are known by the names listed in Table 5.2.

Gasification reactions are generally endothermic, but some of them can be exothermic as well. For example, those of carbon with oxygen and hydrogen (R3, R4, and R5 in Table 5.2) are exothermic, whereas those with carbon dioxide and steam (reactions R1 and R2) are endothermic. The heat of reaction given in Table 5.2 for various reactions refers to a temperature of 25 °C.

Speed of Char Reactions

The rate of gasification of char (comprising of mainly carbon) depends primarily on its reactivity and the reaction potential of the gasifying medium. Oxygen, for example, is the most active, followed by steam and carbon dioxide. The rate of the char–oxygen reaction ($C + 0.5O_2 \rightarrow CO$) is the fastest among the four in Table 5.2 (R1, R2, R3, and R4). It is so fast that it quickly consumes the oxygen, leaving hardly any free oxygen for any other reactions.

The rate of the char–steam reaction ($C + H_2O \rightarrow CO + H_2$) is three to five orders of magnitude slower than that of the char–oxygen reaction. The Boudouard, or char–carbon dioxide, reaction ($C + CO_2 \rightarrow 2CO$) is six to seven orders of magnitude slower (Smoot and Smith, 1985). The rate of the water–gas or water–steam gasification reaction (R2) is about two to five times faster than that of the Boudouard reaction (R1) (Blasi, 2009).

The char–hydrogen reaction that forms methane ($C + 2H_2 \rightarrow CH_4$) is the slowest of all. Walker et al. (1959) estimated the relative rates of the four reactions, at 800 °C temperature and 10 K Pa pressure, as 10^5 for oxygen, 10^3 for steam, 10^1 for carbon dioxide, and 3×10^{-3} for hydrogen. The relative rates, R, may be shown as

$$R_{C+O_2} >> R_{C+H_2O} > R_{C+CO_2} >> R_{C+H_2} \tag{5.5}$$

When steam reacts with carbon it can produce CO and H_2. Under certain conditions the steam and carbon reaction can also produce CH_4 and CO_2.

Boudouard Reaction Model

The gasification of char in carbon dioxide is popularly known as the *Boudouard reaction.*

$$C + CO_2 \leftrightarrow 2CO \text{ (reaction R1 in Table 5.2)} \tag{5.6}$$

Blasi (2009) describes the Boudouard reaction through the following steps. In the first step, CO_2 dissociates at a carbon-free active site (C_{fas}), releasing carbon monoxide and forming a carbon–oxygen surface complex, C(O). This reaction can move in the opposite direction as well, forming a carbon active site and CO_2 in the second step. In the third step, the carbon–oxygen complex produces a molecule of CO.

$$\text{Step 1} \quad C_{fas} + CO_2 \xrightarrow{k_{b1}} C(O) + CO \tag{5.7}$$

$$\text{Step 2} \quad C(O) + CO \xrightarrow{k_{b2}} C_{fas} + CO_2 \tag{5.8}$$

$$\text{Step 3} \quad C(O) \xrightarrow{k_{b3}} CO \tag{5.9}$$

where k_i is the rate of the ith reaction.

The rate of the char gasification reaction in CO_2 is insignificant: below 1000 K.

Water–Gas Reaction Model

The gasification of char in steam, known as the *water–gas reaction*, is perhaps the most important gasification reaction.

$$C + H_2O \leftrightarrow CO + H_2 \text{ (R2 in Table 5.2)} \tag{5.10}$$

The first step involves the dissociation of H_2O on a free active site of carbon (C_{fas}), releasing hydrogen and forming a surface oxide complex of carbon C(O). In the second and third steps, the surface oxide complex produces a new free active site and a molecule of CO.

$$\text{Step 1} \quad C_{fas} + H_2O \xrightarrow{k_{w1}} C(O) + H_2 \tag{5.11}$$

$$\text{Step 2} \quad C(O) + H_2 \xrightarrow{k_{w2}} C_{fas} + H_2O \tag{5.12}$$

$$\text{Step 3} \quad C(O) \xrightarrow{k_{w3}} CO \tag{5.13}$$

Some models (Blasi, 2009) also include the possibility of hydrogen inhibition by C(H) or $C(H)_2$ complexes as here:

$$C_{fas} + H_2 \leftrightarrow C(H)_2 \tag{5.14}$$

$$C_{fas} + 0.5\,H_2 \leftrightarrow C(H) \tag{5.15}$$

The presence of hydrogen has a strong inhibiting effect on the char gasification rate in H_2O. For example, 30% hydrogen in the gasification atmosphere can reduce the gasification rate by a factor as high as 15 (Barrio et al., 2001). So an effective means of accelerating the water–gas reaction is continuous removal of hydrogen from the reaction site.

Shift Reaction Model

The shift reaction is an important gas-phase reaction. It increases the hydrogen content of the gasification product at the expense of carbon monoxide. This reaction is also called the "water–gas shift reaction" in some literature (Klass, 1998, p. 277), though it is much different from the water–gas reaction (R2).

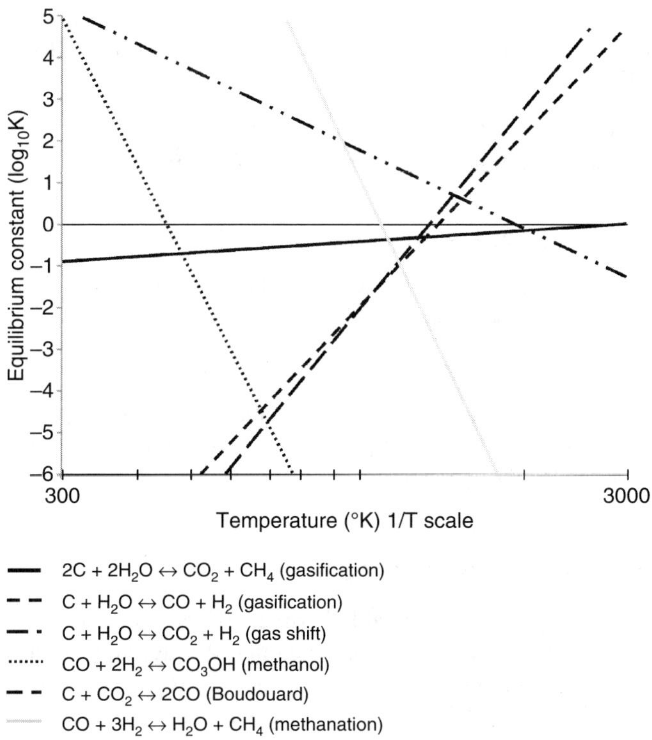

FIGURE 5.4 Equilibrium constants for selected gasification reactions. *(Source: Adapted from Probstein and Hicks, 2006, p. 63.)*

$$CO + H_2O \leftrightarrow CO_2 + H_2 - 41.2\ \text{kJ/mol (reaction R9 in Table 5.2)} \qquad (5.16)$$

This is a prestep in syngas production in the downstream of a gasifier, where the ratio of hydrogen and carbon monoxide in the product gas is critical.

The shift reaction is slightly exothermic, and its equilibrium yield decreases slowly with temperature. Depending on temperature, it may be driven in either direction—that is, products or reactants. However, it is not sensitive to pressure (Petersen and Werther, 2005).

Above 1000 °C the shift reaction (R9) rapidly reaches equilibrium, but at a lower temperature it needs heterogeneous catalysts. Figure 5.4 (Probstein and Hicks, 2006, p. 63) shows that this reaction has a higher equilibrium constant at a lower temperature, which implies a higher yield of H_2 at a lower temperature. With increasing temperature, the yield decreases but the reaction rate increases. Optimum yield is obtained at about 225 °C.

Because the reaction rate at such a low temperature is low, catalysts like chromium-promoted iron, copper-zinc, and cobalt-molybdenum are used (Probstein and Hicks, 2006, p. 124). At higher temperatures (350–600 °C) Fe-based catalysts may be employed. Pressure exerts no appreciable effect on the H_2/CO ratio. Commercial shift conversions of CO use these catalysts (Boerrigter and Rauch, 2005):

- Copper-promoted catalyst, at about 300–510 °C
- Copper-zinc-aluminum oxide catalyst, at about 180–270 °C

Hydrogasification Reaction Model

This reaction involves the gasification of char in a hydrogen environment, which leads to the production of methane.

$$C + 2H_2 \rightarrow CH_4 \text{ (reaction R3 in Table 5.2)} \tag{5.17}$$

The rate of this reaction is much slower than that of the other reactions, and so it is not discussed here. It is of importance only when the production of synthetic natural gas (SNG) is desired.

5.3.4 Char Combustion Reactions

Most gasification reactions are endothermic. To provide the required heat of reaction as well as that required for heating, drying, and pyrolysis, a certain amount of exothermic combustion reaction is allowed in a gasifier. Reaction R5 ($C + O_2 \rightarrow CO_2$) is the best in this regard as it gives the highest amount of heat (394 kJ) per k.mol of carbon consumed. The next best is R4 ($C + 1/2O_2 \rightarrow CO$), which also produces the fuel gas CO, but produces only 111 kJ/mol of heat. The speed of R4 is relatively slow.

When carbon comes in contact with oxygen, both R4 and R5 can take place, but their extent depends on temperature. A partition coefficient, β, may be defined to determine how oxygen will partition itself between the two. R4 and R5 may be combined and written as

$$\beta C + O_2 \rightarrow 2(\beta - 1)CO + (2 - \beta)CO_2 \tag{5.18}$$

The value of the partition coefficient β lies between 1 and 2 and depends on temperature. One of the commonly used expressions (Arthur, 1951) for β is

$$\beta = \frac{[CO]}{[CO_2]} = 2400e^{-\left(\frac{6234}{T}\right)} \tag{5.19}$$

where T is the surface temperature of the char.

Combustion reactions are generally faster than gasification reactions under similar conditions. Table 5.3 compares the rate of combustion and gasification for a biomass char at a typical gasifier temperature of 900 °C. The combustion rates are at least one order of magnitude faster than the gasification reaction rate. Owing to pore diffusion resistance, finer char particles have a much higher reaction rate.

TABLE 5.3 Comparison of the Effect of Pore Diffusion on Char Gasification and Combustion Rates

Particle Size (μm)	Combustion Rate (min^{-1})	Gasification Rate (min^{-1})	Combustion Rate/ Gasification Rate (–)
6350	0.648	0.042	15.4
841	5.04	0.317	15.9
74	55.9	0.975	57.3

Source: Adapted from Reed, 2002, p. II-189.

Another important difference between char gasification and combustion reactions in a fluidized bed is that during gasification the temperature of the char particle is nearly the same as the bed temperature because of simultaneous exothermic and endothermic reactions on it (Gomez-Barea et al., 2008). In combustion, the char particle temperature can be much hotter than the bed temperature (Basu, 1977).

The relative amounts of fuel, oxidant (air or oxygen), and steam (if used) govern the fraction of carbon or oxygen that enters R5 or R4 (Table 5.2). Any more oxidant than that needed for the endothermic reaction will increase the gasifier temperature unnecessarily as well as reduce the quality of the product by diluting it with carbon dioxide. Example 5.1 illustrates how the heat balance works out in a gasifier.

Example 5.1

In an updraft gasifier, the water–gas gasification reaction ($C + H_2O \rightarrow CO + H_2 + 131$ kJ/mol) is to be carried out. Assume that drying and other losses in the system need 50% additional heat. Find a means to adjust the extent of the combustion reaction by controlling the supply of oxygen and carbon such that this need is met.

Solution

The reaction needs 131 kJ of heat for gasification of each mol of carbon. In oxygen-deficient or substoichiometric conditions like that present in a gasifier, the exothermic combustion reaction ($C + 1/2O_2 \rightarrow CO - 111$ kJ/mol) is more likely to take place than the more complete combustion reaction ($C + O_2 \rightarrow CO_2 - 394$ kJ/mol). If we adjust the feedstock such that for every mole of carbon gasified, only p moles of carbon will be partially oxidized using $p/2$ mol of oxygen, the heat released by the combustion reaction will exactly balance the heat needed by the gasification reaction. In that case the reaction is

$$C + H_2O \rightarrow CO + H_2 + 131\,\text{kJ/mol} \qquad \text{(i)}$$

- Heat required for endothermic reaction/k.mol C = 131 kJ
- Heat required for drying, etc. = 0.5 × 131 = 65.5 kJ
- Total heat required = 131 + 65.5 = 196.5 kJ

If p moles of carbon participate in the exothermic reaction, R4,

$$pC + 0.5pO_2 \rightarrow pCO - 111p \quad \text{(ii)}$$

Then we have $111p = 196.5$ or $p = 1.77$

Adding reactions (i) and (ii), we get the net reaction

$$2.77C + H_2O + 0.88O_2 \rightarrow 1.77CO + H_2$$

Thus, for (2.77 × 12) kg of carbon, we need (2 + 16) kg of steam and (0.88 × 32) kg of oxygen. If we add more oxygen, the combustion reaction, R5, may take place and the temperature of the combustion zone may rise further.

5.3.5 Catalytic Gasification

Use of catalysts in the thermochemical conversion of biomass may not be essential, but it can help under certain circumstances. Two main motivations for catalysts are:

- Removal of tar from the product gas, especially if the downstream application or the installed equipment cannot tolerate it (see Chapter 4 for more details).
- Reduction in methane content of the product gas, particularly when it is to be used as syngas (CO, H_2 mixture).

The development of catalytic gasification is driven by the need for tar reforming. When the product gas passes over the catalyst particles, the tar or condensable hydrocarbon can be reformed on the catalyst surface with either steam or carbon dioxide, thus producing additional hydrogen and carbon monoxide. The reactions may be written in simple form as

Steam reforming reaction:

$$C_nH_m + nH_2O \xrightarrow{\text{catalyst}} (n + m/2)H_2 + nCO \quad (5.20)$$

Carbon dioxide (or dry) reforming reaction:

$$C_nH_m + nCO_2 \xrightarrow{\text{catalyst}} 2nCO + (m/2)H_2 \quad (5.21)$$

As we can see, instead of undesirable tar or soot, we get additional fuel gases through the catalytic tar-reforming reactions (Eq. 5.20). Both gas yield and the heating value of the product gas improve.

The other option for tar removal is thermal cracking, but it requires a high (>1100 °C) temperature and produces soot; thus, it cannot harness the lost energy in tar hydrocarbon.

The second motivation for catalytic gasification is removal of methane from the product gas. For this we can use either catalytic steam reforming or catalytic carbon dioxide reforming of methane. Reforming is very important for the production of syngas, which cannot tolerate methane and requires a precise ratio of CO and H_2 in the product gas. In steam reforming, methane reacts with steam in the temperature range of 700 to 1100 °C in the presence of a metal-based catalyst, and thus it is reformed into CO and H_2 (Li et al., 2007):

$$CH_4 + H_2O \xrightarrow{\text{catalyst}} CO + 3H_2 + 206\ \text{kJ/mol} \quad \text{– steam reforming of methane} \tag{5.22}$$

This reaction is widely used in hydrogen production from methane, for which nickel-based catalysts are very effective.

The carbon dioxide reforming of methane is not as widely used commercially as steam reforming, but it has the special attraction of reducing two greenhouse gases (CO_2 and CH_4) in one reaction, and it can be a good option for removal of carbon dioxide from the product gas. The reaction is highly endothermic (Wang and Lu, 1996):

$$CH_4 + CO_2 \xrightarrow{\text{catalyst}} 2CO + 2H_2 + 247\ \text{kJ/mol} \quad \text{– dry reforming of methane} \tag{5.23}$$

Nickel-based catalysts are also effective for the dry-reforming reaction (Liu et al., 2008).

Catalyst Selection

Catalysts for reforming reactions are to be chosen keeping in view their objective and practical use. Some important catalyst selection criteria for the removal of tar are as follows:

- Effective
- Resistant to deactivation by carbon fouling and sintering
- Easily regenerated
- Strong and resistant to attrition
- Inexpensive

For methane removal, the following criteria are to be met in addition to those in the previous list:

- Capable of reforming methane
- Must provide the required CO/H_2 ratio for the syngas process

Catalysts can work in in-situ and post-gasification reactions. The former may involve impregnating the catalyst in the biomass prior to gasification. It can be added directly in the reactor, as in a fluidized bed. Such application is effective in reducing the tar, but it is not effective in reducing methane (Sutton et al., 2001). In post-gasification, catalysts are placed in a secondary reactor downstream of the gasifier to convert the tar and methane formed. This has the additional advantage of being independent of the gasifier operating condition.

The second reactor can be operated at temperatures optimum for the reforming reaction.

The catalysts in biomass gasification are divided into three groups: earth metal, alkali metal, and nickel based.

Earth metal catalysts. Dolomite ($CaCO_3.MgCO_3$) is very effective for disposal of tar, and it is inexpensive and widely available, obviating the need for catalyst regeneration. It can be used as a primary catalyst by mixing with the biomass or as a secondary catalyst in a reformer downstream, which is also called a *guard bed*. Calcined dolomite is significantly more effective than raw dolomite (Sutton et al., 2001). Neither, however, is very useful for methane conversion. The rate of the reforming reaction is higher with carbon dioxide than with steam.

Alkali metal catalysts. Potassium carbonate and sodium carbonate are important in biomass gasification as primary catalysts. K_2CO_3 is more effective than Na_2CO_3. Unlike dolomite, they can reduce methane in the product gas through a reforming reaction. Many biomass types have inherent potassium in their ash, so they can benefit from the catalytic action of the potassium with reduced tar production. However, potassium is notorious for agglomerating in fluidized beds, which offsets its catalytic benefit.

Ni-based catalyst. Nickel is highly effective as a reforming catalyst for reduction of tar as well as for adjustment of the CO/H_2 ratio through methane conversion. It performs best when used downstream of the gasifier in a secondary bed, typically at 780 °C (Sutton et al., 2001). Deactivation of the catalyst with carbon deposits is an issue. Nickel is relatively inexpensive and commercially available though not as cheap as dolomite. Appropriate catalyst support is important for optimum performance.

5.3.6 Gasification Processes in the Reactors

The sequence of gasification reactions depends to some extent on the type of gas–solid contacting reactors used. A brief description of this process as it occurs in some principal reactor types follows.

Moving-Bed Reactor

To explain the reaction process in moving-bed gasifiers, we take the example of a simple updraft gasifier reactor (Figure 5.5).

In a typical updraft gasifier, fuel is fed from the top; the product gas leaves from the top as well. The gasifying agent (air, oxygen, steam, or their mixture), is slightly preheated and enters the gasifier through a grid at the bottom. The gas then rises through a bed of descending fuel or ash in the gasifier chamber.

The air (the gasifying medium), as it enters the bottom of the bed, meets hot ash and unconverted chars descending from the top (Figure 5.5). The

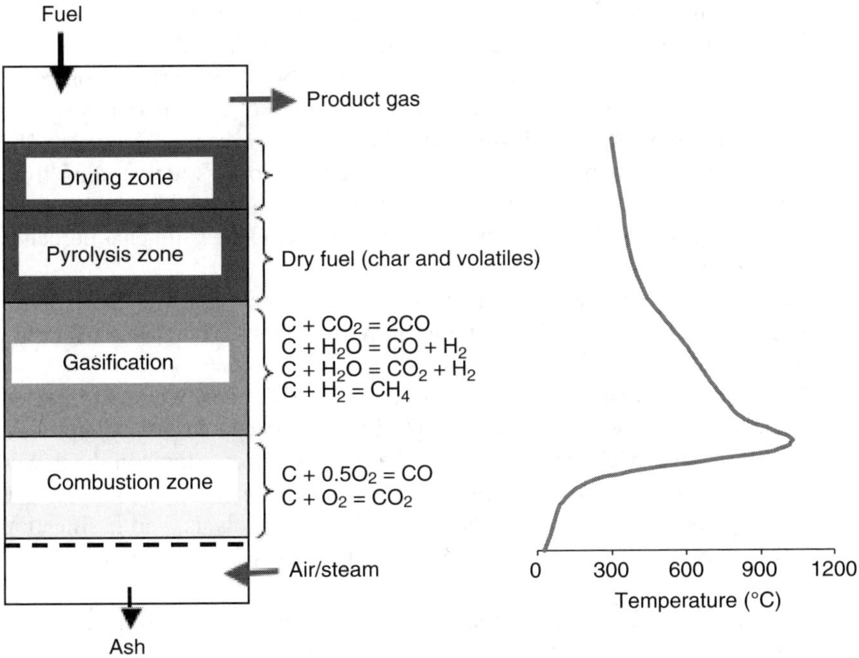

FIGURE 5.5 Stages of gasification in an updraft gasifier.

temperature in the bottom layer well exceeds the ignition temperature of carbon, so the highly exothermic combustion reaction (Eq. 5.24) takes place in the presence of excess oxygen. The released heat heats the upward-moving gas as well as the descending solids.

$$C + O_2 \rightarrow CO_2 - 394\ \text{kJ/mol} \tag{5.24}$$

The combustion reaction (Eq. 5.24), being very fast, rapidly consumes most of the available oxygen. As the available oxygen is reduced further up, the combustion reaction changes into partial combustion, releasing CO and a moderate amount of heat.

$$C + 1/2\,O_2 \rightarrow CO - 111\ \text{kJ/mol} \tag{5.25}$$

The hot gas, a mixture of CO, CO_2, and steam (from the feed and the gasifying medium), moves further up into the gasification zone, where char from the upper bed is gasified by Eq. (5.26). The carbon dioxide concentration increases rapidly in the first combustion zone, but once the oxygen is nearly depleted, the CO_2 enters the gasification reaction (Eq. 5.26) with char, resulting in a decline in CO_2 concentration in the gasification zone.

$$C + CO_2 \rightarrow 2CO + 172\ \text{kJ/mol} \tag{5.26}$$

$$C + H_2O \rightarrow CO + H_2 + 131\ \text{kJ/mol}$$

Sensible heating of the hot gas provides the heat for the two endothermic gasification reactions in Eq. (5.26): R1 and R2 (Table 5.2). These are responsible for most of the gasification products like hydrogen and carbon monoxide. Because of their endothermic nature, the temperature of the gas reduces.

The zone above the gasification zone is for the pyrolysis of biomass. The residual heat of the rising hot gas heats up the dry biomass, descending from above. The biomass then decomposes (pyrolyzed) into noncondensable gases, condensable gases, and char. Both gases move up while the solid char descends with other solids.

The topmost zone dries the fresh biomass fed into it using the balance enthalpy of the hot product gas coming from the bottom. This gas is a mixture of gasification and pyrolysis products.

In an updraft gasifier biomass fed from the top descends, while air injected from the side meets with the pyrolysis product, releasing heat (see Chapter 6). Thereafter, both product gas and solids (char and ash) move down in the downdraft gasifier. Here, a part of the pyrolysis gas may burn above the gasification zone. Thus, the thermal energy required for drying, pyrolysis, and gasification is supplied by the combustion of pyrolysis gas. This phenomenon is called *flaming pyrolysis*.

In downdraft gasifiers, the reaction regions are different from those for updraft gasifiers. Here, steam and oxygen or air are fed into a lower section of the gasifier (Figure 5.6) with the biomass. The pyrolysis and combustion products flow downward. The hot gas then moves downward over the remaining

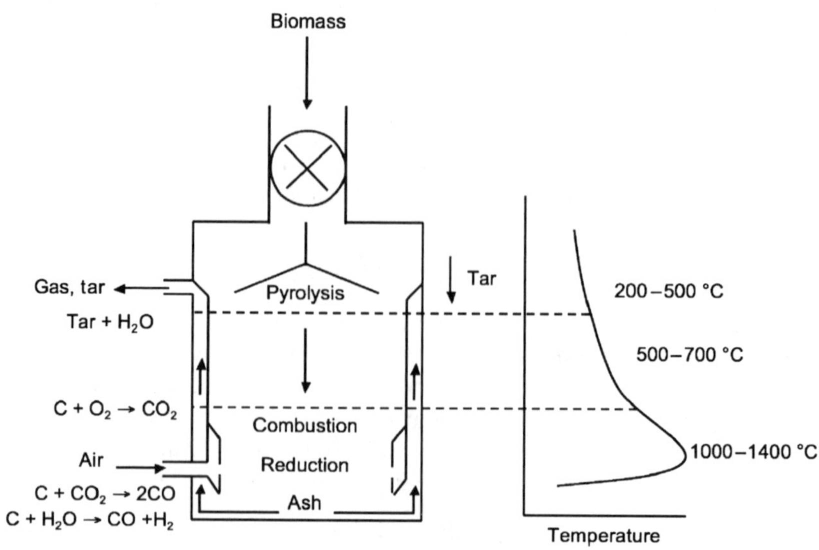

FIGURE 5.6 Gasification reactions in a downdraft gasifier.

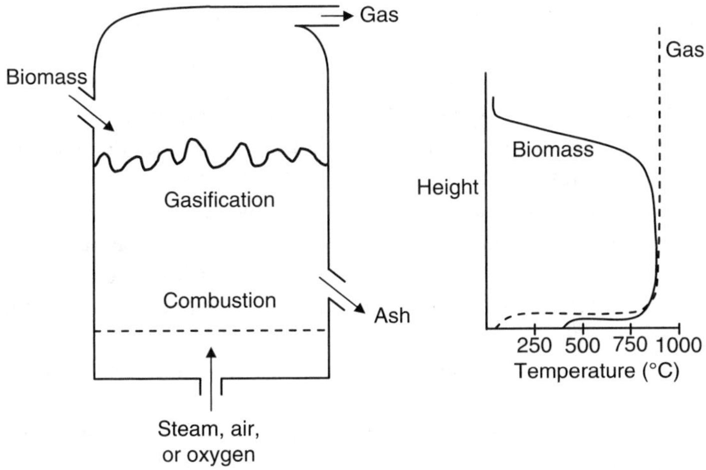

FIGURE 5.7 Schematic of a bubbling fluidized-bed gasifier. *(Source: Higman and van der Burgt, 2008, p. 106.)*

hot char, where gasification takes place. Such an arrangement results in tar-free but low-energy-content gases.

Fluidized-Bed Reactor

In a bubbling fluidized bed, the fuel fed from either the top or the sides mixes relatively fast over the whole body of the fluid bed (Figure 5.7). The gasifying medium (air, oxygen, steam, or their mixture) also serves as the fluidizing gas and so is sent through the bottom of the reactor.

In a typical fluidized-bed gasifier, fresh solid fuel particles are brought into contact with hot bed solids that quickly heat the particles to the bed temperature and make them undergo rapid drying and pyrolysis, producing char and gases.

Though the bed solids are well mixed, the fluidizing gas remains generally in plug-flow mode, entering from the bottom and leaving from the top. Upon entering the bottom of the bed, the oxygen goes into fast exothermic reactions (R4, R5, and R8 in Table 5.2) with char mixed with bed materials. The bed materials immediately disperse the heat released by these reactions to the entire fluidized bed. The amount of heat released near the bottom grid depends on the oxygen content of the fluidizing gas and the amount of char that comes in contact with it. The local temperature in this region depends on how vigorously the bed solids disperse heat from the combustion zone.

Subsequent gasification reactions take place further up as the gas rises. The bubbles of the fluidized bed can serve as the primary conduit to the top. They are relatively solids-free. While they help in mixing, the bubbles can also allow gas to bypass the solids without participating in the gasification reactions. The pyrolysis products coming in contact with the hot solids break down into noncondensable gases. If they escape the bed and rise into the cooler freeboard, tar and char are formed.

A bubbling fluidized bed cannot achieve complete char conversion because of the back-mixing of solids. The high degree of solid mixing helps a bubbling fluidized-bed gasifier achieve temperature uniformity, but owing to the intimate mixing of fully gasified and partially gasified fuel particles, any solids leaving the bed contain some partially gasified char. Char particles entrained from a bubbling bed can also contribute to the loss in a gasifier. The other important problem with fluidized-bed gasifiers is the slow diffusion of oxygen from the bubbles to the emulsion phase. This encourages the combustion reaction in the bubble phase, which decreases gasification efficiency.

In a circulating fluidized bed (CFB), solids circulate around a loop that is characterized by intense mixing and longer solid residence time within its solid circulation loop. The absence of any bubbles avoids the gas-bypassing problem of bubbling fluidized beds.

Fluidized-bed gasifiers typically operate in the temperature range of 800 to 1000 °C to avoid ash agglomeration. This is satisfactory for reactive fuels such as biomass, municipal solid waste (MSW), and lignite. Since fluidized-bed gasifiers operate at relatively low temperatures, most high-ash fuels, depending on ash chemistry, can be gasified without the problem of ash sintering and agglomeration. Owing to the large thermal inertia and vigorous mixing in fluidized-bed gasifiers, a wider range of fuels or a mixture of them can be gasified. This feature is especially attractive for biomass fuels, such as agricultural residues and wood, that may be available for gasification at different times of the year. For these reasons, many developmental activities on large-scale biomass gasification are focused on fluidized-bed technologies.

Entrained-Flow Reactor

Entrained-flow gasifiers are preferred for the integrated gasification combined cycle (IGCC) plants. Reactors of this type typically operate at 1400 °C and 20 to 70 bar pressure, where powdered fuel is entrained in the gasifying medium. Figure 5.8 shows two entrained-flow gasifier types. In the first one, oxygen, the most common gasifying medium, and the powdered fuel enter from the side; in the second one they enter from the top.

In entrained-flow gasifiers, the combustion reaction, R5 (Eq. 5.24), may take place right at the entry point of the oxygen, followed by reaction R4 (Eq. 5.25) further downstream, where the excess oxygen is used up.

Powdered fuel (< 75 micron) is injected into the reactor chamber along with oxygen and steam (air is rarely used). To facilitate feeding into the reactor, especially if it is pressurized, the fuel may be mixed with water to make a slurry. The gas velocity in the reactor is sufficiently high to fully entrain the fuel particles. Slurry-fed gasifiers need additional reactor volume for evaporation of the large amount of water mixed with the fuel. Furthermore, their oxygen

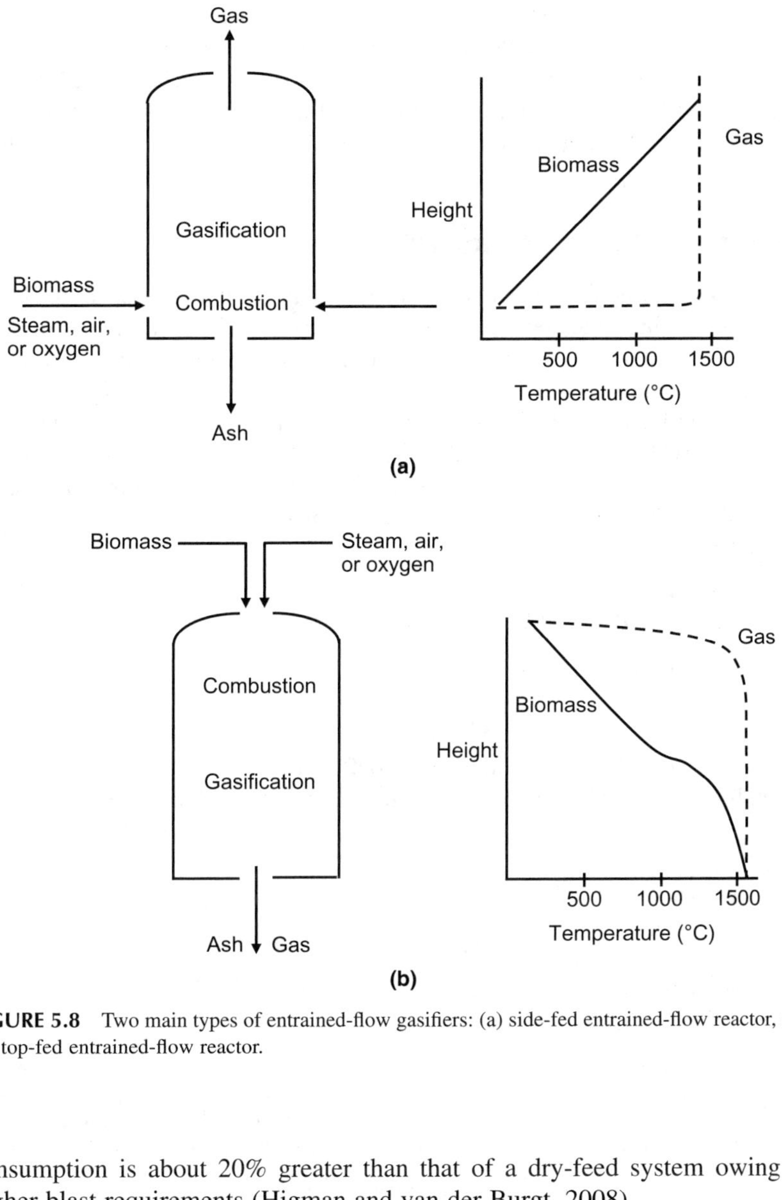

FIGURE 5.8 Two main types of entrained-flow gasifiers: (a) side-fed entrained-flow reactor, and (b) top-fed entrained-flow reactor.

consumption is about 20% greater than that of a dry-feed system owing to higher blast requirements (Higman and van der Burgt, 2008).

Entrained flow gasifiers are of two types depending on how and where the fuel is injected into the reactor. Chapter 6 discusses several types. In all of these designs, oxygen, upon entering the reactor, reacts rapidly with the volatiles and char in exothermic reactions. These raise the reactor temperature well above the melting point of ash, resulting in complete destruction of tar or oil. Such high temperatures should give a very high level of carbon conversion.

An entrained-flow gasifier may be viewed as a plug-flow reactor. Although the gas is heated to the reactor temperature rapidly upon entering, solids heat up less slowly along the reactor length because of the reactor's large thermal capacity and plug-flow nature, as shown in Figure 5.8. Some entrained-flow reactors are modeled as stirred tank reactors because of the rapid mixing of solids.

5.4 KINETICS OF GASIFICATION

Stoichiometric calculations can help determine the products of reaction. Not all reactions are instantaneous and completely convert reactants into products. Many of the chemical reactions discussed in the preceding sections proceed at a finite rate and to a finite extent.

To what extent a reaction progresses is determined by its equilibrium state. Its kinetic rates, on the other hand, determine how fast the reaction products are formed and whether the reaction completes within the gasifier chamber. A review of the basics of chemical equilibrium may be useful before discussing its results.

5.4.1 Chemical Equilibrium

Let us consider the reaction:

$$nA + mB \xrightarrow{k_{for}} pC + qD \tag{5.27}$$

where n, m, p, and q are stoichiometric coefficients. The rate of this reaction, r_1, depends on C_A and C_B, the concentration of the reactants A and B, respectively.

$$r_1 = k_{for}\, C_A^n\, C_B^m \tag{5.28}$$

The reaction can also move in the opposite direction:

$$pC + qD \xrightarrow{k_{back}} nA + mB \tag{5.29}$$

The rate of this reaction, r_2, is similarly written in terms of C_C and C_D, the concentration of C and D, respectively:

$$r_2 = k_{back}\, C_C^p\, C_D^q \tag{5.30}$$

When the reaction begins, the concentration of the reactants A and B is high. So the forward reaction rate r_1 is initially much higher than r_2, the reverse reaction rate, because the product concentrations are relatively low. The reaction in this state is not in equilibrium, as $r_1 > r_2$. As the reaction progresses, the forward reaction increases the buildup of products C and D. This increases the reverse reaction rate. Finally, a stage comes when the two rates are equal to each other ($r_1 = r_2$). This is the equilibrium state. At equilibrium,

- There is no further change in the concentration of the reactants and the products.

- The forward reaction rate is equal to the reverse reaction rate.
- The Gibbs free energy of the system is at minimum.
- The entropy of the system is at maximum.

Under equilibrium state, we have

$$r_1 = r_2$$

$$k_{for}\, C_A^n\, C_B^m = k_{back}\, C_C^p\, C_D^q \tag{5.31}$$

Reaction Rate Constant

A rate constant, k_i, is independent of the concentration of reactants but is dependent on the reaction temperature, T. The temperature dependency of the reaction rate constant is expressed in Arrhenius form as

$$k = A_0 \exp\left(-\frac{E}{RT}\right) \tag{5.32}$$

where A_0 is a pre-exponential constant, R is the universal gas constant, and E is the activation energy for the reaction.

The ratio of rate constants for the forward and reverse reactions is the equilibrium constant, K_e. From Eq. (5.31) we can write

$$K_e = \frac{k_{for}}{k_{back}} = \frac{C_C^p C_D^q}{C_A^n C_B^q} \tag{5.33}$$

The equilibrium constant, K_e, depends on temperature but not on pressure. Table 5.4 gives values of equilibrium constants and heat of formation of some gasification reactions (Probstein and Hicks, 2006, pp. 62–64).

TABLE 5.4 Equilibrium Constants and Heats of Formation for Five Gasification Reactions

	Equilibrium Constant ($\log_{10}K$)			Heat of Formation (kJ/mol)	
Reaction	*298 K*	*1000 K*	*1500 K*	*1000 K*	*1500 K*
$C + \frac{1}{2}O_2 \rightarrow CO$	24.065	10.483	8.507	−111.9	−116.1
$C + O_2 \rightarrow CO_2$	69.134	20.677	13.801	−394.5	−395.0
$C + 2H_2 \rightarrow CH_4$	8.906	−0.999	−2.590	−89.5	−94.0
$2C + 2H_2 \rightarrow C_2H_4$	−11.940	−6.189	−5.551	38.7	33.2
$H_2 + \frac{1}{2}O_2 \rightarrow H_2O$	40.073	10.070	5.733	−247.8	−250.5

Source: Data compiled from Probstein and Hicks, 2006, p. 64.

Gibbs Free Energy

Gibbs free energy, G, is an important thermodynamic function. Its change in terms of a change in entropy, ΔS, and enthalpy, ΔH, is written as

$$\Delta G = \Delta H - T\Delta S \tag{5.34}$$

The change in enthalpy or entropy for a reaction system is computed by finding the enthalpy or entropy changes of individual gases in the system. It is explained in Example 5.2. An alternative approach uses the empirical equations given by Probstein and Hicks (2006). It expresses the Gibbs function (Eq. 5.35) and the enthalpy of formation (Eq. 5.36) in terms of temperature, T, the heat of formation at the reference state at 1 atmosphere and 298 K, and a number of empirical coefficients, a', b', and so forth.

$$\Delta G^{0}_{f,T} = \Delta h^{0}_{298} - a'T\ln(T) - b'T^2 - \left(\frac{c'}{2}\right)T^3 - \left(\frac{d'}{3}\right)T^4 + \left(\frac{e'}{2T}\right) + f' + g'T \quad \text{kJ/mol} \tag{5.35}$$

$$\Delta H^{0}_{f,T} = \Delta h^{0}_{298} + a'T + b'T^2 + c'T^3 + d'T^4 + \left(\frac{e'}{T}\right) + f' \quad \text{kJ/mol} \tag{5.36}$$

The values of the empirical coefficients for some common gases are given in Table 5.5.

The equilibrium constant of a reaction occurring at a temperature T may be known using the value of Gibbs free energy.

$$K_e = \exp\left(-\frac{\Delta G}{RT}\right) \tag{5.37}$$

Here, ΔG is the standard Gibbs function of reaction or free energy change for the reaction, R is the universal gas constant, and T is the gas temperature.

Example 5.2

Find the equilibrium constant at 2000 K for the reaction

$$CO_2 \rightarrow CO + \tfrac{1}{2}O_2$$

Solution

Enthalpy change is written by taking the values for it from the NIST-JANAF thermochemical tables (Chase, 1998) for 2000 K:

$$\begin{aligned}\Delta H &= (h_f^0 + \Delta h)_{CO} + (h_f^0 + \Delta h)_{O_2} - (h_f^0 + \Delta h)_{CO_2} \\ &= 1\,\text{mol}\,(-110{,}527 + 56{,}744)\ \text{J/mol} + 1/2\ \text{mol}\,(0 + 59{,}175)\ \text{J/mol} \\ &\quad - 1\,\text{mol}\,(-393{,}522 + 91{,}439)\ \text{J/mol} = 277{,}887\ \text{J}\end{aligned}$$

The change in entropy, ΔS, is written in the same way as for taking the values of entropy change from the NIST-JANAF tables (see list that follows on page 140).

TABLE 5.5 Heat of Combustion, Gibbs Free Energy, and Heat of Formation at 298 K, 1 Atm, and Empirical Coefficients from Eqs. 5.35 and 5.36

Product	*HHV* (kJ/mol)	$\Delta_f G_{298}$ (kJ/mol)	$\Delta_f H_{298}$ (kJ/mol)	Empirical Coefficients						
				a'	b'	c'	d'	e'	f'	g'
C	393.5	0	0							
CO	283	−137.3	−110.5	5.619×10^{-3}	-1.19×10^{-5}	6.383×10^{-9}	-1.846×10^{-12}	-4.891×10^{2}	0.868	-6.131×10^{-2}
CO_2	0	−394.4	−393.5	-1.949×10^{-2}	3.122×10^{-5}	-2.448×10^{-8}	6.946×10^{-12}	-4.891×10^{2}	5.27	−0.1207
CH_4	890.3	−50.8	−74.8	-4.62×10^{-2}	1.13×10^{-5}	1.319×10^{-8}	-6.647×10^{-12}	-4.891×10^{2}	14.11	0.2234
C_2H_4	1411	68.1	52.3	-7.281×10^{-2}	5.802×10^{-5}	-1.861×10^{-8}	5.648×10^{-13}	-9.782×10^{2}	20.32	−0.4076
CH_3OH	763.9	−161.6	−201.2	-5.834×10^{-2}	2.07×10^{-5}	1.491×10^{-8}	-9.614×10^{-12}	-4.891×10^{2}	16.88	−0.2467
H_2O (steam)	0	−228.6	−241.8	-8.95×10^{-3}	-3.672×10^{-6}	5.209×10^{-9}	-1.478×10^{-12}	0	2.868	−0.0172
H_2O (water)	0	−237.2	−285.8							
O_2	0	0	0							
H_2	285.8	0	0							

Source: Adapted from Probstein and Hicks, 2006, pp. 55, 61.

$$\begin{aligned}\Delta S &= 1\times S_{CO} + \tfrac{1}{2}\times S_{O_2} - 1\times S_{CO_2}\\ &= (1\,\text{mol}\times 258.71\ \text{J/mol K}) + (1/2\ \text{mol}\times 268.74\ \text{J/mol K})\\ &\quad - (1\,\text{mol}\times 309.29\ \text{J/mol K})\\ &= 83.79\ \text{J/K}\end{aligned}$$

From Eq. (5.34), the change in the Gibbs free energy can be written as

$$\begin{aligned}\Delta G &= \Delta H - T\Delta S\\ &= 277.887\ \text{kJ} - (2{,}000\ \text{K}\times 83.79\ \text{J/K}) = 110.307\ \text{kJ}\end{aligned}$$

The equilibrium constant is calculated using Eq. (5.37):

$$K_{2000K} = e^{-\frac{\Delta G}{RT}} = e^{-\left(\frac{110.307}{0.008314*2000}\right)} = 0.001315 \tag{5.38}$$

Kinetics of Gas–Solid Reactions

The rate of gasification of char is much slower than the rate of pyrolysis of the biomass that produces the char. Thus, the volume of a gasifier is more dependent on the rate of char gasification than on the rate of pyrolysis. The char gasification reaction therefore plays a major role in the design and performance of a gasifier.

Typical temperatures of the gasification zone in downdraft and fluidized-bed reactors are in the range of 700 to 900 °C. The three most common gas–solid reactions that occur in the char gasification zone are

$$\text{Boudouard reaction:}\quad (\text{R1: C} + \text{CO}_2 \rightarrow 2\text{CO}) \tag{5.39}$$

$$\text{Water–gas reaction:}\quad (\text{R2: C} + \text{H}_2\text{O} \leftrightarrow \text{CO} + \text{H}_2) \tag{5.40}$$

$$\text{Methanation reaction:}\quad (\text{R3: C} + 2\text{H}_2 \leftrightarrow \text{CH}_4). \tag{5.41}$$

The water–gas reaction, R2, is dominant in a steam gasifier. In the absence of steam, when air or oxygen is the gasifying medium, the Boudouard reaction, R1, is dominant. However, the steam gasification reaction rate is higher than the Boudouard reaction rate.

Another important gasification reaction is the shift reaction, R9 ($CO + H_2O \leftrightarrow CO_2 + H_2$), which takes place in the gas phase. It is discussed in the next section. A popular form of the gas–solid char reaction, r, is the nth-order expression:

$$r = \frac{1}{(1-X)^m}\frac{dX}{dt} = A_0 e^{-\frac{E}{RT}} P_i^n\ \text{s}^{-1} \tag{5.42}$$

where X is the fractional carbon conversion, A_0 is the apparent pre-exponential constant (1/s), E is the activation energy (kJ/mol), m is the reaction order with respect to the carbon conversion, T is the temperature (K), and n is the reaction order with respect to the gas partial pressure, P_i. The universal gas constant, R, is 0.008314 kJ/mol.K.

Boudouard Reaction

Referring to the Boudouard reaction (R1) in Eq. (5.6), we can use the Langmuir–Hinshelwood rate, which takes into account CO inhibition (Cetin et al., 2005) to express the apparent gasification reaction rate, r_b:

$$r_b = \frac{k_{b1}P_{CO_2}}{1+(k_{b2}/k_{b3})P_{CO}+(k_{b1}/k_{b3})P_{CO_2}}\ s^{-1} \tag{5.43}$$

where P_{CO} and P_{CO_2} are the partial pressure of CO and CO_2, respectively, on the char surface (bar). The rate constants, k_i, are given in the form, $A\ \exp(-E/RT)$ $bar^{-n}s^{-n}$, where A is the pre-exponential factor ($bar^{-n}.s^{-n}$). Barrio and Hustad (2001) gave some values of the pre-exponential factor and the activation energy for Birch wood (Table 5.6).

When the concentration of CO is relatively small, and when its inhibiting effect is not to be taken into account, the kinetic rate of gasification by the Boudouard reaction may be expressed by a simpler nth-order equation as

$$r_b = A_b e^{-\frac{E}{RT}} P_{CO_2}^n\ s^{-1} \tag{5.44}$$

For the Boudouard reaction, the values of the activation energy, E, for biomass char are typically in the range of 200 to 250 kJ/mol, and those of the exponent, n, are in the range of 0.4 to 0.6 (Blasi, 2009). Typical values of A, E, and n for birch, poplar, cotton, wheat straw, and spruce are given in Table 5.7.

The reverse of the Boudouard reaction has a major implication, especially in catalytic reactions, as it deposits carbon on its catalyst surfaces, thus deactivating the catalyst.

$$2CO \rightarrow CO_2 + C - 172\ kJ/mol \tag{5.45}$$

TABLE 5.6 Activation Energy and Pre-Exponential Factors for Birch Char Using the Langmuir-Hinshelwood Rate Constants for CO_2 Gasification

Langmuir-Hinshelwood Rate Constants (s^{-1} bar^{-1})	Activation Energy, E (kJ/mol)	Pre-Exponential Actor, A (s^{-1} bar^{-1})
k_{b1}	165	1.3×10^5
k_{b2}	20.8	0.36
k_{b3}	236	3.23×10^7

Source: Adapted from Barrio and Hustad, 2001.

TABLE 5.7 Typical Values for Activation Energy, Pre-Exponential Factor, and Reaction Order for Char in the Boudouard Reaction

Char Origin	Activation Energy, E (kJ/mol)	Pre-Exponential Factor, A (s^{-1} bar^{-1})	Reaction Order, n (–)	Reference
Birch	215	3.1×10^6 s^{-1} $bar^{-0.38}$	0.38	Barrio and Hustad, 2001
Dry poplar	109.5	153.5 s^{-1} bar^{-1}	1.2	Barrio and Hustad, 2001
Cotton wood	196	4.85×10^8 s^{-1}	0.6	DeGroot and Shafizadeh, 1984
Douglas fir	221	19.67×10^8 s^{-1}	0.6	DeGroot and Shafizadeh, 1984
Wheat straw	205.6	5.81×10^6 s^{-1}	0.59	Risnes et al., 2001
Spruce	220	21.16×10^6 s^{-1}	0.36	Risnes et al., 2001

The preceding reaction becomes thermodynamically feasible when (P_{CO}^2/P_{CO_2}) is much greater than that of the equilibrium constant of the Boudouard reaction (Littlewood, 1977).

Water–Gas Reaction

Referring to the water–gas reaction, the kinetic rate, r_w, may also be written in Langmuir-Hinshelwood form to consider the inhibiting effect of hydrogen and other complexes (Blasi, 2009).

$$r_w = \frac{k_{w1}P_{H_2O}}{1+(k_{w1}/k_{w3})P_{H_2O}+(k_{w2}/k_{w3})P_{H_2}}\ s^{-1} \tag{5.46}$$

where P_i is the partial pressure of gas i in bars.

Typical rate constants according to Barrio et al. (2001) for beech wood are

$$k_{w1} = 2.0\times10^7 \exp(-199/RT);\ bar^{-1}s^{-1}$$

$$k_{w2} = 1.8\times10^6 \exp(-146/RT);\ bar^{-1}s^{-1}$$

$$k_{w3} = 8.4\times10^7 \exp(-225/RT)\ bar^{-1}s^{-1}$$

Most kinetic analysis, however, uses a simpler nth-order expression for the reaction rate:

$$r_w = A_w e^{-\frac{E}{RT}} P_{H_2O}^n\ s^{-1} \tag{5.47}$$

Typical values for the activation energy, E, for steam gasification of char for some biomass types are given in Table 5.8.

TABLE 5.8 Activation Energy, Pre-Exponential Factor, and Reaction Order for Char for the Water–Gas Reaction

Char Origin	Activation Energy, *E* (kJ/mol)	Pre-Exponential Factor, A_w (s^{-1} bar^{-1})	Reaction Order, *n* (–)	Reference
Birch	237	2.62×10^8 s^{-1} bar^{-n}	0.57	Barrio et al., 2001
Beech	211	0.171×10^8 s^{-1} bar^{-n}	0.51	Barrio et al., 2001
Wood	198	0.123×10^8 s^{-1} atm^{-n}	0.75	Hemati and Laguerie, 1988
Various biomass	180–200		0.04–1.0	Blasi, 2009

Hydrogasification Reaction (Methanation)

The hydrogasification reaction is as follows:

$$C + 2H_2 \Leftrightarrow CH_4 \tag{5.48}$$

With freshly devolatilized char, this reaction progresses rapidly, but graphitization of carbon soon causes the rate to drop to a low value. The reaction involves volume increase, and so pressure has a positive influence on it. High pressure and rapid heating help this reaction. Wang and Kinoshita (1993) measured the rate of this reaction and obtained values of $A = 4.189 \times 10^{-3}$ s^{-1} and $E =$ 19.21 kJ/mol.

Steam Reforming of Hydrocarbon

For production of syngas (CO, H_2) direct reforming of hydrocarbon is an option. Here, a mixture of hydrocarbon and steam is passed over a nickel-based catalyst at 700 to 900 °C. The final composition of the product gas depends on the following factors (Littlewood, 1977):

- H/C ratio of the feed
- Steam/carbon (S/C) ratio
- Reaction temperature
- Operating pressure

The mixture of CO and H_2 produced can be subsequently synthesized into required liquid fuels or chemical feedstock. The reactions may be described as

$$C_nH_m + \frac{4m-n}{4}H_2O \Leftrightarrow \frac{4m+n}{8}CH_4 + \frac{4m-n}{8}CO_2 \tag{5.49}$$

$$CH_4 + H_2O \Leftrightarrow CO + 3H_2 \tag{5.50}$$

$$CO + H_2O \Leftrightarrow CO_2 + H_2 \tag{5.51}$$

The first reaction (Eq. 5.48) is favorable at high pressure, as it involves an increase in volume in the forward direction. The equilibrium constant of the first reaction increases with temperature while that of the third reaction (Eq. 5.51), which is also known as the shift reaction, decreases.

Kinetics of Gas-Phase Reactions

Several gas-phase reactions play an important role in gasification. Among them, the shift reaction (R9), which converts carbon monoxide into hydrogen, is most important.

$$R9: CO + H_2O \xrightarrow{k_{for}} CO_2 + H_2 - 41.1\ \text{kJ/mol} \tag{5.52}$$

This reaction is mildly exothermic. Since there is no volume change, it is relatively insensitive to changes in pressure.

The equilibrium yield of the shift reaction decreases slowly with temperature. For a favorable yield, the reaction should be conducted at low temperature, but then the reaction rate will be slow. For an optimum rate, we need catalysts. Below 400 °C, a chromium-promoted iron formulation catalyst ($Fe_2O_3 - Cr_2O_3$) may be used (Littlewood, 1977).

Other gas-phase reactions include CO combustion, which provides heat to the endothermic gasification reactions:

$$R6: CO + 1/2\,O_2 \xrightarrow{k_{for}} CO_2 - 284\ \text{kJ/mol} \tag{5.53}$$

These homogeneous reactions are reversible. The rate of forward reactions is given by the rate coefficients of Table 5.9.

TABLE 5.9 Forward Reaction Rates, r, for Gas-Phase Homogeneous Reactions

Reaction	Reaction Rate (r)	Heat of Formation ($m^3.mol^{-1}.s^{-1}$)	Reference
$H_2 + ½O_2 \rightarrow H_2O$	$K\,C_{H2}^{1.5}C_{O_2}$	$51.8\,T^{1.5}$ exp ($-3420/T$)	Vilienskii and Hezmalian, 1978
$CO + ½O_2 \rightarrow CO_2$	$K\,C_{CO}C_{O2}^{0.5}C_{H2O}^{0.5}$	2.238×10^{12} exp ($-167.47/RT$)	Westbrook and Dryer, 1981
$CO + H_2O \rightarrow CO_2 + H_2$	$K\,C_{CO}C_{H_2O}$	0.2778 exp ($-12.56/RT$)	Petersen and Werther, 2005

Note: Here, the gas constant, R, is in kJ/mol.K.

For the backward CO oxidation reaction ($CO + \frac{1}{2} O_2 \xleftarrow{k_{back}} CO_2$), the rate, k_{back}, is given by Westbrook and Dryer (1981) as

$$k_{back} = 5.18 \times 10^8 \exp(-167.47/RT)\, C_{CO_2} \tag{5.54}$$

For the reverse of the shift reaction ($CO + H_2O \xleftarrow{k_{back}} CO_2 + H_2$), the rate is given as

$$k_{back} = 126.2 \exp(-47.29/RT)\, C_{CO_2} C_{H_2}\ \text{mol.m}^{-3} \tag{5.55}$$

If the forward rate constant is known, then the backward reaction rate, k_{back}, can be determined using the equilibrium constant from the Gibbs free energy equation:

$$K_{equilibrium} = \frac{k_{for}}{k_{back}} = \exp\left(\frac{-\Delta G^0}{RT}\right) \text{at 1 atm pressure} \tag{5.56}$$

ΔG^0 for the shift reaction may be calculated (see Callaghan, 2006) from a simple correlation of

$$\Delta G^0 = -32.197 + 0.031T - (1774.7/T),\ \text{kJ/mol} \tag{5.57}$$

where T is in K.

Example 5.3

For shift reaction $CO + H_2O \rightarrow CO_2 + H_2$, the equilibrium constant at 625 K is given as 20 and that at 1667 K as 0.368. Assume that the reaction begins with 1 mole of CO, 1 mole of H_2O, and 1 mole of nitrogen. Find:

- The equilibrium constant at 1100 K and 1 atm.
- The equilibrium mole fraction of carbon dioxide.
- Whether the reaction is endothermic or exothermic.
- If pressure is increased to 100 atm, the impact of the equilibrium constant at 1100 K.

Solution

Part (a). For the shift reaction, the Gibbs free energy at a certain temperature can be calculated from Eq. (5.57):

$$\Delta G^0 = -32.197 + 0.031\text{T} - (1774.7/T)$$

at 1100 K, ΔG^0 = 0.2896 kJ/mol.

The equilibrium constant can be calculated from Eq. (5.56):

$$K_{equilibrium} = \frac{k_{for}}{k_{back}} = \exp\left(\frac{-\Delta G^0}{RT}\right)$$

$$K_{equilibrium} = \exp\left(\frac{-0.2896}{0.008314 * 1100}\right)$$

$$K_{equilibrium} = 0.9688$$

Part (b). At equilibrium, the rate of the forward reaction will be equal to the rate of the backward reaction, or $K_{equilibrium} = 1$. So, using the definition of the equilibrium constant, we have

$$K_{equilibrium} = \frac{p_{CO_2} p_{H_2}}{p_{CO} p_{H_2O}} = 1$$

where p denotes the partial pressure of the various species. In this reaction, nitrogen stays inert and does not react. Thus, 1 mole of nitrogen comes out from it. If x moles of CO and H_2O react to form x moles of CO_2 and H_2, then at equilibrium, $(1 - x)$ moles of CO and H_2O remain unreacted. We can list the component mole fraction as:

Species	Mole	Mole fraction
CO	$(1 - x)$	$(1 - x) / 3$
H_2O	$(1 - x)$	$(1 - x) / 3$
CO_2	x	$x/3$
H_2	x	$x/3$
N_2	1	1/3

The mole fraction y is related to the partial pressure, p, by the relation $yP = p$, where P stands for total pressure.

Substituting the values for the partial pressures of the various species,

$$1 = \frac{\left(\frac{x}{3}P\right)\left(\frac{x}{3}P\right)}{\left(\frac{1-x}{3}P\right)\left(\frac{1-x}{3}P\right)}$$

Solving for x, we get $x = 0.5$. Thus, the mole fraction of CO_2 at equilibrium = $(1 - x)/3 = 0.5/3 = 0.1667$.

Part (c). To determine if this reaction is exothermic or endothermic, the standard heats of formation of the individual components are taken from the NIST-JANAF thermochemical tables (Chase, 1998).

$$\Delta H = (h_f^0)_{CO_2} + (h_f^0)_{H_2} - [(h_f^0)_{CO} + (h_f^0)_{H_2O}]$$

$$\Delta H = -393.52 \text{ kJ/mol} - 0 \text{ kJ/mol} - [-110.53 \text{ kJ/mol} - 241.82 \text{ kJ/mol}]$$

$$\Delta H = -41.17 \text{ kJ/mol}$$

Since 41.17 kJ/mol of heat is given out, the reaction is exothermic.

Part (d). This reaction does not depend on pressure, as there is no volume change. The equilibrium constant changes only with temperature, so the equilibrium constant at 100 atm is the same as that at 1 atm, for 1100 K. The equilibrium constant is 0.9688 at 100 atm, for 1100 K.

5.4.2 Char Reactivity

Reactivity, generally a property of a solid fuel, is the value of the reaction rate under well-defined conditions of gasifying agent, temperature, and pressure.

Proper values or expressions of char reactivity are necessary for all gasifier models. This topic has been studied extensively for more than 60 years, and a large body of information is available, especially for coal. These studies unearthed important effects of char size, surface area, pore size distribution, catalytic effect, mineral content, pretreatment, and heating. The origin of the char and the extent of its conversion also exert some influence on reactivity.

Char can originate from any hydrocarbon—coal, peat, biomass, and so forth. An important difference between chars from biomass and those from fossil fuels like coal or peat is that the reactivity of biomass chars increases with conversion while that of coal or peat char decreases. Figure 5.3 plots the reactivity for hardwood and peat against their conversion (Liliedahl and Sjostrom, 1997). It is apparent that, while the conversion rate (at conversion 0.8) of hardwood char in steam is 9% per minute, that of peat char under similar conditions is only 1.5% per minute.

Effect of Pyrolysis Conditions

The pyrolysis condition under which the char is produced also affects the reactivity of the char. For example, van Heek and Muhlen (1990) noted that the reactivity of char (in air) is much lower when produced above 1000 °C compared to that when produced at 700 °C. High temperatures reduce the number of active sites of reaction and the number of edge atoms. Longer residence times at peak temperature during pyrolysis also reduce reactivity.

Effect of Mineral Matter in Biomass

Inorganic materials in fuels can act as catalysts in the char–oxygen reaction (Zolin et al., 2001). In coal, inorganic materials reside as minerals, whereas in biomass they generally remain as salts or are organically bound. Alkali metals, potassium, and sodium are active catalysts in reactions with oxygen-containing species. Dispersed alkali metals in biomass contribute to the high catalytic activity of inorganic materials in biomass. In coal, CaO is also dispersed, but at high temperatures it sinters and vaporizes, blocking micropores.

Inorganic matter also affects pyrolysis, giving char of varying morphological characteristics. Potassium and sodium catalyze the polymerization of volatile matter, increasing the char yield; at the same time they produce solid materials that deposit on the char pores, blocking them. During subsequent oxidation of the char, the alkali metal catalyzes this process. Polymerization of volatile matter dominates over the pore-blocking effect. A high pyrolysis temperature may result in thermal annealing or loss of active sites and thereby loss of char reactivity (Zolin et al., 2001).

Intrinsic Reaction Rate

Char gasification takes place on the surface of solid char particles, which is generally taken to be the outer surface area. However, char particles are highly porous, and the surface areas of the inner pore walls are several orders of magnitude higher than the external surface area. For example, the actual surface area (BET) of an internal pore of a 1-mm-diameter beechwood char is 660 cm^2 while its outer surface is only 3.14 cm^2. Thus, if there is no physical restriction, the reacting gas can potentially enter the pores and react on their walls, resulting in a high overall char conversion rate. For this reason, two char particles with the same external surface area (size) may have widely different reaction rates because of their different internal structure.

From a scientific standpoint, it is wise to express the surface reaction rate on the basis of the actual surface on which the reaction takes place rather than the external surface area. The rate based on the actual pore wall surface area is the *intrinsic reaction rate*; the rate based on the external surface area of the char is the *apparent reaction rate*. The latter is difficult to measure, so sometimes it is taken as the reactive surface area determined indirectly from the reaction rate instead of the total pore surface area measured by the physical adsorption of nitrogen. This is known as the BET area (Klose and Wolki, 2005).

Mass Transfer Control

For the gasification reaction to take place within the char's pores, the reacting gas must enter the pores. If the availability of the gas is so limited that it is entirely consumed by the reaction on the outer surface of the char, gasification is restricted to the external surface area. This can happen because of the limitation of the mass transfer of gas to the char surface. We can illustrate using the example of char gasification in CO_2:

$$C + CO_2 \rightarrow 2CO \qquad (5.58)$$

Here, the CO_2 gas has to diffuse to the char surface to react with the active carbon sites. The diffusion, however, takes place at a finite rate. If the kinetic rate of this reaction is much faster than the diffusion rate of CO_2 to the char surface, all of the CO_2 gas molecules transported are consumed on the external surface of the char, leaving none to enter the pores and react on their surfaces. As the overall reaction is controlled by diffusion, it is called the *diffusion-* or *mass-transfer-controlled* regime of reaction.

On the other hand, if the kinetic rate of reaction is slow compared to the transport rate of CO_2 molecules, then the CO_2 will diffuse into the pores and react on their walls. The reaction in this situation is "kinetically controlled."

$$\begin{array}{ll} \text{Diffusion rate} \gg \text{kinetic rate} & [\text{Kinetic control reaction}] \\ \text{Diffusion rate} \ll \text{kinetic rate} & [\text{Diffusion control reaction}] \end{array} \qquad (5.59)$$

Between the two extremes lie intermediate regimes. The relative rates of chemical reaction and diffusion determine the gas concentration profile in the vicinity of the char particle; how the reaction progresses; and how char size, pore distribution, reaction temperature, char gas relative velocity, and so forth,

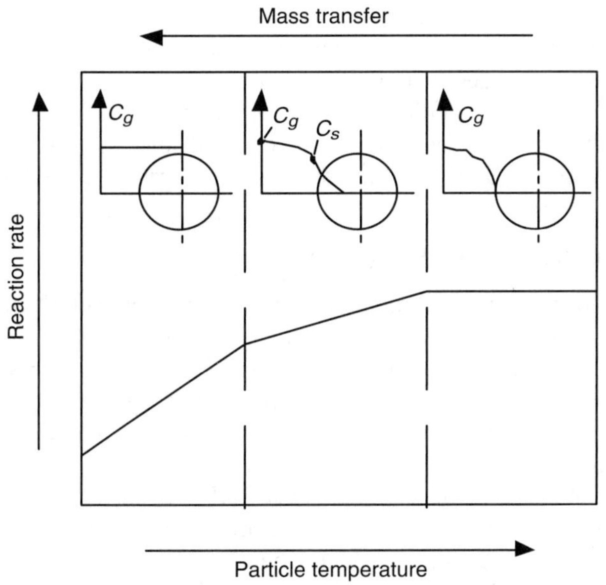

FIGURE 5.9 Char gasification regimes in a porous biomass char particle.

influence overall char conversion. Figure 5.9 shows how the concentration profile of CO_2 around the particle changes with temperature. With a rise in the surface temperature, the kinetic rate increases and therefore the overall reaction moves from the kinetic to the diffusion-controlled regime, resulting in less reaction within the pores.

The overall gasification rate of char particles, Q, when both mass transfer and kinetic rates are important, may be written as

$$Q = \frac{P_g}{\dfrac{1}{h_m} + \dfrac{1}{R_c}} \text{ kg Carbon/m}^2\text{.s} \tag{5.60}$$

where P_g is the concentration in partial pressure (bar) of the gasifying agent outside the char particle, h_m is the mass transfer rate (kg carbon/(m^{2}bar.s)) to the surface, and R_c is the kinetic rate of reaction: kg carbon/(m^{2}bar.s).

5.5 GASIFICATION MODELS

Optimal conversion of chemical energy of the biomass or other solid fuel into the desired gas depends on proper configuration, sizing, and choice of gasifier operating conditions. In commercial plants, optimum operating conditions are often derived through trials on the unit or by experiments on pilot plants. Even though expensive, experiments can give more reliable design data than can be obtained through modeling or simulation. There is, however, one major limitation with experimental data. If one of the variables of the original process changes, the optimum operating condition chosen from the specific experimental condition is no longer valid. Furthermore, an experimentally found optimum parameter can be size-specific; that is, the optimum operating condition for one size of gasifier is not necessarily valid for any other size. The right choice between experiment and modeling, then, is necessary for a reliable design.

5.5.1 Simulation versus Experiment

Simulation, or mathematical modeling, of a gasifier may not give a very accurate prediction of its performance, but it can at least provide qualitative guidance on the effect of design and operating or feedstock parameters. Simulation allows the designer or plant engineer to reasonably optimize the operation or the design of the plant using available experimental data for a pilot plant or the current plant.

Simulation can also identify operating limits and hazardous or undesirable operating zones, if they exist. Modern gasifiers, for example, often operate at a high temperature and pressure and are therefore exposed to extreme operating conditions. To push the operation to further extreme conditions to improve the gasifier performance may be hazardous, especially if it is done with no prior idea of how the gasifier might behave at those conditions. Modeling may provide a less expensive means of assessing the benefits and the associated risk.

Simulation can never be a substitute for good experimental data, especially in the case of gas–solid systems such as gasifiers. A mathematical model, however sophisticated, is useless unless it can reproduce real operation with an acceptable degree of deviation (Souza-Santos, 2004). Still, a good mathematical model can

- Find optimum operating conditions or a design for the gasifier.
- Identify areas of concern or danger in operation.
- Provide information on extreme operating conditions (high temperature, high pressure) where experiments are difficult to perform.
- Provide information over a much wider range of conditions than one can obtain experimentally.
- Better interpret experimental results and analyze abnormal behavior of a gasifier, if that occurs.
- Assist scale-up of the gasifier from one successfully operating size to another, and from one feedstock to another.

5.5.2 Gasifier Simulation Models

Gasifier simulation models may be classified into the following groups:

- Thermodynamic equilibrium
- Kinetic

- Computational fluid dynamics (CFD)
- Artificial neural network

The thermodynamic equilibrium model predicts the maximum achievable yield of a desired product from a reacting system (Li et al., 2001). In other words, if the reactants are left to react for an infinite time, they will reach equilibrium yield. The yield and composition of the product at this condition is given by the equilibrium model, which concerns the reaction alone without taking into account the geometry of the gasifier.

In practice, only a finite time is available for the reactant to react in the gasifier. So, the equilibrium model may give an ideal yield. For practical applications, we need to use the kinetic model to predict the product from a gasifier that provides a certain time for reaction. A kinetic model studies the progress of reactions in the reactor, giving the product compositions at different positions along the gasifier. It takes into account the reactor's geometry as well as its hydrodynamics.

CFD models (Euler type) solve a set of simultaneous equations for conservation of mass, momentum, energy, and species over a discrete region of the gasifier. Thus, they give distribution of temperature, concentration, and other parameters within the reactor. If the reactor hydrodynamics is well known, a CFD model provides a very accurate prediction of temperature and gas yield around the reactor.

Neural network analysis is a relatively new simulation tool for modeling a gasifier. It works somewhat like an experienced operator, who uses his or her years of experience to predict how the gasifier will behave under a certain condition. This approach requires little prior knowledge about the process. Instead, the neural network *learns* by itself from sample experimental data (Guo et al., 1997).

Thermodynamic Equilibrium Models

Thermodynamic equilibrium calculation is independent of gasifier design and so is convenient for studying the influence of fuel and process parameters. Though chemical or thermodynamic equilibrium may not be reached within the gasifier, this model provides the designer with a reasonable prediction of the maximum achievable yield of a desired product. However, it cannot predict the influence of hydrodynamic or geometric parameters, like fluidizing velocity, or design variables, like gasifier height.

Chemical equilibrium is determined by either of the following:

- The equilibrium constant
- Minimization of the Gibbs free energy

Prior to 1958 all equilibrium computations were carried out using the equilibrium constant formulation of the governing equations (Zeleznik and Gordon,

1968). Later, computation of equilibrium compositions by Gibbs free energy minimization became an accepted alternative.

This section presents a simplified approach to equilibrium modeling of a gasifier based on the following overall gasification reactions:

$$\text{R1: } CO_2 + C \rightarrow 2\,CO \tag{5.61}$$

$$\text{R2: } C + H_2O \rightarrow H_2 + CO \tag{5.62}$$

$$\text{R3: } C + 2H_2 \rightarrow CH_4 \tag{5.63}$$

$$\text{R9: } CO + H_2O \rightarrow CO_2 + H_2 \tag{5.64}$$

From a thermodynamic point of view, the equilibrium state gives the maximum conversion for a given reaction condition. The reaction is considered to be zero dimensional and there are no changes with time (Li et al., 2001). An equilibrium model is effective at higher temperatures (>1500 K), where it can show useful trends in operating parameter variations (Altafini et al., 2003). For equilibrium modeling, one may use stoichiometric or nonstoichiometric methods (Basu, 2006).

Stoichiometric Equilibrium Models

In the stoichiometric method, the model incorporates the chemical reactions and species involved. It usually starts by selecting all species containing C, H, and O, or any other dominant elements. If other elements form a minor part of the product gas, they are often neglected.

Let us take the example of 1 mole of biomass being gasified in *d* moles of steam and *e* moles of air. The reaction of the biomass with air (3.76 moles of nitrogen, 1 mole of oxygen) and steam may then be represented by

$$\begin{aligned} CH_aO_bN_c + dH_2O + e(O_2 + 3.76N_2) \rightarrow n_1C + n_2H_2 + n_3CO \\ + n_4H_2O + n_5CO_2 + n_6CH_4 + n_7N_2 \end{aligned} \tag{5.65}$$

where $n_1 \ldots n_7$ are stoichiometric coefficients. Here, $CH_aO_bN_c$ is the chemical representation of the biomass and *a*, *b*, and *c* are the mole ratios (H/C, O/C, and N/C) determined from the ultimate analysis of the biomass. With *d* and *e* as input parameters, the total number of unknowns is seven.

An atomic balance of carbon, hydrogen, oxygen, and nitrogen gives

$$\text{C: } n_1 + n_3 + n_5 + n_6 = 1 \tag{5.66}$$

$$\text{H: } 2n_2 + 2n_4 + 4n_6 = a + 2d \tag{5.67}$$

$$\text{O: } n_3 + n_4 + 2n_5 = b + d + 2e \tag{5.68}$$

$$\text{N: } n_7 = c + 7.52e \tag{5.69}$$

During the gasification process, reactions R1, R2, R3, and R9 (see Table 5.2) take place. The water–gas shift reaction, R9, can be considered a result of the subtraction of the steam gasification and Boudouard reactions, so we consider the equilibrium of reactions R1, R2, and R3 alone. For a gasifier pressure, P, the equilibrium constants for reactions R_1, R_2, and R_3 are given by

$$K_{e1} = \frac{y_{CO}^2 P}{y_{CO_2}} \quad R1 \tag{5.70}$$

$$K_{e2} = \frac{y_{CO} y_{H_2} P}{y_{H_2O}} \quad R2 \tag{5.71}$$

$$K_{e3} = \frac{y_{CH_4}}{y_{H_2}^2 P} \quad R3 \tag{5.72}$$

where y_i is the mole fraction for species i of CO, H_2, H_2O, and CO_2.

The two sets of equations (stoichiometric and equilibrium) may be solved simultaneously to find the coefficients, $(n_1 \ldots n_7)$, and hence the product gas composition in an equilibrium state. Thus, by solving seven equations (Eqs. 5.66–5.72) we can find seven unknowns $(n_1 \ldots n_7)$, which give both the yield and the product of the gasification for a given air/steam-to-biomass ratio. The approach is based on the simplified reaction path and the chemical formula of the biomass.

This is a greatly simplified example of the stoichiometric modeling of a gasification reaction. The complexity increases with the number of equations considered. For a known reaction mechanism, the stoichiometric equilibrium model predicts the maximum achievable yield of a desired product or the possible limiting behavior of a reacting system.

Nonstoichiometric Equilibrium Models

In nonstoichiometric modeling, no knowledge of a particular reaction mechanism is required to solve the problem. In a reacting system, a stable equilibrium condition is reached when the Gibbs free energy of the system is at the minimum. So, this method is based on minimizing the total Gibbs free energy. The only input needed is the elemental composition of the feed, which is known from its ultimate analysis. This method is particularly suitable for fuels like biomass, the exact chemical formula of which is not clearly known.

The Gibbs free energy, G_{total} for the gasification product comprising N species ($i = 1 \ldots N$) is given by

$$G_{total} = \sum_{i=1}^{N} n_i \Delta G_{f,i}^0 + \sum_{i=1}^{N} n_i RT \ln\left(\frac{n_i}{\sum n_i}\right) \tag{5.73}$$

where $\Delta G_{f,i}^0$ is the Gibbs free energy of formation of species i at standard pressure of 1 bar.

Equation (5.73) is to be solved for unknown values of n_i to minimize G_{total}, bearing in mind that it is subject to the overall mass balance of individual elements. For example, irrespective of the reaction path, type, or chemical formula of the fuel, the amount of carbon determined by ultimate analysis must be equal to the sum total of all carbon in the gas mixture produced. Thus, for each jth element we can write

$$\sum_{i=1}^{N} a_{i,j} n_i = A_j \tag{5.74}$$

where $a_{i,j}$ is the number of atoms of the jth element in the ith species, and A_j is the total number of atoms of element j entering the reactor. The value of n_i should be found such that G_{total} will be minimum. We can use the Lagrange multiplier methods to solve these equations.

The Lagrange function (L) is defined as

$$L = G_{total} - \sum_{j=1}^{K} \lambda_j \left(\sum_{i=1}^{N} a_{ij} n_i - A_j \right) \text{kJ/mol} \tag{5.75}$$

where λ is the Lagrangian multiplier for the jth element.

To find the extreme point, we divide Eq. (5.75) by RT and take the derivative,

$$\left(\frac{\partial L}{\partial n_i} \right) = 0 \tag{5.76}$$

Substituting the value of G_{total} from Eq. (5.73) in Eq. (5.75), and then taking its partial derivative, the final equation is of the form given by

$$\left(\frac{\partial L}{\partial n_i} \right) = \frac{\Delta G^0_{f,i}}{RT} + \sum_{i=1}^{N} \ln \left(\frac{n_i}{n_{total}} \right) + \frac{1}{RT} \sum_{j=1}^{K} \lambda_j \left(\sum_{i=1}^{N} a_{ij} n_i \right) = 0 \tag{5.77}$$

Kinetic Models

Gas composition measurements for gasifiers often vary significantly from those predicted by equilibrium models (Peterson and Werther, 2005; Li et al., 2001; Kersten, 2002). This shows the inadequacy of equilibrium models and underscores the need of kinetic models to simulate gasifier behavior.

A kinetic model gives the gas yield and product composition a gasifier achieves after a finite time (or in a finite volume in a flowing medium). Thus, it involves parameters such as reaction rate, residence time of particles, and reactor hydrodynamics. For a given operating condition and gasifier configuration, the kinetic model can predict the profiles of gas composition and temperature inside the gasifier and overall gasifier performance.

The model couples the hydrodynamics of the gasifier reactor with the kinetics of gasification reactions inside the gasifier. At low reaction temperatures, the reaction rate is very slow, so the residence time required for complete conversion is long. Therefore, kinetic modeling is more suitable and accurate at relatively low operating temperatures (<800 °C) (Altafini et al., 2003). For higher temperatures, where the reaction rate is faster, the equilibrium model may be of greater use.

Kinetic modeling has two components: (1) reaction kinetics and (2) reactor hydrodynamics.

Reaction Kinetics

Reaction kinetics must be solved simultaneously with bed hydrodynamics and mass and energy balances to obtain the yields of gas, tar, and char at a given operating condition.

As the gasification of a biomass particle proceeds, the resulting mass loss is manifested either through reduction in size with unchanged density or reduction in density with unchanged size. In both cases the rate is expressed in terms of the external surface area of the biomass char. Some models, where the reaction is made up of char alone, can define a reaction rate based on reactor volume. There are thus three ways of defining the char gasification reaction for biomass: (1) shrinking core model, (2) shrinking particle model, and (3) volumetric reaction rate model.

Reactor Hydrodynamics

The kinetic model considers the physical mixing process and therefore requires knowledge of reactor hydrodynamics. The hydrodynamics may be defined in terms of the following types with increasing sophistication and accuracy:

- Zero dimensional (stirred tank reactor)
- One dimensional (plug flow)
- Two dimensional
- Three dimensional

Unlike other models, the kinetic model is sensitive to the gas–solid contacting process involved in the gasifier. Based on this process, the model may be divided into three groups: (1) moving or fixed bed, (2) fluidized bed, and (3) entrained flow. Short descriptions of these are given in Section 5.6.

Neural Network Models

An alternative to the sophisticated modeling of a complex process, especially for one not well understood, is an artificial neural network (ANN). An ANN model mimics the working of the human brain and provides some human characteristics in solving models (Abdulsalam, 2005). It cannot produce an analytical solution, but it can give numerical results. This technique has been used with reasonable success to predict gas yield and composition from gasification of bagasse, cotton stem, pine sawdust, and poplar in fluidized beds (Guo et al., 1997); in municipal solid waste; and also in a fluidized bed (Xiao et al., 2009).

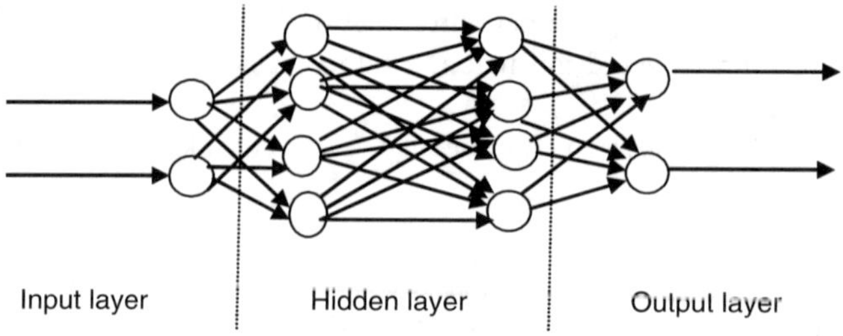

FIGURE 5.10 Schematic of a multilayer feed-forward neural network. *(Source: Adapted from Kalogirou, 2001.)*

The ANN model can deal with complex gasification problems. It uses a high-speed architecture of three hidden layers of neurons (Kalogirou, 2001): one to receive the input(s), one to process them, and one to deliver output(s). Figure 5.10 shows the arrangement of neuron layers and the connection patterns between them. Kalogirou (2001) suggested the following empirical formula to estimate the number of hidden neurons:

$$\text{Number of hidden neurons} = \tfrac{1}{2}(\text{inputs} + \text{outputs}) + \sqrt{\text{number of training patterns}} \tag{5.78}$$

The input layer has two values associated with it: inputs and weights. Weights are used to transfer data from layer to layer. In the first step, the information is processed at the nodes and then added up (summation); the result is passed through an activation function. The outcome is the node's "activation value," which is multiplied by the specific weight and transferred to the next node.

Network Training

Training modifies the connection weights in some orderly fashion using learning methods (Guo et al., 2001). It begins with a set of data (with inputs and outputs targeted); the weights are adjusted until the difference between the neural network output and the corresponding target is minimum (Kalogirou et al., 1999). When the training process satisfies the required tolerance, the network holds the weights constant and uses the network to make output predictions. After training, the weights contain meaningful information. A back-propagation algorithm is used to train the network. Multilayer feed-forward neural networks are used to approximate the function.

A neural network may return poor results for data that differ from the original data it was trained with. This happens sometimes when limited data are available to calibrate and evaluate the constants of the model (Hajek and Judd, 1995). After structuring the neural network, information starts to flow from the input layer to the output layer according to the concepts described here.

CFD Models

Computational fluid dynamics can have an important role in the modeling of a fluidized-bed gasifier. A CFD-based code involves a solution of conservation of mass, momentum, species, and energy over a defined domain or region. The equations can be written for an element, where the flux of the just-mentioned quantities moving in and out of the element is considered with suitable boundary conditions.

A CFD code for gasification typically includes a set of submodels for the sequence of operations such as the vaporization of a biomass particle, its pyrolysis (devolatilization), the secondary reaction in pyrolysis, and char oxidation (Di Blasi, 2008; Babu and Chaurasia, 2004). Further sophistications such as a subroutine for fragmentation of fuels during gasification and combustion are also developed (Syred et al., 2007). These subroutines can be coupled with the transport phenomenon, especially in the case of a fluidized-bed gasifier.

The hydrodynamic or transport phenomenon for a laminar flow situation is completely defined by the Navier-Stokes equation, but in the case of turbulent flow a solution becomes difficult. A complete time-dependent solution of the instantaneous Navier-Stokes equation is beyond today's computation capabilities (Wang and Yan, 2008), so it is necessary to assume some models for the turbulence. The Reynolds-averaged Navier-Stokes (k-ε) model or large eddy simulation filters are two means of accounting for turbulence in the flow.

For a fluidized bed, the flow is often modeled using the Eulerian-Lagrange concept. The discrete phase is applied to the particle flow; the continuous phase, to the gas. Overmann and associates (2008) used the Euler-Euler and Euler-Lagrange approaches to model wood gasification in a bubbling fluidized bed. Their preliminary results found both to have comparable agreement with experiments. If the flow is sufficiently dilute, the particle–particle interaction and the particle volume in the gas are neglected.

A two-fluid model is another computational fluid dynamics approach. Finite difference, finite element, and finite volume are three methods used for discretization. Commercial software such as ANSYS, ASPEN, Fluent, Phoenics, and CFD2000 are available for solution (Miao et al., 2008). A review and comparison of these codes is given in Xia and Sun (2002) and Norton et al. (2007).

Recent progress in numerical solution and modeling of complex gas–solid interactions has brought CFD much closer to real-life simulation. If successful, it will be a powerful tool for optimization and even design of thermochemical reactors like gasifiers (Wang and Yan, 2008). CFD models are most effective in modeling entrained-flow gasifiers, where the gas–solid flows are less complex than those in fluidized beds and the solid concentration is low.

Models developed by several investigators employ sophisticated reaction kinetics and complex particle–particle interaction. Most of them, however, must use some submodels, fitting parameters or major assumptions into areas where precise information is not available. Such weak links in the long array make the final result susceptible to the accuracy of those "weak links." If the final results are known, we can use them to back-calculate the values of the unknown parameters or to refine the assumptions used.

The CFD model can thus predict the behavior of a given gasifier over a wider range of parameters using data for one situation, but this prediction might not be accurate if the code is used for a different gasifier with input parameters that are substantially different from the one for which experimental data are available.

5.6 KINETIC MODEL APPLICATIONS

This section briefly discusses how kinetic models can be applied to the three major gasifier types.

5.6.1 Moving-Bed Gasifiers

A basic moving-bed or fixed-bed gasifier can use the following assumptions:

- The reactor is uniform radially (i.e., no temperature or concentration gradient exists in the radial direction).
- The solids flow downward (in a updraft gasifier) as a plug flow.
- The gas flows upward as a plug flow.
- The interchange between two phases takes place by diffusion.

The mass balance of a gas species, j, can be written (Souza-Santos, 2004, p. 134) as

$$u_g \frac{d\rho_{g,j}}{dz} = D_{g,j} \frac{d^2 \rho_{g,j}}{dz^2} + R_{m,j} \tag{5.79}$$

where u_g is the superficial gas velocity, z is the distance, $\rho_{g,j}$ is the density of the jth gas, and $D_{g,j}$ is the diffusivity of the jth gas. $R_{m,,j}$, the production or consumption of the jth gas element, is related to $Q_{gasification}$ heat generation or absorption.

Similarly, an energy balance equation can be written for a dz element as

$$\rho_g C_{pg} u_g \frac{dT}{dz} = \lambda_g \frac{d^2T}{dz^2} + Q_{gasification} + Q_{conv} + Q_{rad} + Q_{mass} \tag{5.80}$$

where, $Q_{gasification}$, Q_{conv}, Q_{rad}, and Q_{mass} are the net heat flow into the element due to gasification, convection, radiation, and mass transfer, respectively. These terms can be positive or negative. ρ_g, C_{pg}, and λ_g are the density, specific heat, and thermal conductivity of the bulk gas, respectively.

Equations (5.79) and (5.80) can be solved simultaneously with appropriate expression for the reaction rate, $R_{m,j}$.

5.6.2 Fluidized-Bed Gasifiers

The kinetic modeling of fluidized-bed gasifiers requires several assumptions or submodels. It takes into account how the fluidized-bed hydrodynamics is viewed in terms of heat and mass transfer, and gas flow through the fluidized bed. The bed hydrodynamics defines the transport of the gasification medium through the system, which in turn influences the chemical reaction on the biomass surface. Each of these is subject to some assumptions or involves submodels.

One can use several versions of the fluidization model:

- Two-phase model of bubbling fluidized bed: bubbling and emulsion phases
- Three-phase model of bubbling fluidized bed: bubbling, cloud, and emulsion phases
- Fluidized bed divided into horizontal sections or slices
- Core-annulus structure

Gas flow through the bed can be modeled as:

- Plug flow in the bubbling phase; ideally mixed gas in the emulsion phase
- Ideally mixed gases in both phases
- Plug flow in both phases (there is exchange between phases)
- Plug flow through the bubble and emulsion phases without mass transfer between phases
- Plug flow of gas upward in the core and solid backflow in the annulus

The following sections present the essentials of a model for a circulating fluidized-bed combustor and one for a bubbling fluidized-bed gasifier (Kaushal et al., 2008). A typical one-dimensional steady-state model of a circulating fluidized-bed combustor, as shown in Figure 5.11, assumes gases as ideal and in the plug-flow regime. The riser is divided into three hydrodynamic zones: lower dense bed zone, intermediate middle zone, and top dilute zone. The solids are assumed uniform in size with no attrition. Char is a homogeneous matrix of carbon, hydrogen, and oxygen.

A bubbling fluidized-bed gasifier is divided into several zones with different hydrodynamic characteristics: dense zone and freeboard zone for bubbling beds; core-annulus for circulating beds. The dense zone additionally deals with the drying and devolatilization of the introduced feed. Superheated steam is introduced at the lower boundary of the dense zone. Each zone is further divided into cells, which individually calculate their local hydrodynamic and thermodynamic state using chosen equations or correlations. The cells are solved sequentially from bottom to top, with the output of each considered the input for the next. The conservation equations for carbon, bed material, and energy are evaluated not in each cell but across the entire zone. Therefore, each

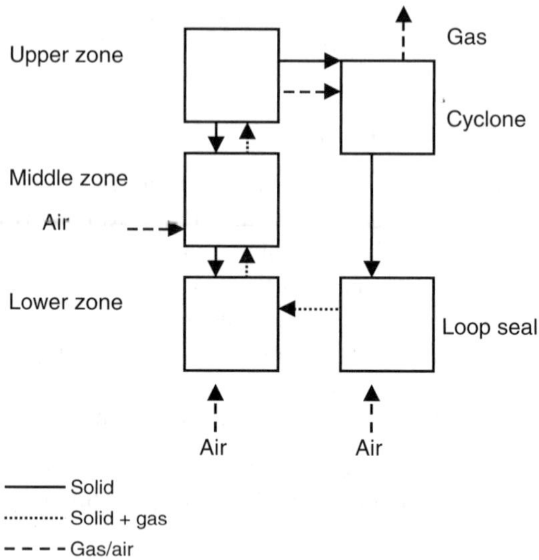

FIGURE 5.11 Model of a circulating fluidized-bed gasifier.

zone shows a homogeneous char concentration in the bed material and a uniform temperature. Additional input parameters to the model are geometric data, particle properties, and flow rates.

Hydrodynamic Submodel (Bubbling Bed)

The dense zone (assumed to be the bubbling bed) is modeled according to the modified two-phase theory. Bubble size is calculated as a function of bed height (Darton and LaNauze, 1977), and it is assumed that all bubbles at any cross-section are of uniform size:

$$d_b = 0.54\frac{(U-U_{mf})^{0.4}}{g^{0.2}}\left(z+4\sqrt{\frac{A}{N_{or}}}\right)^{0.8} \tag{5.81}$$

where A/N_{or} is the number of orifices per unit of cross-section area of the bed.

The interphase mass transfer between bubbles and emulsion, essential for the gas–solid reactions, is modeled semi-empirically using the specific bubble surface as the exchange area, the concentration gradient, and the mass-transfer coefficient. The mass-transfer coefficient, K_{BE}, based on the bubble–emulsion surface area (Sit and Grace, 1978), is

$$K_{BE} = \frac{U_{mf}}{4} + \sqrt{\frac{4\varepsilon_{mf} D_r U_B}{\pi d_B}} \tag{5.82}$$

where U_{mf} and ε_{mf} are, respectively, fluidization velocity and voidage at a minimum fluidizing condition, D_r is the bed diameter, and U_B is the rise velocity of a bubble of size d_B.

The axial mean voidage in the freeboard is calculated using an exponential decay function.

Reaction Submodel

Gasification reactions proceed at a finite speed; this process is divided into three steps: drying, devolatilization, and gasification. The time taken for drying and devolatilization of the fuel is much shorter than the time taken for gasification of the remaining char. Some models assume instantaneous drying and devolatilization because the rate of reaction of the char, which is the slowest, largely governs the overall process.

The products of devolatilization are CO_2, CO, H_2O, H_2, and CH_4. The gases released during drying and devolatilization are not added instantaneously to the upflowing gas stream, but are added along the height of the gasifier in a predefined pattern. The total mass devolatilized, $m_{volatile}$, is therefore the sum of the carbon, hydrogen, and oxygen volatilized from the solid biomass.

$$m_{volatile} = m_{char} + m_{hydrogen} + m_{oxygen} \tag{5.83}$$

Char gasification, the next critical step, may be assumed to move simultaneously through reactions R1, R2, and R3 (Table 5.2). As these three reactions occur simultaneously on the char particle, reducing its mass, the overall rate is given as

$$m_{char} = m_{Boudouard} + m_{steam} + m_{methanation} \tag{5.84}$$

The conversion of the porous char particle may be modeled assuming that the process follows shrinking particle (diminishing size), shrinking core (diminishing size of the unreacted core), or progressive conversion (diminishing density). The shift reaction is the most important homogenous reaction followed by steam reforming. The bed materials may catalyze the homogeneous reactions, but only in the emulsion phase, because the bubble phase is assumed to be free of solids.

5.6.3 Entrained-Flow Gasifiers

Extensive work on the modeling of entrained-flow gasifiers is available in the literature. Computational fluid dynamics (CFD) has been successfully applied to this gasifier type. This section presents a simplified approach to entrained-flow gasification following the work of Vamvuka et al. (1995).

The reactor is considered to be a steady-state, one-dimensional plug–flow reactor in the axial direction and well mixed radially—similar to that shown in Figure 5.12. Fuel particles shrink as they are gasified. Five gas–solid reactions (R1–R5 in Table 5.2) can potentially take place on the char particle surface. The reduction in the mass of char particles is the sum of these individual reactions, so if there are N_c char particles in the unit gas volume, the total reduction, W_c, in the plug flow is as shown in the equation that follows the figure.

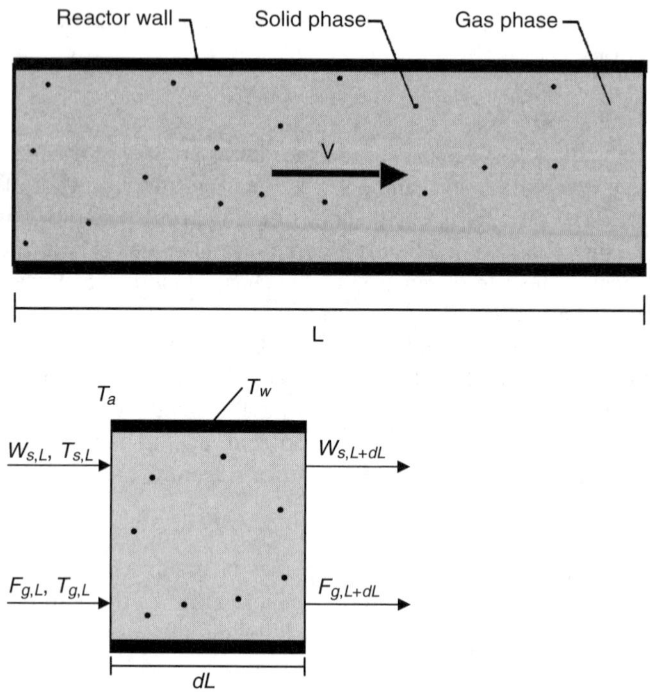

FIGURE 5.12 One-dimensional entrained-flow model.

$$dW_c = -(N_c A dz)\sum_{k=1}^{5} r_k(T_s, L_r) \tag{5.85}$$

where $r_k(T_s, L_r)$ is the surface reaction rate of the kth reaction (one of R1–R5) at the reactor's surface temperature, T_s, and length, L_r. A is its cross-section area.

Gaseous reactants diffuse to the char surface to participate in k reactions. Thus, if a_{jk} is the mass of the jth gas, required for the kth reaction, the overall diffusion rate of this gas from free stream concentration, y_j, to the char surface, y_{js}, may be related to the total of all reactions consuming the jth gas as follows:

$$\sum_{k=1}^{5} a_{jk} r_k(T_s, L_r) = 4\pi r_c^2\left[\frac{D_{gj}P}{RT_g r_c}(y_j - y_{js})\right] \tag{5.86}$$

where y_{js} and y_j are mole fractions of gas on the char surface and in the bulk gas, respectively; P is the reactor pressure; and D_{gj} is the diffusion coefficient of the jth gas in the mixture of gases.

The surface reaction rate, $r_k(T_s, L_r)$, may be written in nth-order form as

$$r_k(T_s, L_r) = 4\pi r_c^{\,2} K_{sk}(T_s)(P y_{js})^n \text{ mol/s} \tag{5.87}$$

where n is the order or reaction, and $K_{sk}(T_s)$ is the surface reaction rate constant at temperature T_s.

For conversion of gaseous species, we can write

$$\frac{dF_{gj}}{dZ} = \pm N_c A \sum_{k=1}^{5} a_{jk} r_k(T_s, L_r) \tag{5.88}$$

where $a_{j,k}$ is the stoichiometric coefficient for the jth gas in the kth reaction.

The total molar flow rate of the jth gas is found by adding the contribution of each of nine gas–solid and gas–gas reactions:

$$F_{gj} = F_{gj0} + \sum a_{jk} \xi_k \tag{5.89}$$

where F_{gj0} is the initial flow rate of the gas.

Energy Balance

Some of the five equations (reactions R1–R5) are endothermic while some are exothermic. The overall heat balance of reacting char particles is known from a balance of a particle's heat generation and heat loss to the gas by conduction and radiation.

$$\begin{aligned}\frac{d(W_c C_{pc} T_s)}{dZ} &= -N_c A \left[\sum_{k=1}^{5} r_k(T_s, L_r)\right] \Delta H_k(T_s) \\ &+ 4\pi r_c^2 \left[\frac{\lambda_g}{r_c}(T_s - T_g) + \sigma_p \sigma (T_c^4 - T_g^4)\right]\end{aligned} \tag{5.90}$$

where C_{pc} is the specific heat of the char, ΔH_k is the heat of reaction of the kth reaction at the char surface at temperature T_s, e_p is the emissivity of the char particle, λ_g is the thermal conductivity of the gas, and σ is the Stefan-Boltzmann constant.

A similar heat balance for the gas in an element dz in length can be carried out as

$$\begin{aligned}\frac{d\left(\sum_j F_{gj} C_{pg} T_g\right)}{dZ} &= -A \left[\sum_{k=6}^{9} \xi_k \Delta H_k(T_g)\right] \\ &- 4\pi r_c^2 N_C A \left[\frac{\lambda_g}{r_c}(T_g - T_c) + e_p \sigma (T_g^4 - T_c^4)\right] \\ &- \left[h_{conv}(T_g - T_w) + e_w \sigma (T_g^4 - T_w^4)\right] \pi D_r\end{aligned} \tag{5.91}$$

where $\Delta\xi_k$ is the extent of the gas-phase kth reaction with the heat of reaction, $\Delta H_k\ (T_g)$; h_{conv} is the gas-wall convective heat transfer coefficient; and D_r is the reactor's internal diameter.

The first term on the right of Eq. (5.91) is the net heat absorption by the gas-phase reaction, the second is the heat transfer from the gas to the char particles, and the third is the heat loss by the gas at temperature T_g to the wall at temperature T_w.

The equations are solved for an elemental volume, $A_r dL_r$, with boundary conditions from the previous upstream cell. The results are then used to solve the next downstream cell.

Symbols and Nomenclature

A = cross-sectional area of bed or reactor (m^2)
A_0 = pre-exponential coefficient in Eq. (5.42) (s^{-1})
A_b, A_w = pre-exponential coefficients in Eqs. (5.44) and (5.47), respectively ($bar^{-n}\ s^{-1}$)
A_j = total number of atoms of element j entering the reactor (–)
$a_{i,j}$ = number of atoms of jth element in ith species (–)
a_{jk} = mass of jth gas, required for the kth reaction (kg)
C_i = molar concentration of ith gas (mol/m^3)
C_{pc} = specific heat of char (kJ/kg.K)
C_{pg} = specific heat of the bulk gas
D_r = internal diameter of the reactor (m)
$D_{g,j}$ = diffusion coefficient of the jth gas in the mixture of gases (m^2/s)
d_b = diameter of the bubble (m)
E = activation energy (kJ/mol)
e_p = emissivity of char particle (–)
F_{gl0} = initial flow rate of the gas (mol/s)
F_{gl} = molar flow rate of the lth gas (mol/s)
G_{total} = total Gibbs free energy (kJ)
g = acceleration due to gravity, 9.81 (m/s^2)
ΔH_k = heat of reaction of kth reaction at char surface (kJ/mol)
ΔH = enthalpy change (kJ)
h_i^0, h_f^0 = heat of formation at reference state (kJ)
h_{conv} = gas-wall convective heat transfer coefficient (kW/m^2K)
h_m = mass-transfer coefficient (kg carbon/$m^2.bar^2.s$)
k = first-order reaction rate constant (s^{-1})
k_0 = pre-exponential factor (s^{-1})
k_{liq} = rate constant for the liquid yield of pyrolysis (s^{-1})
k_{BE} = bubble-emulsion mass exchange coefficient (m/s)
k_c = rate constant for the char yield of pyrolysis (s^{-1})
k_g = rate constant for the gas yield of pyrolysis (s^{-1})
k_I = rate constant of three primary pyrolysis reactions taken together (s^{-1})
K = number of element in Eq (5.77)
k_{w1}, k_{w2}, k_{w3} = rate constants in Eq (5.47) ($bar^{-1}\ s^{-1}$)
K_{sk} = surface reaction rate constant for kth reaction, $mol/m^2.bar^n$
K_e, $K_{equilibrium}$ = equilibrium constant (–)
l = number of gaseous reactants (–)
L_r = length of the reactor (m)
L = Lagrangian function (–)
m_b = mass of the biomass in the primary pyrolysis process (kg)
m_0 = initial mass of the biomass (kg)
m_c = mass of the biomass remaining after complete conversion (kg)
m = reaction order with respect to carbon conversion in Eq. 5.42 (–)

m, n, p, q = stoichiometric coefficients in Eqs. 5.27–5.29
n = reaction order with respect to the gas partial pressure, Eq. 5.44 (–)
N = number of species present (–)
N_c = number of char particles in unit gas volume (–)
N_{or} = number of orifices in a bed of area (A_r)
P_i = partial pressure of the species i (bar)
P = total pressure of the species (bar)
Q = char gasification rate (kg carbon/m^2.s)
$Q_{gasification}$, Q_{conv}, Q_{rad}, and Q_{mass} = energy transfer due to gasification, convection, radiation, respectively (kW/m^3 of bed)
R = gas constant (8.314 J/mol.K, or 8.314×10^{-5} m^3.bar/mol.K)
R_c = chemical kinetic reaction rate (kg carbon/m^2.bar^{2}s)
$R_{m,g,j}$ = rate of production or consumption of gas species j (kg/m^{3}s)
r_i = reaction rate of the ith reaction (s^{-1})
r_c = char particle radius (m)
T = temperature (K)
T_s = surface temperature of char particles (K)
T_g = gas temperature (K)
T_w = wall temperature (K)
t = time (s)
u_g = superficial gas velocity in Eq. 5.80 (m/s)
U = fluidization velocity (m/s)
U_B = bubble rise velocity (m/s)
U_{mf} = minimum fluidization velocity (m/s)
X = fractional change in the carbon mass of the biomass (kg)
y = mole fraction of a species (–)
y_l = mole fraction of gas in the bulk (–)
y_{ls} = mole fraction of gas on the char surface (–)
z = height above grid or distance along a reactor from fuel entry (m)
α_{lk} = stoichiometric coefficient for lth gas in kth reaction (–)
β = partition coefficient (–)
λ = Lagrangian multiplier (–)
λ_g = thermal conductivity of gas (kJ/m.K)
σ = Stefan-Boltzmann constant (5.67×10^{-8} W m^{-2} K^{-4})
$\Delta G, \Delta G^0$ = change in Gibbs free energy (kJ)
ΔG^0_{fi} = change in Gibbs free energy of formation of species i (kJ)
$\Delta\xi_k$ = extent of gas-phase kth reaction (–)
ρ_j = density of jth gas (kg/m^3)
ε_{mf} = voidage at minimum fluidization condition
ρ_g = density of the bulk gas
ΔS = entropy change (kJ/K)

Chapter 6

Design of Biomass Gasifiers

6.1 INTRODUCTION

The design of a gasification plant includes the gasifier reactor as well as its auxiliary or support equipment. A typical biomass gasification plant design comprises the following systems:

- Gasifier reactor
- Biomass-handling system
- Biomass-feeding system
- Gas-cleanup system
- Ash or solid residue-removal system

This chapter deals with the design of the gasifier reactor alone. Chapter 8 discusses the design of the handling and feeding systems. Gas-cleaning systems are briefly discussed in Chapters 4 and 9.

As with most process plant equipment, the design of a gasifier may be divided into three major phases:

Phase 1. Process design and preliminary sizing
Phase 2. Optimization of design
Phase 3. Detailed mechanical design

For cost estimation and/or for submission of initial bids, most manufacturers use the first step of sizing the gasifier. The second step is considered only for a confirmed project—that is, when an order is placed and the manufacturer is ready for the final stage of detailed mechanical or manufacturing design.

This chapter mainly concerns the first phase and, briefly, the second phase (design optimization). To set the ground for design methodologies, a short description of different gasifier types is presented, followed by a discussion of design considerations and design methodologies.

Biomass Gasification and Pyrolysis. DOI: 10.1016/B978-0-12-374988-8.00006-4

TABLE 6.1 Comparison of Some Commercial Gasifiers

Parameters	Fixed/Moving Bed	Fluidized Bed	Entrained Bed
Feed size	<51 mm	<6 mm	<0.15 mm
Tolerance for fines	Limited	Good	Excellent
Tolerance for coarse	Very good	Good	Poor
Exit gas temperature	450–650 °C	800–1000 °C	>1260 °C
Feedstock tolerance	Low-rank coal	Low-rank coal and excellent for biomass	Any coal including caking but unsuitable for biomass
Oxidant requirements	Low	Moderate	High
Reaction zone temperature	1090 °C	800–1000 °C	1990 °C
Steam requirement	High	Moderate	Low
Nature of ash produced	Dry	Dry	Slagging
Cold-gas efficiency	80%	89%	80%
Application	Small capacities	Medium-size units	Large capacities
Problem areas	Tar production and utilization of fines	Carbon conversion	Raw-gas cooling

Source: Data compiled from Basu, 2006.

6.1.1 Gasifier Types

Gasifiers are classified mainly on the basis of their gas–solid contacting mode and gasifying medium. Based on the gas–solid contacting mode, gasifiers are broadly divided into three principal types (Table 6.1): (1) fixed or moving bed, (2) fluidized bed, and (3) entrained flow. Each is further subdivided into specific types as shown in Figure 6.1. Major western technology providers, as listed in the figure, supply their gasification technologies as per one of these.

One gasifier type is not necessarily suitable for the full range of gasifier capacities. There is an appropriate range of application for each. For example, the moving-bed (updraft and downdraft) type is used for smaller units (10 kWth–10 MWth); the fluidized-bed type is more appropriate for intermediate units (5 MWth–100 MWth); entrained-flow reactors are used for large-capacity units (>50 MWth). Figure 6.2 shows the overlapped range of application for different

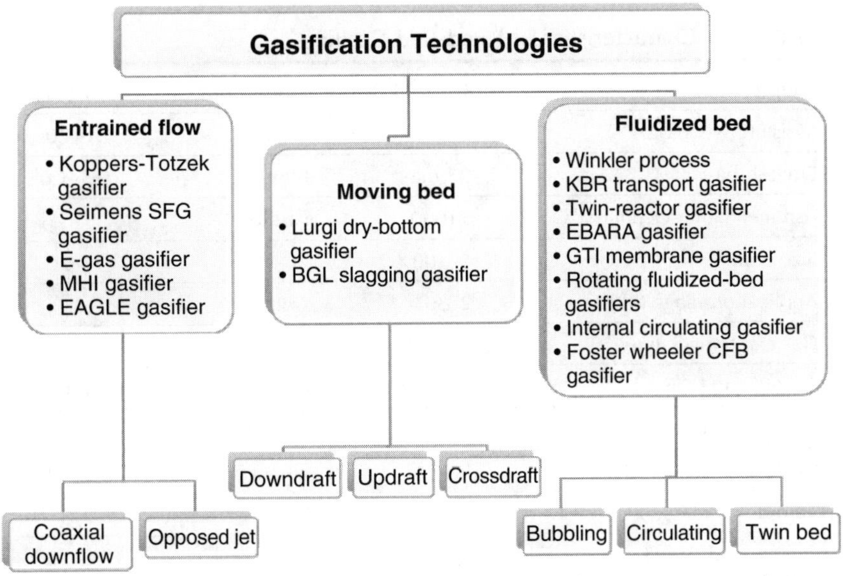

FIGURE 6.1 Gasification technologies and their commercial suppliers.

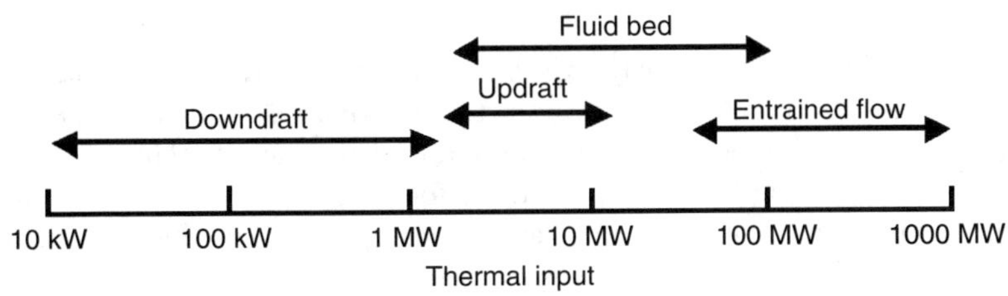

FIGURE 6.2 Range of applicability for biomass gasifier types.

types of gasifiers developed with data from Maniatis (2001) and Knoef (2005). Crossdraft gasifiers are for the smallest size while entrained flow are the largest size gasifiers.

6.2 FIXED-BED/MOVING-BED GASIFIERS

In entrained-flow and fluidized-bed gasifiers, the gasifying medium conveys the fuel particles through the reactor, but in a *fixed-bed* (also known as *moving-bed*) gasifier the fuel is supported on a grate (hence its name). This type is also called *moving-bed* because the fuel moves down in the gasifier as a plug. Fixed-bed gasifiers can be built inexpensively in small sizes, which is one of their major attractions. For this reason, large numbers of small-scale moving-bed biomass gasifiers are in use around the world.

TABLE 6.2 Characteristics of Fixed-Bed Gasifiers

Fuel (wood)	Updraft	Downdraft	Crossdraft
Moisture wet basis (%)	60 max	25 max	10–20
Dry-ash basis (%)	25 max	6 max	0.5–1.0
Ash melting temperature (°C)	>1000	>1250	
Size (mm)	5–100	20–100	5–20
Application range (MW)	2–30	1–2	
Gas exit temperature (°C)	200–400	700	1250
Tar (g/Nm^3)	30–150	0.015–3.0	0.01–0.1
Gas LHV (MJ/Nm^3)	5–6	4.5–5.0	4.0–4.5
Hot-gas efficiency (%)	90–95	85–90	75–90
Turn-down ratio (–)	5–10	3–4	2–3
Hearth load (MW/m^2)	<2.8		

Source: Adapted from Knoef, 2005, p. 26.

Both mixing and heat transfer within the moving (fixed) bed are rather poor, which makes it difficult to achieve uniform distribution of fuel, temperature, and gas composition across the cross-section of the gasifier. Thus, fuels that are prone to agglomeration can potentially form agglomerates during gasification. This is why fixed-bed gasifiers are not very effective for biomass fuels or coal with a high caking index in large-capacity units.

There are three main types of fixed- or moving-bed gasifier: (1) updraft, (2) downdraft, and (3) crossdraft. Table 6.2 compares their characteristics.

6.2.1 Updraft Gasifiers

An updraft gasifier is one of the oldest and simplest of all designs. Here, the gasification medium (air, oxygen, or steam) travels upward while the bed of fuel moves downward, and thus the gas and solids are in countercurrent mode. The product gas leaves from the top as shown in Figure 6.3. The gasifying medium enters the bed through a grate or a distributor, where it meets with the hot bed of ash. The ash drops through the grate, which is often made moving (rotating or reciprocating), especially in large units to facilitate ash discharge. Chapter 5 describes this process in more detail.

Updraft gasifiers are suitable for high-ash (up to 25%), high-moisture (up to 60%) biomass. They are also suitable for low-volatile fuels such as charcoal. Tar production is very high (30–150 g/nm^3) in an updraft gasifier, which makes

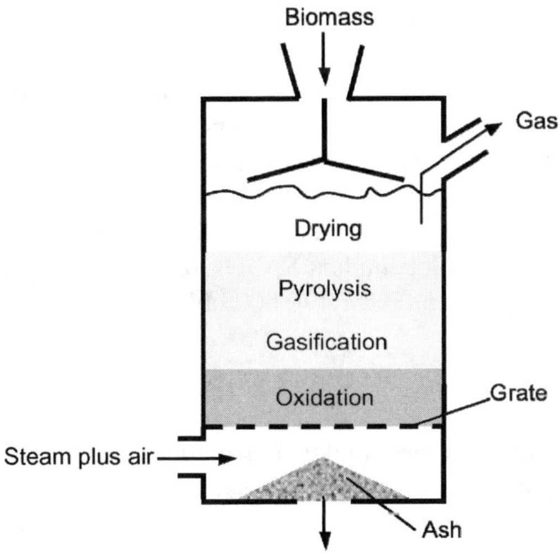

FIGURE 6.3 Schematic of an updraft gasifier.

it unsuitable for high-volatility fuels. On the other hand, as a countercurrent unit, an updraft gasifier utilizes combustion heat very effectively and achieves high cold-gas efficiency (Section 6.11.1). Therefore, it is more suitable for direct firing, where the gas produced is burnt in a furnace or boiler with no cleaning or cooling required. Since the gas is not fired in an engine or stored, the tar produced does not have to be cleaned.

Updraft gasifiers find commercial use in small units like cooking stoves in villages and in large units like South African Synthetic Oils (SASOL) for production of gasoline from coal. The following is a brief description of two important large-scale commercial updraft gasifier technologies.

Dry-Ash Gasifier

Lurgi, a process development company, developed a pressurized dry-ash updraft gasifier. It is called *dry ash* because the ash produced is not molten. One that produces molten ash is called a *slagging* gasifier.

Though the peak temperature (in the combustion zone) is 1200 °C, the maximum gasification temperature is 700 to 900 °C. The reactor pressure is in the neighborhood of 3 MPa, and the residence time of coal in the gasifier is between 30 and 60 minutes (Ebasco, 1981). The gasification medium is a mixture of steam and oxygen, steam and air, or steam and oxygen-enriched air. It uses a relatively high steam/fuel carbon ratio (~1.5).

The coal is first screened to between 3 and 40 mm (Probstein and Hicks, 2006, p. 162) and then fed into a lock hopper. The gasifying agent moves upward in the gasifier while the solids descend. The reactor is a double-walled pressure vessel. Between the two walls lies water that quickly boils into steam under pressure, utilizing the heat loss from the reactor. As the coal travels down the reactor, it undergoes drying, devolatilization, gasification, and combustion. Typical residence time in the gasifier is about an hour (Probstein and Hicks, 2006, p. 162). In a dry-ash gasifier, the temperature is lower than the melting point of ash, so the solid residue dries and is removed from the reactor by a rotating grate.

The dry-ash technology has been used at SASOL in South Africa, the world's biggest gasification complex. SASOL produces 55 million Nm^3/day of syngas, which is used to produce 170,000 bbl/day of Fischer-Tropsch liquid fuel.

Slagging Gasifier

The British Gas/Lurgi consortium developed a moving-bed gasifier that works on the same principle as the dry-ash gasifier, except a much higher temperature (1500–1800 °C) is used in the combustion zone to melt the ash (hence its name, *slagging gasifier*). Such a high temperature requires a lower steam-to-fuel ratio (~0.58) than that used in dry-ash units (Probstein and Hicks, 2006, p. 169).

Coal crushed to 5 to 80 mm is fed into the gasifier through a lock hopper system (Minchener, 2005). The gasifier's tolerance for coal fines is limited, so briquetting is used in places where the coal carries too many of them. Gasification agents, oxygen and steam, are introduced into the pressurized (~3 MPa) gasifier vessel through sidewall-mounted tuyers (lances) at the elevation where combustion and slag formation occur.

The coal introduced at the top gradually descends through several process zones. The feed is first dried in the top zone and then devolatilized. The descending coal is transformed into char and then passes into the gasification (reaction) zone. Below this zone, any remaining carbon is oxidized, and the ash content is liquefied, forming slag. Slag is withdrawn from the slag pool through an opening in the hearth plate at the bottom of the gasifier vessel. The product gas leaves from the top, typically at 400 to 500 °C (Minchener, 2005).

6.2.2 Downdraft Gasifiers

A downdraft gasifier is a co-current reactor where air enters the gasifier at a certain height below the top. The product gas flows downward (giving the name *downdraft*) and leaves through a bed of hot ash (Figures 6.4 and 6.5). Since it passes through the high-temperature zone of hot ash, the tar in the product gas finds favorable conditions for cracking (see Chapter 4). For this reason, a downdraft gasifier, of all types, has the lowest tar production rate.

Air from a set of nozzles, set around the gasifier's periphery, flows downward and meets with pyrolyzed char particles, developing a combustion zone (zone III shown schematically in Figure 6.5 and described in the discussion of throatless downdraft gasifiers that follows) of about 1200 to 1400 °C. Then the

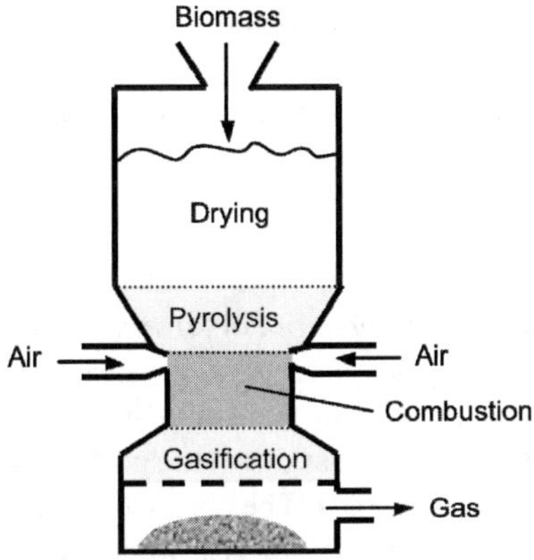

FIGURE 6.4 Schematic of a throated-type downdraft gasifier.

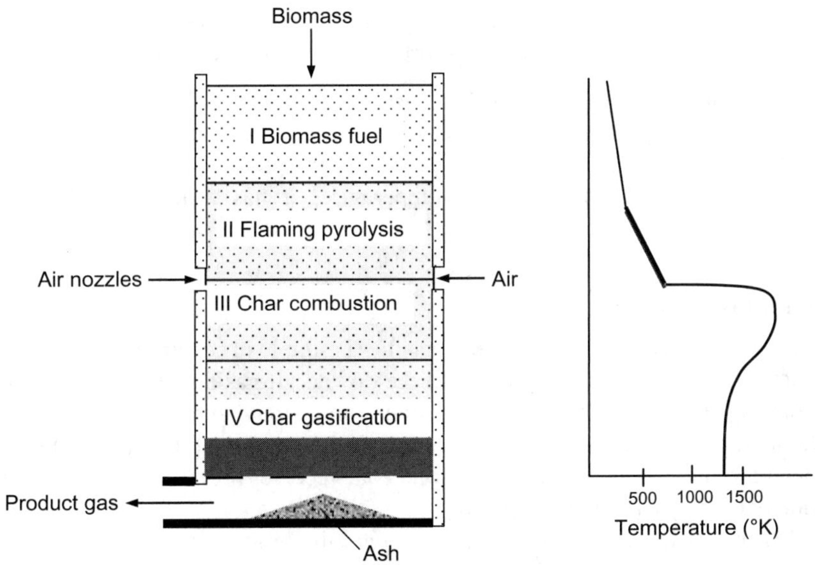

FIGURE 6.5 Schematic of the operation of a throatless downdraft gasifier. Temperature gradient along the height shown at the right.

gas descends further through the bed of hot char particles (zone IV), gasifying them. The ash produced leaves with the gas, dropping off at the bottom of the reactor.

Downdraft gasifiers work well with internal-combustion engines. The engine suction draws air through the bed of fuel, and gas is produced at the end. Low tar content (0.015–3 g/nm^3) in the product gas is another motivation for their use with internal-combustion engines. A downdraft gasifier requires a shorter time (20–30 minutes) to ignite and bring the plant up to working temperature compared to the time required by an updraft gasifier.

There are two principal types of downdraft gasifier. The throatless (or open core) type is illustrated in Figure 6.5. Reactions in different zones and at different temperatures are plotted on the right. The throated (or constricted) type is shown in Figure 6.4.

Throatless Gasifier

This gasifier type is also called *open top*, or *stratified throatless*. Here, the top is exposed to the atmosphere, and there is no constriction in the gasifier vessel because the walls are vertical. Figure 6.5 shows that a throatless design allows unrestricted movement of the biomass down the gasifier, which is not possible in the throated type shown in Figure 6.4. The absence of a throat avoids bridging or channeling. Open-core is another throatless design, but here air is not added from the middle as in other types of downdraft gasifiers. Air is drawn into the gasifier from the top by the suction created downstream of the gasifier. Such gasifiers are suitable for finer fuels—for example, lighter biomass such as rice husk.

The following are some of the shortcomings of a downdraft gasifier:

- It operates best on pelletized fuel instead of fine light biomass.
- The moisture in the fuel must not exceed 25%.
- A large amount of ash and dust remains in the product gas.
- As a result of its high exit temperature, it has a lower gasification temperature.

Operating Principle

Because an open-top, or a throatless, gasifier is simple in construction, it is used to describe the gasification process in the downdraft gasifier (Figure 6.5). The throatless process can be divided into four zones (Reed and Das, 1988, p. 39). The first, or uppermost, zone receives raw fuel from the top that is dried in air drawn through the first zone. The second zone receives heat from the third zone principally by thermal conduction.

During its journey through the first zone, the biomass heats up (zone I in Figure 6.5). Above 350 °C, it undergoes pyrolysis, breaking down into charcoal, noncondensable gases (CO, H_2, CH_4, CO_2, and H_2O), and tar vapors (condensable gases). The pyrolysis product in zone II receives only a limited supply of air from below and burns in a fuel-rich flame. This is called *flaming pyrolysis*. Most of the tar and char produced burn in zone III, where they generate heat for pyrolysis and subsequent endothermic gasification reactions (Reed and Das, 1988, p. 28).

Zone III contains ash and pyrolyzed char produced in zone II. While passing over the char, hot gases containing CO_2 and H_2O undergo steam gasification and Boudouard reactions, producing CO and H_2. The temperature of the downflowing gas reduces modestly, owing to the endothermic gasification reactions, but it is still above 700 °C.

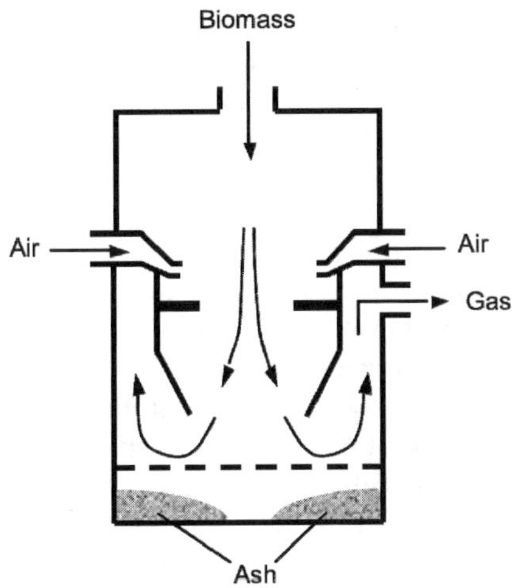

FIGURE 6.6 Constricted downdraft gasifier (Imbert type). Air/oxygen is added through nozzles around the vessel just above the constriction. *(Source: Adapted from Reed and Das, 1988, p. 39.)*

The bottommost layer (zone IV) consists of hot ash and/or unreacted charcoal, which crack any unconverted tar in this layer. Figure 6.5 shows the reactions and temperature distribution along the gasifier height. In one version of the throatless downdraft gasifier, the *open-core* type, the air enters from the top along with the feed. This type is free from some of the problems of other downdraft gasifiers.

Throated Gasifier

The cross-sectional area of a throated (also called *constricted*) gasifier is reduced at the throat and then expanded, as shown in Figure 6.4. The purpose is for the oxidation (combustion) zone to be at the narrowest part of the throat and to force all of the pyrolysis gas to pass through this narrow passage. Air is injected through nozzles just above the constriction. The height of the injection is about one-third of the way up from the bottom (Reed and Das, 1988, p. 33).

The movement of the entire mass of pyrolysis product through this hot and narrow zone results in a uniform temperature distribution over the cross-section and allows most of the tar to crack there. In the 1920s, a French inventor, Georges Imbert, developed the original design, which is popularly known as an *Imbert gasifier* (Figure 6.6).

The fuel, fed at the top, descends along a cylindrical section that serves as storage. The air pyrolyzes the biomass and burns the pyrolysis product or some charcoal. The hot char and the pyrolysis product pass through the throat, where most of the tar is cracked and the char is gasified. Figure 6.6 showed a flat-type throat construction, but it can be a V-type like in Figure 6.4.

Throated downdraft gasifiers are not suitable for scale-up to larger sizes because they do not allow for uniform distribution of flow and temperature in the constricted area. Beyond 1 MWth capacity, an annular constriction can be employed, but this has not been the practice to date.

6.2.3 Crossdraft Gasifiers

A crossdraft gasifier is a co-current moving-bed reactor, in which the fuel is fed from the top and air is injected through a nozzle from the side (Figure 6.7). It is primarily used for gasification of charcoal with very low ash content. Unlike the downdraft and updraft types, it releases the product from its side wall opposite to the entry point of the air for gasification. Because of this configuration, the design is also referred to as *sidedraft*. High-velocity air enters the gasifier through a nozzle set at a certain height above the grate. Excess oxygen in front of the nozzles facilitates combustion (oxidation) of part of the char, creating a very-high-temperature (>1500 °C) zone. The remaining char is then gasified to CO downstream in the next zone (Figure 6.7). The product gas exits from the opposite side of the gasifier. Heat from the combustion zone is conducted around the pyrolysis zone, so the fresh biomass is pyrolyzed while passing through it.

This type of gasifier is generally used in small-scale biomass units. One of its important features is a relatively small reaction zone with low thermal capacity, which gives a faster response time than that of any other moving-bed type. Moreover, startup time (5–10 minutes) is much shorter than in downdraft and updraft units. These features allow a sidedraft gasifier to respond well to load changes when used directly to run an engine. Because its tar production is low (0.01–0.1 g/nm^3), a crossdraft gasifier requires a relatively simple gas-cleaning system.

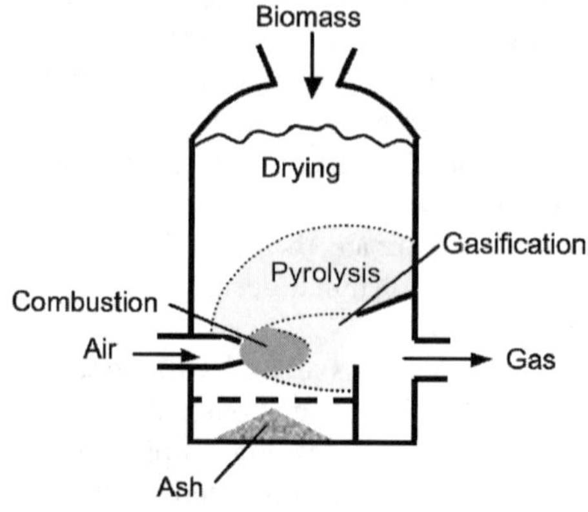

FIGURE 6.7 Schematic of a crossdraft gasifier.

Crossdraft gasifiers can be very light and small (<10 kWe). Since layers of fuel and ash insulate the walls from the high-temperature zone, the gasifier vessel can be constructed of ordinary steel with refractory linings on the nozzle and gas exit zone.

The crossdraft design is less suitable for high-ash or high-tar fuels, but it can handle high-moisture fuels if the top is open so that the moisture can escape. Particle size should be controlled, as unscreened fuel runs the risk of bridging and channeling. Crossdraft gasifiers work better with charcoal or pyrolyzed fuels. For unpyrolyzed fuels, the height of the air nozzle above the grate becomes critical (Reed and Das, 1988, p. 32).

6.3 FLUIDIZED-BED GASIFIERS

Fluidized-bed gasifiers are noted for their excellent mixing and temperature uniformity. A fluidized bed is made of granular solids, called *bed materials*, that are kept in a semi-suspended condition (*fluidized state*) by the passage of the gasifying medium through them at the appropriate velocities. The excellent gas–solid mixing and the large thermal inertia of the bed make this type of gasifier relatively insensitive to the fuel's quality (Basu, 2006). Along with this, the temperature uniformity greatly reduces the risk of fuel agglomeration.

The fluidized-bed design has proved to be particularly advantageous for gasification of biomass. Its tar production lies between that for updraft (~50 g/nm^3) and downdraft gasifiers (~1 g/nm^3), with an average value of around 10 g/nm^3 (Milne et al., 1998, p. 14). There are two principal fluidized-bed types: bubbling and circulating.

6.3.1 Bubbling Fluidized-Bed Gasifier

The bubbling fluidized-bed gasifier, developed by Fritz Winkler in 1921, is perhaps the oldest commercial application of fluidized beds; it has been in commercial use for many years for the gasification of coal (Figure 6.8); for biomass gasification, it is one of the most popular options. A fairly large number of bubbling fluidized-bed gasifiers of varying designs have been developed and are in operation (Lim and Alimuddin, 2008; Narváez et al., 1996).

Because they are particularly suitable for medium-size units (<25 MWth), many biomass gasifiers operate on the bubbling fluidized-bed regime. Depending on operating conditions, bubbling-bed gasifiers can be grouped as low-temperature and high-temperature types. They can also operate at atmospheric or elevated pressures.

In the most common type of fluidized bed, biomass crushed to less than 10 mm is fed into a bed of hot materials. These bed materials are fluidized with steam, air, or oxygen, or their combination, depending on the choice of

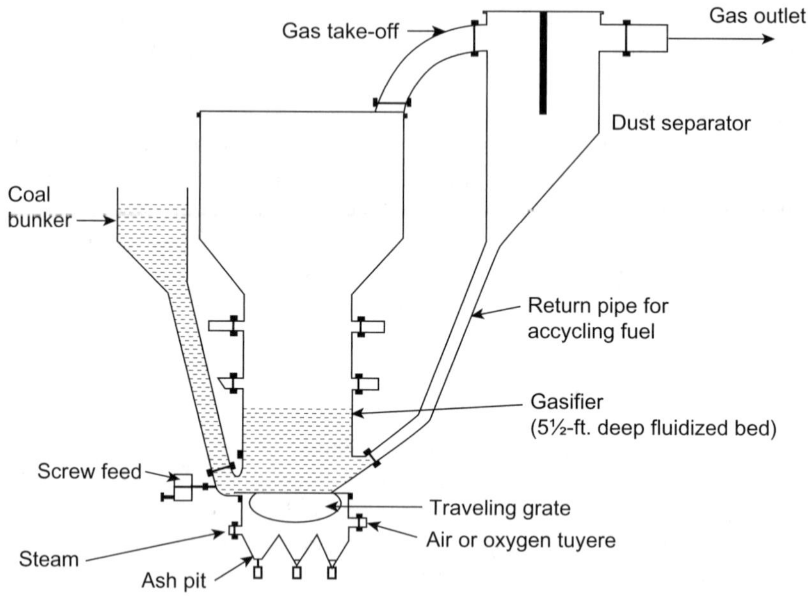

FIGURE 6.8 A sketch of the original Winkler bubbling fluidized-bed gasifier.

gasification medium. The ash generated from either the fuel or the inorganic materials associated with it is drained easily from the bottom of the bed. The bed temperature is normally kept below 980 °C for coal and below 900 °C for biomass to avoid ash fusion and consequent agglomeration.

The gasifying medium may be supplied in two stages. The first-stage supply is adequate to maintain the fluidized bed at the desired temperature; the second-stage supply, added above the bed, converts entrained unreacted char particles and hydrocarbons into useful gas.

High-temperature Winkler (HTW) gasification is an example of high-temperature, high-pressure bubbling fluidized-bed gasification for coal and lignite. Developed by Rheinbraun AG of Germany, the process employs a pressurized fluidized bed operating below the ash-melting point. To improve carbon conversion efficiency, small char particles in the raw gas are separated by a cyclone and returned to the bottom of the main reactor (Figure 6.9).

The gasifying medium (steam and oxygen) is introduced into the fluidized bed at different levels as well as above it. The bed is maintained at a pressure of 10 bars while its temperature is maintained at about 800 °C to avoid ash fusion. The overbed supply of the gasifying medium raises the local temperature to about 1000 °C to minimize production of methane and other hydrocarbons.

The HTW process produces a better-quality gas compared with the gas that is produced by traditional low-temperature fluidized beds. Though originally

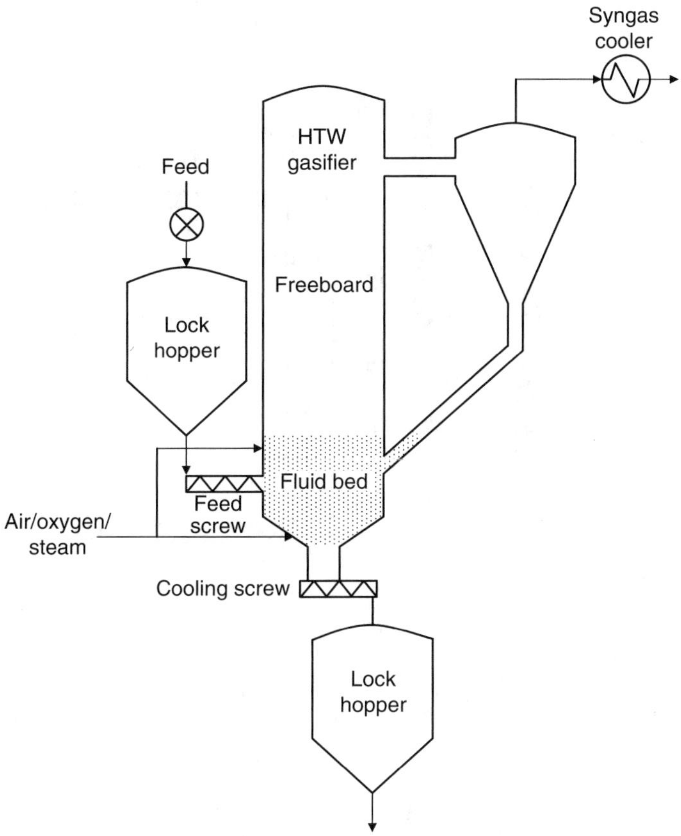

FIGURE 6.9 High-temperature Winkler (HTW) bubbling fluidized-bed gasifier.

developed for coal, it is suitable for lignite and other reactive fuels like biomass and treated municipal solid waste (MSW).

6.3.2 Circulating Fluidized-Bed Gasifier

A circulating fluidized-bed (CFB) gasifier has a special appeal for biomass gasification because of the long gas residence time it provides. It is especially suitable for fuels with high volatiles. A CFB typically comprises a riser, a cyclone, and a solid recycle device (Figure 6.10). The riser serves as the gasifier reactor.

Although the HTW process (Figure 6.9) appears similar to a CFB, it is only a bubbling bed with limited solid recycle. The circulating and bubbling fluidized beds are significantly different in their hydrodynamic. In a CFB, the solids are dispersed all over the tall riser, allowing a long residence time for the gas as well as for the fine particles. The fluidization velocity in a CFB is much

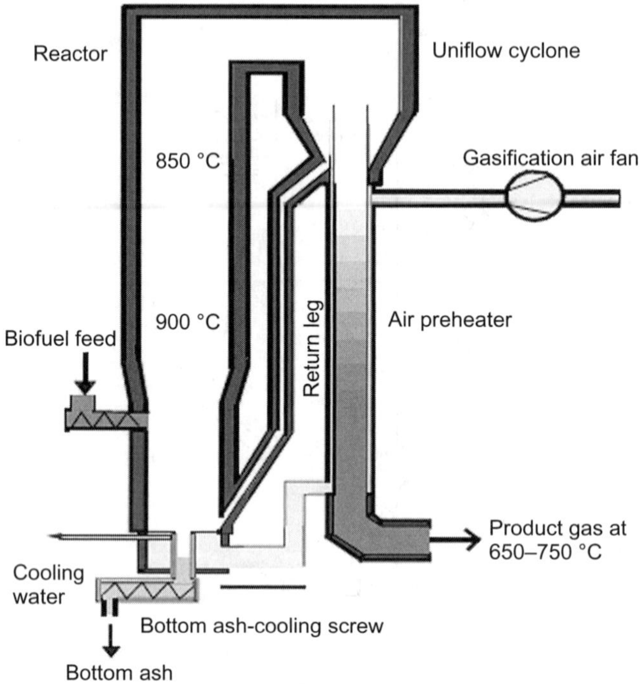

FIGURE 6.10 Circulating fluidized-bed gasifier. *(Source: Adapted from Foster Wheeler.)*

higher (3.5–5.5 m/s) than that in a bubbling bed (0.5–1.0 m/s). Also, there is large-scale migration of solids out of the CFB riser. These are captured and continuously returned to the riser's base. The recycle rate of the solids and the fluidization velocity in the riser are sufficiently high to maintain the riser in a special hydrodynamic condition, known as *fast fluidized bed*. Depending on the fuel and the application, the riser operates at a temperature of 800 to 1000 °C.

The hot gas from the gasifier passes through a cyclone, which separates most of the solid particles associated with it, and the loop seal returns the particles to the bottom of the gasifier. Foster Wheeler developed a CFB gasifier where an air preheater is located in the standpipe below the cyclone to raise the temperature of the gasification air and indirectly raise the gasifier temperature (Figure 6.10).

Many commercial gasifiers of this type have been installed in different countries. At the time of writing, the biggest among these is a 60-MWth unit in a coal-fired and natural-gas-fired power plant in Lahti, Finland, to provide a cheap supplementary fuel by gasifying waste wood and refuse-derived fuels (RDFs). Several manufacturers around the world have developed versions of the CFB gasifier that work on the same principle and vary only in engineering details.

TABLE 6.3 Comparison of Hydrodynamic Operating Conditions of Commercial Transport Gasifier and Circulating Fluidized Bed of Fluid Catalyst Cracking Units

Parameter	Smith et al., 2002	Peterson and Werther, 2005	Zhu and Venderbosch, 2005
Particle size (μm)	200–350	180–230	20–150
Riser velocity (m/s)	12–18	3.5–5.0	6–28
Circulation rate (kg/m^2·s)	730–3400	2.5–9.2*	400–1200
Riser temperature (°C)	910–1010	800–900	500–550
Riser pressure (bar)	140–270 psig	1 bar	150–300 kPa
Reactor	KBR gasifier	CFB gasifier	FCC cracker

*Computed from comparable units.

Transport Gasifier

This type of gasifier has the characteristics of both entrained-flow and fluidized-bed reactors. The hydrodynamics of a transport gasifier is similar to that of a fluid catalytic cracking reactor. A transport gasifier operates at circulation rates, velocities, and riser densities considerably higher than those of a conventional circulating fluidized bed. This results in higher throughput, better mixing, and higher mass and heat-transfer rates. The fuel particles are also very fine (Basu, 2006) and as such it requires a pulverizer or a hammer mill. A comparison of typical hydrodynamic operating conditions in a transport gasifier and in a fluid catalytic cracking unit is given in Table 6.3.

A transport gasifier consists of a mixing zone, a riser, a disengager, a cyclone, a standpipe, and a J-leg. Coal, sorbent (for sulfur capture), and air are injected into the reactor's mixing zone. The disengager removes the larger carried-over particles, and the separated solids return to the mixing section through the J-valve located at the base of the standpipe (Figure 6.11). Most of the remaining finer particles are removed by a cyclone located downstream, from which the gas exits the reactor. The reactor can use either air or oxygen as the gasification medium.

Use of oxygen as the gasifying medium avoids nitrogen, the diluting agent in the product gas. For this purpose, air is more suitable for power generation, while oxygen is more suitable for chemicals production. The transport gasifier has proved to be effective for gasification of coal, but it is yet to be proven for biomass.

Twin Reactor System

One of the major problems in air gasification of coal or biomass is the dilution of product gas by the nitrogen in the air used for the exothermic combustion

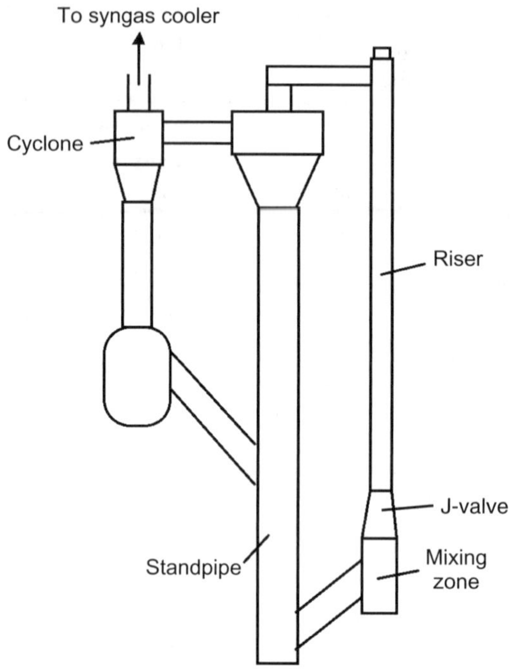

FIGURE 6.11 A sketch of a typical transport fluidized-bed gasifier.

reaction necessary in a self-sustained gasifier. To avoid this, oxygen is used instead, but oxygen gasification is expensive and highly energy intensive (see Example 6.5 later in chapter). A twin reactor (e.g., a dual fluidized bed) overcomes this problem by separating the combustion reactor from the gasification reactor such that the nitrogen released in the air combustion does not dilute the product gas. Twin reactor systems are used for coal and biomass. They are either externally circulating or internally circulating.

This type of system has some limitations; for example, Corella et al. (2007) identified two major design issues with the dual fluidized-bed system:

- Biomass contains less char than coal contains; however, if this char is used for gasification the amount of char available may not be sufficient to provide the required endothermic heat to the gasifier reactor to maintain a temperature above 900 °C. External heating may be necessary.
- Though the gasifier runs on steam, only a small fraction (<10%) of the steam participates in the gasification reaction; the rest of it simply leaves the gasifier, consuming a large amount of heat and diluting the gas.

The Technical University of Vienna used the externally circulating system to gasify various types of biomass in an industrial plant in Gussing, Austria.

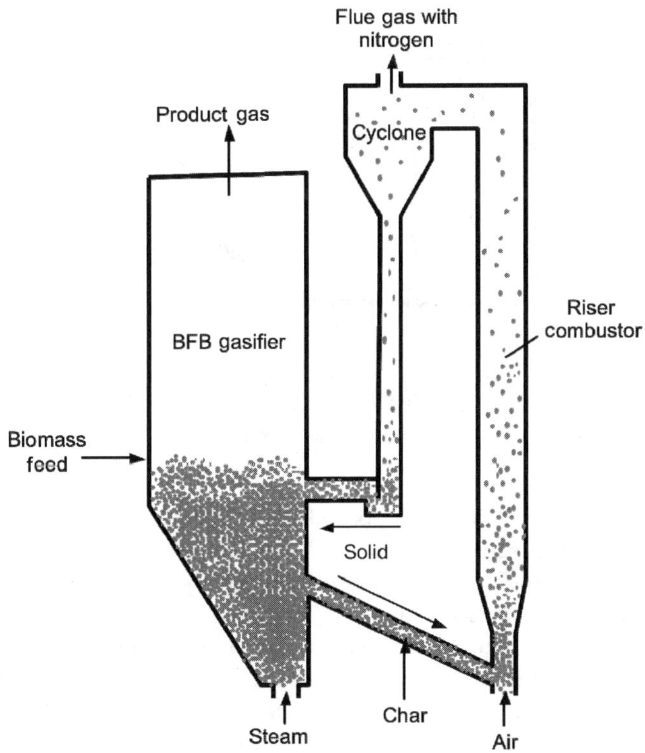

FIGURE 6.12 Twin reactor (dual fluidized-bed) gasifier.

The system is comprised of a bubbling fluidized-bed gasifier and a circulating fluidized-bed combustor (Figure 6.12). The riser in a CFB operates as a combustor; the bubbling fluidized bed in the return leg operates as a gasifier. Pyrolysis and gasification take place in the bubbling fluidized bed, which is fluidized by superheated steam. Unconverted char and tar move to the riser through a nonmechanical valve. The riser is fluidized by air.

Tar and gas produced during pyrolysis are combusted in the riser's combustion zone. Heat generated by combustion raises the temperature of the inert bed material to around 900 °C. This material leaves the riser and is captured by the cyclone at the riser exit. The collected solids drop into a standpipe and are then circulated into the bubbling fluidized-bed reactor to supply heat for its endothermic reactions. The char is gasified in the bubbling bed in the presence of steam, producing the product gas. This system overcomes the problem of tar by burning it in the combustor. In this way, a product gas relatively free of tar can be obtained.

The Rentech-Silvagas process is also based on the externally circulating principle. Here, both the combustor and the gasifier work on circulating fluidized-bed principles. In the internally circulating design, the fluidized-bed

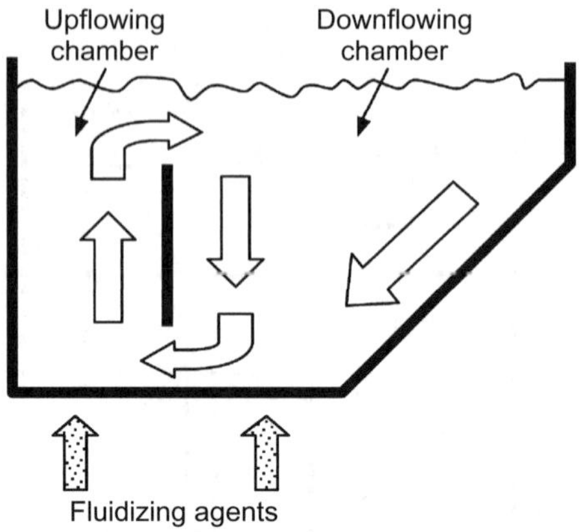

FIGURE 6.13 Internally circulating dual fluidized-bed gasifier.

reactor is divided into two chambers and connected by a window at the bottom of the division wall separating them. The chambers are fluidized at different velocities (Figure 6.13), which result in their having varying bed densities. As the bed height is the same in both, the hydrostatic pressure at the bottom of the two chambers is different. The biomass and sand thus flow from the higher-density chamber to the lower-density chamber, creating a continuous circulation of bed materials similar to the natural circulation in a boiler. This helps increase the residence time of solids in the fluidized bed.

Such an arrangement can provide a more uniform distribution of biomass particles in the reactor, with increased gasification yield and decreased tar and fine solids (char) in the syngas (Freda et al., 2008). A special feature of the twin reactor is that more air or oxygen can be added in one part of the bed to encourage combustion, and more steam can be added in another part to encourage gasification.

Chemical Looping Gasifier

Chemical looping is a relatively new concept. Its primary motivation is production of two separate streams of gases—a product gas rich in hydrogen and a gas stream rich in carbon dioxide—such that the CO_2 can be sequestrated while the hydrogen can be used for applications that require hydrogen-rich gas. The system uses calcium oxide as a carrier of carbon dioxide between two reactors: a gasifier (bubbling fluidized bed) and a regenerator (circulating fluidized bed). The CO_2 produced during gasification is captured by the CaO and released in a second reactor during sorbent regeneration.

Figure 6.14 is a schematic of the chemical looping process. Biomass is fed into the gasifier that receives calcium oxide from the regenerator and superheated steam from an external source. During gasification, the carbon dioxide produced is captured by the calcium oxide that makes up the bubbling fluidized bed (Acharya et al., 2009), as follows:

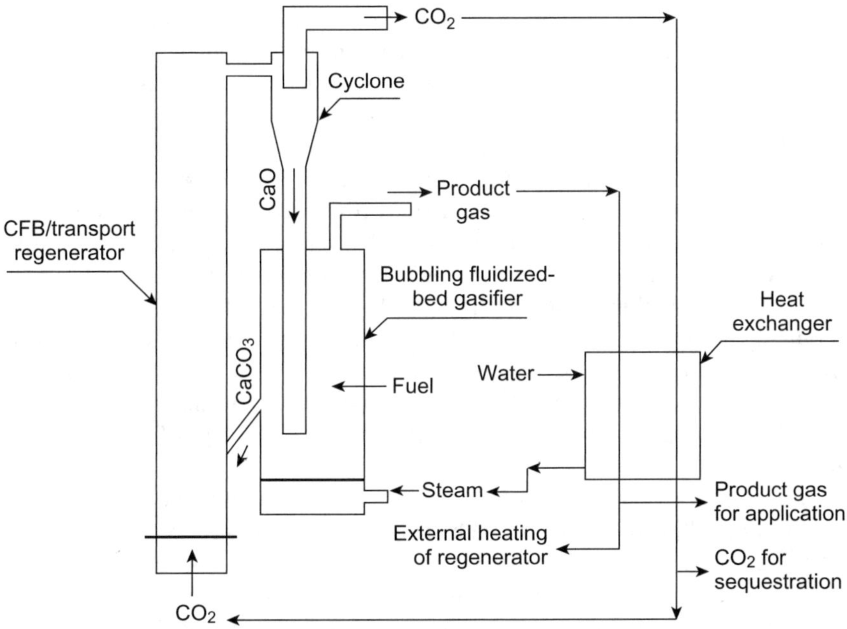

FIGURE 6.14 Chemical looping gasification with CaO as the carrier of CO_2 between the gasifier and the regenerator.

$$\text{Gasification reaction: } C_nH_hO_o + (2n-p)H_2O + nCaO \leftrightarrow nCaCO_3 + \left(\frac{h}{2}+2n-o\right)H_2 \quad (6.1)$$

$$CO + H_2O \leftrightarrow CO_2 + H_2 \quad (6.2)$$

$$CO_2 \text{ removal reaction: } CaO + CO_2 \rightarrow CaCO_3 \quad (6.3)$$

Immediate removal of the reaction product, CO_2, from the system increases the rate of forward reaction (Eq. 6.2), enhancing the water–gas shift reaction, therefore yielding more hydrogen in the product gas. The calcium carbonate formed in the gasifier (Eq. 6.3) is transferred to a circulating/transport regenerator, where it is calcined into calcium oxide and carbon dioxide.

$$\text{Regeneration: } CaCO_3 \rightarrow CaO + CO_2 + 178.3\ \text{kJ/mol} \quad (6.4)$$

The carbon dioxide and the product gas leave the regenerator and gasifier, respectively, at a high temperature. The hot product can be used for generation of steam needed for gasification.

6.4 ENTRAINED-FLOW GASIFIERS

Entrained flow is the most successful and widely used gasifier type for large-scale gasification of coal, petroleum coke, and refinery residues. It is ideally suited to most types of coal except low-rank coal, which, like lignite and biomass, is not attractive because of its large moisture content. High-ash coal is also less suitable because cold-gas efficiency decreases with increasing ash content. For slurry-fed coal, the economic limit is 20% ash; for dry feed it is 40% (Higman and Burgt, 2008, p. 122).

The suitability of entrained-flow gasification for biomass is questionable for a number of reasons. Owing to a short residence time (a few seconds) in entrained-flow reactors, the fuel needs to be very fine, and grinding fibrous biomass into such fine particles is difficult. For biomass with CaO but no alkali, the ash-melting point is high, and therefore it has a higher oxygen requirement. The melting point of biomass ash with a high alkali content is much lower than that of coal. This reduces the oxygen required to raise the temperature of the ash above its melting point. However, molten biomass ash is highly aggressive, which greatly shortens the life of the gasifier's refractory lining.

For these reasons entrained-flow reactors are not preferred for biomass gasification. Still, they have the advantage of easily destroying tar, which is very high in biomass and is a major problem in biomass gasification.

Entrained-flow gasifiers are essentially co-current plug-flow reactors, where gas and fuel travel. The hydrodynamics is similar to that of the well-known pulverized-coal (PC) boiler, where the coal is ground in a pulverizing mill to sizes below 75 micron and then conveyed by part of the combustion air to a set of burners suitably located around the furnace. The reactor geometry of the entrained-flow gasifier is much different from the furnace geometry of a PC boiler. Additionally, an entrained-flow gasifier works in a substoichiometric supply of oxygen, whereas a PC boiler requires excess oxygen.

The gasification temperature of an entrained-flow gasifier generally well exceeds 1000 °C. This allows production of a gas that is nearly tar-free and has a very low methane content. A properly designed and operated entrained-flow gasifier can have a carbon conversion rate close to 100%. The product gas, being very hot, must be cooled in downstream heat exchangers that produce the superheated steam required for gasification.

Figure 6.15 describes the working principle of an entrained-flow gasifier by means of a simplified sketch. The high-velocity jet forms a recirculation zone near the entry point. Fine fuel particles are rapidly heated by radiative heat from the hot walls of the reactor chamber and from the hot gases downstream, and start burning in excess oxygen. The bulk of the fuel is consumed near the entrance zone through devolatilization; here the temperature may rise to as high as 2500 °C.

The combustion reaction consumes nearly all of the oxygen feed, so the residual char undergoes gasification reactions in CO_2 and H_2O environments downstream of this zone. These reactions are relatively slow compared to the devolatilization reaction, so the char takes much longer to complete its conversion to gases. For this reason, a large reactor length is required.

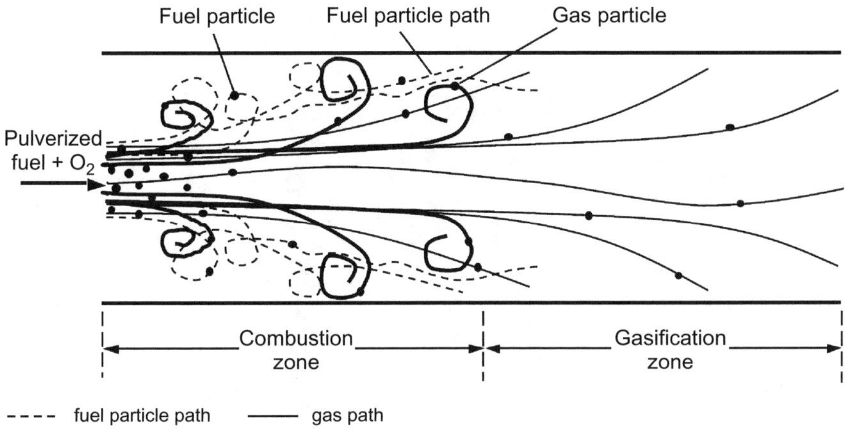

FIGURE 6.15 Simplified sketch of gas–solid flow in an entrained-flow gasifier.

Entrained-flow gasifier design may be classified into two broad groups: (1) the top-fed downflow (used by GE Energy and Siemens SFG), shown in Figure 6.16; and (2) the side-fed upflow (used by Koppers-Totzek, the Shell gasification process, Prenflo, and the Lurgi multipurpose), shown in Figure 6.17.

6.4.1 Top-Fed Gasifier

Top-fed gasifiers use a vertically cylindrical reactor vessel, into which pulverized fuel (biomass or coal) and gasifying agent(s) are conveyed by oxygen and injected from the top. This vessel resembles a vertical furnace with a downward burner (Figure 6.16). The fuel and the gasifying agent(s) are injected into the reactor through a jet that generally sits at the reactor's middle section.

The fuel gasifier (SFG) process of Siemens uses a top-fired reactor design, in which the reactants are introduced through the single centrally mounted burner. This has several advantages. First, it is of an axisymmetrical construction, reducing equipment costs; second, the flow of reactant occurs from a single burner, reducing the number of burners to be controlled; finally, the product gas and the slag flow in the same direction, which reduces any potential blockage in a slag trap (Higman and van der Burgt, 2008, p. 132).

6.4.2 Side-Fed Gasifier

In side-fed gasifiers, powdered fuel is injected through horizontal nozzles set opposite each other in the reactor's lower section (Figure 6.17). Jets of fuel and gasifying agents form a stirred-tank reactor characterized by a high degree of mixing. The product gas moves upward and exits through the top. Because of the high oxygen availability in this mixing zone, rapid exothermic reactions take place, raising the gas temperature to well above the ash-melting point

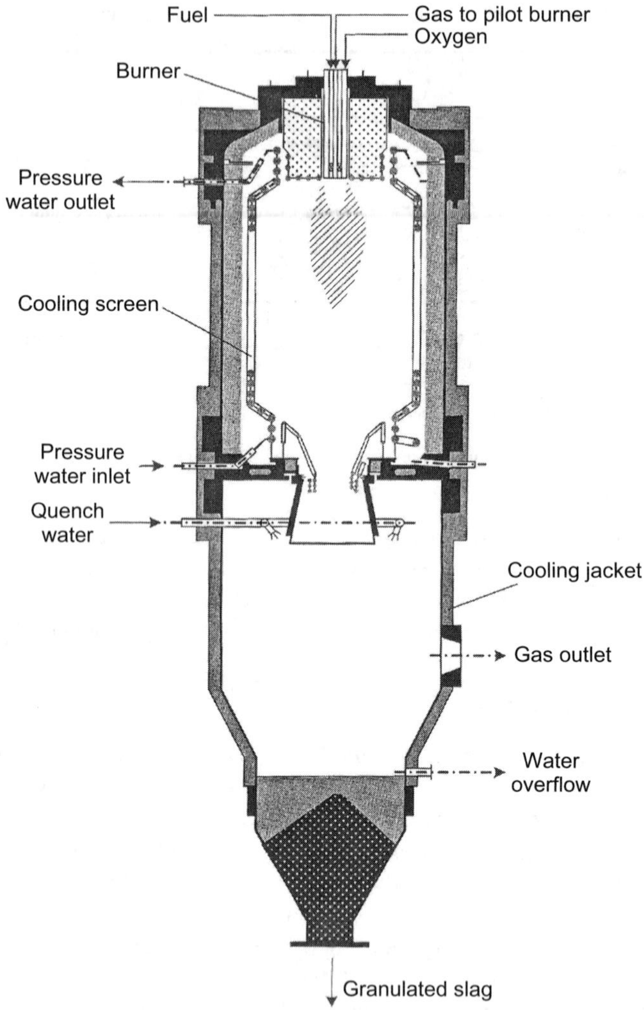

FIGURE 6.16 A schematic of a top-fed downflow entrained-flow gasifier.

(>1400 °C). Thus, the ash, instead of traveling up, is separated in this zone as slag from the fuel, and drained. Some gasifier designs (e.g., E-gas, MHI, Eagle) inject additional fuel further downstream from the main reaction zone.

The Koppers-Totzek atmospheric pressure gasifier also uses side feeding. It consists of two side-mounted burners where a mixture of coal and oxygen is injected. The gas leaves from the top of the gasifier at temperatures around 1500 °C and is quenched with water downstream. The reactor has a steam jacket to protect its shell from high temperatures (Higman and van der Burgt, 2008, p. 129).

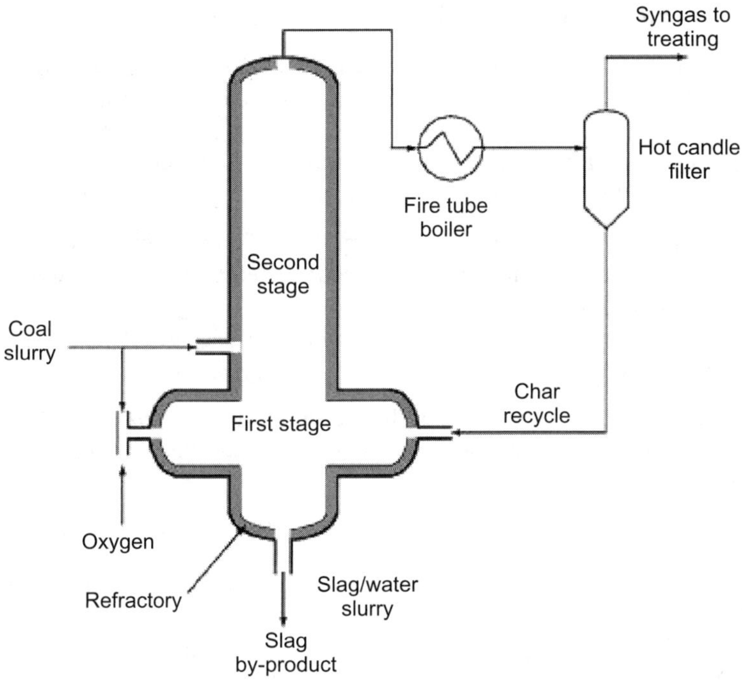

FIGURE 6.17 A schematic of a side-fed entrained-flow gasifier.

The E-gas gasifier is a side-fed two-stage entrained-flow slagging gasifier with a coal–water slurry feed. It is designed to use sub-bituminous coal (Figure 6.17). The coal slurry is fed at the nonslagging stage, where the upflowing gas heats it. Thus, the gas exits at a lower temperature and then passes through a fire-tube boiler and is filtered in a hot candle filter. The char, separated out by the filter, is taken back to the slagging zone. The slag is quenched in a water bath at the bottom of the slagging reactor.

6.4.3 Advantages of Entrained-Flow Gasifiers

Entrained-flow gasifiers have several advantages over other types:

- Low tar production
- A range of acceptable feed
- Ash produced as slag
- High-pressure, high-temperature operation
- Very high conversion of carbon
- Low methane content well suited for synthetic gas production

6.4.4 Entrained-Flow Gasification of Biomass

For thermal gasification of the refractory components of biomass (those difficult to gasify) such as lignin, the minimum temperature requirement is similar to that for coal (~900 °C) (Higman and van de Burgt, 2008, p. 147). Entrained-flow gasification of biomass is therefore rather limited and has not been seen on a commercial scale for the following reasons:

- The residence time in the reactor is very short. For the reactions to complete, the biomass particles must be finely ground. Being fibrous, biomass cannot be pulverized easily.
- Molten ash from biomass is highly aggressive because of its alkali compounds and can corrode the gasifier's refractory or metal lining.

Given these shortcomings, entrained-flow gasifiers are not popular for biomass. However, there is at least one successful entrained-flow biomass gasifier, known as the Choren process.

Choren Process

The Choren entrained-flow biomass gasifier is comprised of three stages (Figure 6.18). The first stage receives biomass in a horizontal stirred-type low-temperature reactor for pregasification at 400 to 500 °C in a limited supply of air. This produces solid char and a tar-rich volatile product. The latter flows into the second chamber (stage 2), an entrained-flow combustor where oxygen and the product gas from the first stage are injected downward into the reactor. Combustion raises the temperature to 1300 to 1500 °C and completely cracks the tar. The hot combustion product flows into the third chamber (stage 3), where the char is gasified.

The solid char received from the first stage is pulverized and fed into the third stage of the Choren process. It is gasified in the hot gasification medium produced in the second stage. Endothermic gasification reactions reduce the

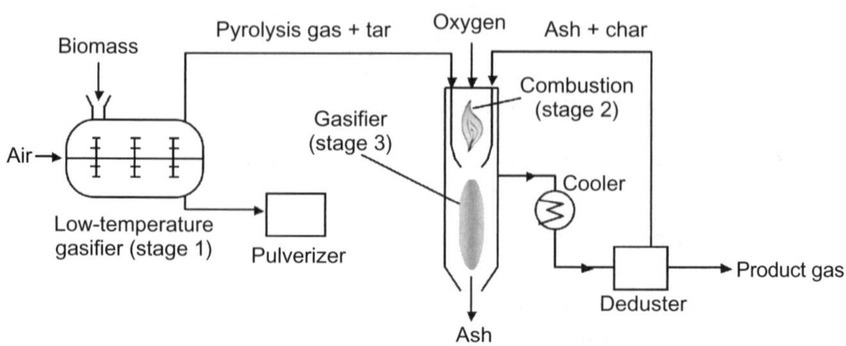

FIGURE 6.18 Choren process. The biomass is gasified in an entrained-flow gasifier, facilitated by a rotary-type partial gasifier (stage 1).

temperature to about 800 °C. Char and ash from the product gas are separated and recycled into the second-stage combustor. The ash melts at the high temperature in the combustor and is drained from the bottom. Now molten, the ash freezes, forming a layer on the membrane wall that protects the wall against the corrosive action of fresh molten biomass ash. The product gas is processed downstream for Fisher-Tropsch synthesis or other applications.

6.5 PLASMA GASIFICATION

In plasma gasification, high-temperature plasma helps gasify biomass hydrocarbons. It is especially suitable for MSW and other waste products. This process may also be called "plasma pyrolysis" because it essentially involves thermal disintegration of carbonaceous material into fragments of compounds in an oxygen-starved environment. The heart of the process is a plasma gun, where an intense electric arc is created between two electrodes spaced apart in a closed vessel through which an inert gas is passed (Figure 6.19).

Though the temperature of the arc is extremely high (~13,000 °C), the temperature downstream, where waste products are brought in contact with it, is much lower (2700–4500 °C). The downstream temperature is still sufficiently high, however, to pyrolyze complex hydrocarbons into simple gases such as CO and H_2. Simultaneously, all inorganic components (e.g., glass, metals,

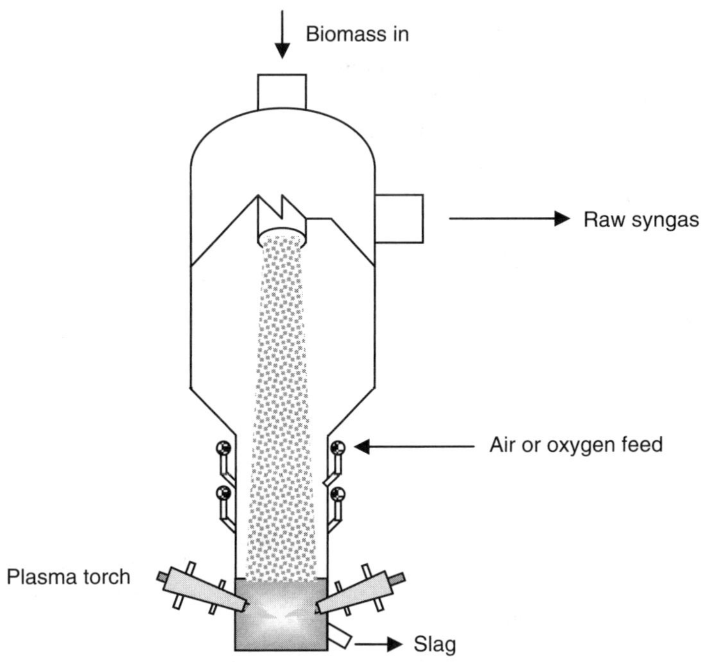

FIGURE 6.19 Plasma gasification of solid waste.

silicates, heavy metals) are fused into a volcanic-type lava, which after cooling forms a basaltic slag. The product gas leaves the gasifier at very high temperatures (1000–1200 °C).

A typical plasma reactor provides a relatively long residence time for the gas in the gasifier. This and the high temperature cause the tar products to be cracked and harmful products like dioxin and furan to be destroyed.

Owing to the high reactor temperature and the presence of chlorine in wastes, the life of the reactor liner is an issue. However, an attractive feature is that plasma gasification is relatively insensitive to the quality of the feedstock. This is the result of an independent energy source run by electricity instead of partial combustion of the gasification product.

6.6 PROCESS DESIGN

The design of a gasifier involves both process and hardware. The process design gives the type and yield of the product, operating conditions, and the basic size of the reactor. The hardware design involves structural and mechanical components, such as grate, main reactor body, insulation, cyclone, and others, that are specific to the reactor type. This section focuses on gasifier process design.

6.6.1 Design Specification

For any design, specification of the plant is very important. The input includes the specification of the fuel, gasification medium, and product gas. A typical fuel specification will include proximate and ultimate analysis, operating temperatures, and ash properties. The specification of the gasifying medium is based on the selection of steam, oxygen, and/or air and their proportions.

These parameters could influence the design of the gasifier, as follows:

- The desired heating value of the product gas dictates the choice of gasification medium. Table 6.4 gives typical ranges of heating value for different mediums.

TABLE 6.4 LHV of Product Gas Ranges and Choice of Gasifying Medium

Gasification Medium	Range of Heating Value of Product Gas (MJ/Nm^3)
Air gasification	4–7
Steam gasification	10–18
Oxygen gasification	12–28

- Hydrogen can be maximized with steam, but if it is not a priority, oxygen or air is a better option, as it reduces the energy used in generating steam and the energy lost through unutilized steam.
- If nitrogen in the product gas is not acceptable, air cannot be chosen.
- Capital cost is lower for air, followed by steam. A much larger investment is needed for an oxygen plant, which also consumes a large amount of auxiliary power.
- Equivalence ratio

For the product gas, the specification includes:

- Desired gas composition
- Desired heating value
- Desired production rate (Nm^3/s or MWth produced)
- Yield of the product gas per unit fuel consumed
- Required power output of the gasifier, Q

The design outputs of process design include geometry and operating and performance parameters.

Basic size includes reactor configuration, cross-section area, and height (hardware design). Important operating parameters are: (1) reactor temperature; (2) preheat temperature of the steam, air, or oxygen; and (3) amount (i.e., steam/biomass ratio) and relative proportion of the gasifying medium (i.e., steam/oxygen ratio). Performance parameters of a gasifier include carbon conversion and cold-gas efficiency.

A typical process design starts with a mass balance followed by an energy balance. The following subsections describe the calculation procedures for these.

6.6.2 Mass Balance

Basic mass and energy balance is common to all types of gasifiers. It involves calculations for product gas flow and fuel feed rate.

Product Gas Flow Rate

The gasifier's required power output, Q (MWth), is an important input parameter specified by the client. Based on this, the designer makes a preliminary estimation of the amount of fuel to be fed into the gasifier and the amount of gasifying medium. The volume flow rate of the product gas, V_g (Nm^3/s), from its desired lower heating value, LHV_g (MJ/Nm^3), is found by

$$V_g = \frac{Q}{LHV_g} \, Nm^3/s \tag{6.5}$$

The net heating value or lower heating value (LHV) can be calculated from its composition. The composition may be predicted by the equilibrium calculations, described later, or by sophisticated kinetic modeling of the gasifier, as discussed in Chapter 5. In the absence of these, a reasonable guess can be made. either from published data on similar fuels in similar gasification conditions or from the designer's experience.

For example, for air-blown fluidized-bed biomass gasifiers, the LHV is in the range 3.5 to 6 MJ/Nm^3 (Enden and Lora, 2004). For oxygen gasification, it is in the range 10 to 15 MJ/Nm^3 (Ciferno and Marano, 2002). So, for an air-blown gasifier, we start with a value of 5 MJ/Nm^3 as a reasonable guess (Quaak et al., 1999).

Fuel Feed Rate

To find the biomass feed rate, M_f , the required power output is divided by the LHV of the biomass (LHV_{bm}) and by the gasifier efficiency, η_{gef}.

$$M_f = \frac{Q}{LHV_{bm}\eta_{gef}} \tag{6.6}$$

The LHV may be related to the higher heating value (HHV) and its hydrogen and moisture contents (Quaak et al., 1999) as

$$LHV_{bm} = HHV_{daf} - 20{,}300 \times H_{daf} - 2{,}260 \times M_{daf} \tag{6.7}$$

Here, H_{daf} is the hydrogen mass fraction in the fuel, M_{daf} is the moisture mass fraction, and HHV_{daf} is the HHV in kJ/kg on a moisture-ash-free basis. By using the definition of these one can relate the HHV on moisture-ash-free basis to that on only dry-basis value as

$$HHV_{daf} = HHV_d\left(\frac{1-M}{1-ASH-M}\right) \tag{6.8}$$

where the subscripts d and daf refer to dry and moisture-ash-free basis, respectively; M is the moisture fraction; and ASH is the ash fraction in fuel on a raw-fuel basis.

On a dry basis, the HHV, HHV_d, is typically in the range 18 to 21 MJ/kg (Van Loo and Koppejan, 2003, p. 48). It may be calculated from the ultimate analysis for the biomass using the following equation (Van Loo and Koppejan, 2003, p. 29):

$$\begin{aligned} HHV_d = {} & 0.3491\,C + 1.1783\,H + 0.1005\,S - 0.0151\,N \\ & - 0.1034\,O - 0.0211\,ASH \end{aligned} \tag{6.9}$$

where C, H, S, N, O, and ASH are the mass fraction of carbon, hydrogen, sulfur, nitrogen, oxygen, and ash in the fuel on a dry basis.

Flow Rate of Gasifiying Medium

The amount of gasification medium has a major influence on yield and composition of the product gas. This section discusses methods for choosing that amount.

Air

The theoretical air requirement for complete combustion of a unit mass of a fuel, m_{th}, is an important parameter. It is known as the *stoichiometric air* requirement. Its calculation is shown in Eq. 2.34. For an air-blown gasifier operating, the amount of air required, M_a, for gasification of unit mass of biomass is found by multiplying it by another parameter ER:

$$M_a = m_{th} ER \tag{6.10}$$

Here, *ER* is the equivalence ratio.

For a fuel feed rate of M_f, the air requirement of the gasifier, M_{fa}, is

$$M_{fa} = m_{th} ER \cdot M_f \tag{6.11}$$

For a biomass gasifier, 0.25 may be taken as a first-guess value for the equivalence ratio, *ER*. A more detailed discussion of this is presented next.

Equivalence Ratio The equivalence ratio is an important gasifier design parameter. It is the ratio of the actual air–fuel ratio to the stoichiometric air–fuel ratio. This term is generally used for air-deficient situations, such as those found in a gasifier.

$$ER\,(<1.0)_{gasification} = \frac{\text{Actual air}}{\text{Stoichiometric air}} = EA\,(>1.0)_{combustion} \tag{6.12}$$

where *EA* is the excess air coefficient.

In a combustor, the amount of air supplied is determined by the stoichiometric (or theoretical) amount of air and its *excess air coefficient.* In a gasifier, the air supply is only a fraction of the stoichiometric amount. The stoichiometric amount of air may be calculated based on the ultimate analysis of the fuel.

The equivalence ratio, *ER*, dictates the performance of the gasifier. For example, pyrolysis takes place in the absence of air and hence the *ER* is zero; for gasification of biomass, it lies between 0.2 and 0.3.

Downdraft gasifiers give the best yield for *ER,* 0.25 (Reed and Das, 1988, p. 25). With a lower *ER* value, the char is not fully converted into gases. Some units deliberately operate with a low *ER* to maximize their charcoal production. A lower *ER* gives rise to higher tar production, however, so updraft gasifiers, which typically operate with an *ER* of less than 0.25, have higher tar content. With an *ER* above 0.25, some product gases are also burnt, increasing the temperature.

The quality of gas obtained from a gasifier strongly depends on the *ER* value, which must be significantly below 1.0 to ensure that the fuel is gasified rather than combusted. However, an excessively low *ER* value (<0.2) results in several problems, including incomplete gasification, excessive char formation, and a low heating value of the product gas. On the other hand, too high an *ER* (>0.4) results in excessive formation of products of complete combustion,

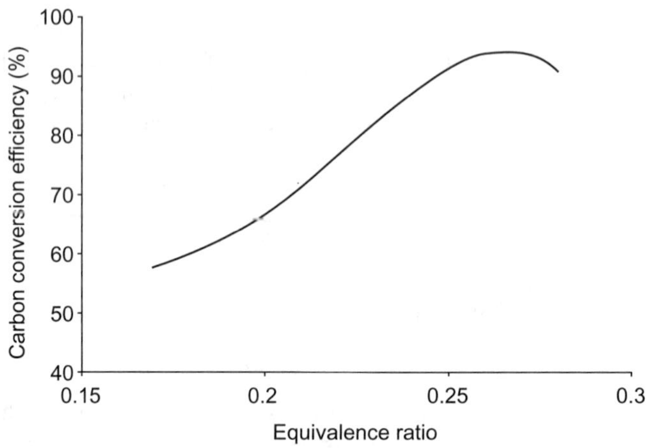

FIGURE 6.20 Effect of equivalence ratio on carbon conversion in a fluidized-bed gasifier.

such as CO_2 and H_2O, at the expense of desirable products, such as CO and H_2. This causes a decrease in the heating value of the gas. In practical gasification systems, the *ER* value is normally maintained within the range of 0.20 to 0.30. Figure 6.20 shows the variation in carbon conversion efficiency of a circulating fluidized-bed gasifier for wood dust against the equivalence ratio. The efficiency increases with ER and then it starts declining. The optimum value here is 0.26, but it may change depending on many factors.

The bed temperature of a fluidized-bed gasifier increases with the *ER* because the higher the amount of air, the greater the extent of the combustion reaction and the higher the amount of heat released (Figure 6.21). Example 6.1 illustrates the calculation procedure for *ER*.

Oxygen

Oxygen is used primarily to provide the thermal energy needed for the endothermic gasification reactions. The bulk of this heat is generated through the following partial and/or complete oxidation reactions of carbon:

$$C + 0.5\ O_2 \rightarrow CO - 111\ \text{kJ/mol} \tag{6.13}$$

$$C + O_2 \rightarrow CO_2 - 394\ \text{kJ/mol} \tag{6.14}$$

It can be seen that for the oxidation of 1 mol of carbon to CO_2, the oxygen requirement is $(2 \times 16)/12 = 2.66$ mol, while that for carbon to CO is $(16/12) = 1.33$ mol. Thus, the reaction in Eq. (6.13) is more likely to take place in oxygen-deficient regions.

Besides supplying the energy for the endothermic gasification reactions, the gasifier must provide energy to raise the feed and gasification medium to the reaction temperature, as well as to compensate for the heat lost to the reactor walls. For a self-sustained gasifier, part of the chemical energy in the biomass

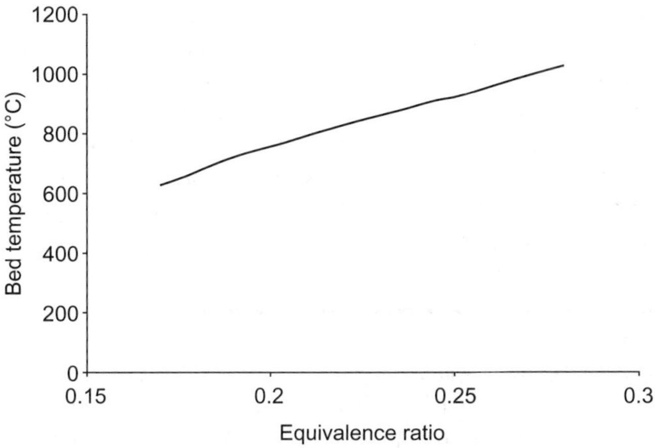

FIGURE 6.21 Gasifier temperature in a CFB riser increases with equivalence ratio.

provides the heat required. The total heat necessary comes from the oxidation reactions. The energy balance of the gasifier is thus the main consideration in determining the oxygen-to-carbon (O/C) ratio.

Equilibrium calculations can show that as the ratio of oxygen to carbon in the feed increases, CH_4, CO, and hydrogen in the product decreases but CO_2 and H_2O in the product increases. Beyond a ratio of 1.0, hardly any CH_4 is produced.

When air is the gasification medium, as is the case for 70% of all gasifiers (Ciferno and Marano, 2002), the nitrogen in it dilutes the product gas. The heating value of the gas is therefore relatively low (4–6 MJ/m^3). When pure oxygen from an air-separation unit is used, the heating value is higher, in the range 10 to 15 MJ/m^3, but a large amount of energy (~2.18 MJ/kg O_2) is spent in separating the oxygen from the air (Grezin and Zakharov, 1988).

The oxygen requirement of a gasifier can be met by either air supply or an air-separation unit that extracts oxygen from air.

Steam

Superheated steam as a gasification medium is used either alone, with air, or with oxygen. It contributes to the generation of hydrogen.

$$C + H_2O \rightarrow CO + H_2 \tag{6.15}$$

The quantity of steam, M_{fh}, is known from the steam-to-carbon (*S/C*) molar ratio.

$$\text{Steam flow rate, } M_{fh} = 18\frac{M_f C}{12}(S/C) \text{ kg steam/kg fuel} \tag{6.16}$$

where M_f is the fuel feed rate, and C is the carbon fraction in the fuel.

The *S/C* mole ratio has an important influence on product composition, as the *ER* has. Both hydrogen and CO increase with an increasing *S/C* ratio for a given temperature and oxygen-to-carbon molar ratio. The production of these two gases increases with decreasing pressure, decreasing oxygen, and decreasing *S/C* ratio. However, there is only a marginal gain in increasing the *S/C* molar ratio above 2 to 3, as the excess steam simply leaves the gasifier unreacted (Probstein and Hicks, 2006, p. 119). So a value in the range of 2.0 to 2.5 can give a reasonable starting value.

Example 6.1

A moving-bed gasifier 4 m in diameter operates at 25 bars of pressure and consumes 750 kg/min (dry-ash–free basis) of bituminous coal, 1930 kg/min of steam, and 280 Nm^3/min of oxygen to produce a product gas that contains 1000 Nm^3 of syngas (a mixture of H_2 and CO). The mean gasifier temperature is 1000 °C. The volumetric composition of the product gas is

CO_2–32%
H_2S–0.4%
CO–15.2%
H_2–42.3%
CH_4–8.6%
C_2H_4–0.8%
N_2–0.7%

The ultimate analysis of the coal on a moisture-ash-free basis is

C–77.3%
H–5.9%
S–4.3%
N–1.4%
O–11.1%

Find

- The *S/C* molar ratio
- The *O/C* molar ratio
- The *ER*
- The hearth load in energy produced per unit of grate area and space velocity

The heating values of the product gas constituents may be taken from Table C.2 in Appendix C.

Solution

From the feed rate of coal, steam, and oxygen, we can find the molar feed rate by dividing the mass rate by the molecular weight as here:

Carbon moles: 750 × 0.773/12 = 48.31 kmol/min
Steam moles: 1980/18 = 107.22 kmol/min
Oxygen moles: 280/22.4 = 12.5 kmol/min

From these we can calculate the following:

S/C molar ratio = 107.22/48.31 = **2.22**
O/C molar ratio = 12.5/48.31 = **0.26**

To find the stoichiometric oxygen requirement, the oxygen required to oxidize carbon to CO_2, hydrogen to H_2O, and sulfur to SO_2 has be to calculated.

- Twelve kg of carbon (1 mol) react with 32 kg of oxygen (1 mol) to produce 1 mol of CO_2:

$$C + O_2 = CO_2$$

Therefore, the oxygen required for 1 kg of carbon is 32/12.

- Thirty-two kg of sulfur (1 mol) react with 32 kg of oxygen (1 mol) to produce 1 mol of SO_2:

$$S + O_2 = SO_2$$

Therefore, the oxygen required for 1 kg of sulfur is 32/32 = 1.

- Similarly, 4 kg of hydrogen react with 32 kg of oxygen to produce H_2O:

$$2H_2 + O_2 = 2H_2O$$

Therefore, the oxygen required for 1 kg of hydrogen is 32/4 = 8.

$$\text{Stoichiometric oxygen requirement} = \frac{32C}{12} + 8H + S - O = \frac{32 \times 0.773}{12}$$
$$+ 8 \times 0.059 + 0.043 - 0.111 = 2.465 \text{ kg of } O_2/\text{kg of fuel}$$

The total O_2 required is

$$750 \times 2.465 = 1848.75 \text{ kg of } O_2/\text{min}$$

The O_2 supplied is

$$\text{moles of } O_2 \times 32 = 12.5 \times 32 = 400 \text{ kg of } O_2/\text{min}$$

From this we can calculate

$$ER = 400/1848.75 = \mathbf{0.22}$$

The syngas constituents in the total product gas are CO (15.2%) and H_2 (42.3%). So, to produce 1000 Nm^3/min of syngas, the amount of product gas, Q_{pr}, is

$$Q_{pr} = 1000/(0.152 + 0.423) = \mathbf{1739\ Nm^3/min}$$

The cross-sectional area of the gasifier reactor, A, is

$$A = \pi\, 4^2/4 = 12.56 \text{ m}^2$$

Assuming the operating temperature to be 1000 °C and the pressure to be 25 bars, the volumetric flow rate of product gas is

$$Q'_{pr} = Q_{pr}\left(\frac{1}{25}\right)\left(\frac{1273}{273}\right) = 324 \text{ m}^3/\text{min}$$

The space velocity of the gas flow Vg is Q'_{pr}/A = 324/ (12.56 × 60) = **0.43 m/s.**

The energy produced per Nm^3 of product gas is found by multiplying the volume fraction by the heating value of each constituent, which is taken from Table C.2 in Appendix C. Adding together the contribution of all product gas constituents gives the total heating value, *HHV*, as

$$HHV = 0.004 \times 25.1 + 0.152 \times (282.99/22.4) + 0.423 \times (285.84/22.4) + 0.086 \times (890.36/22.4) + 0.008 \times 63.4 = 11.33 \text{ MJ/Nm}^3$$

Thus, the total energy produced, E_{total}, is $Q_{pr} \times HHV$

$$= 1739 \times 11.33/60$$
$$= \mathbf{328.3\ MWth}$$

The hearth load is

$$E_{total}/A = 328.3/12.56 = \mathbf{26.14\ MW/m^2}$$

6.6.3 Energy Balance

Unlike combustion reactions, most gasification reactions are endothermic. Thus, heat must be supplied to the gasifier for these reactions to take place at the designed temperature. In laboratory units, this is not an issue because the heat is generally supplied externally. In commercial units, it is a major issue, and it must be calculated and provided for. The amount of external heat supplied to the gasifier depends on the heat requirement of the endothermic reactions as well as on the gasification temperature. The latter is a design choice, and it is discussed next.

Gasification Temperature

Because lignin, a refractory component of biomass, does not gasify well at lower temperatures, thermal gasification of ligno-cellulosic biomass prefers a minimum gasification temperature in the range 800 to 900 °C. For biomass, an entrained-flow gasifier typically maintains a gasification temperature well exceeding 900 °C. For coal, the minimum is 900 °C for most gasifier types (Higman and van der Burgt, 2008, p. 163).

A higher peak gasification temperature is chosen for an entrained-flow gasifier. The higher the ash-melting temperature, the higher the design value of the gasifier temperature. This temperature is raised through the gasifier's exothermic oxidation reactions, so a high reaction temperature also means a high oxygen demand.

In entrained-flow gasifiers, the peak gasification temperature is typically in the range 1400 to 1700 °C, as it is necessary to melt the ash; however, the exit gas temperature is much lower. The peak temperature of a fluidized-bed gasifier is in the range of 700 to 900 °C to avoid softening of bed materials. It is about the same as the gas exit temperature in a fluidized-bed gasifier. In a crossdraft gasifier the gasification temperature is about 1250 °C, whereas the peak gasification temperature is about 1500 °C. The exit-gas temperature of a downdraft gasifier is about 700 °C, but its peak gasifier temperature at the throat is 1000 °C. The updraft gasifier has the lowest gas-exit temperature (200–400 °C), while its gasification temperature may be up to 900 °C (Knoef, 2005). Once the gasification temperature is known, the designer can turn to the heat balance on this basis.

Heat of Reaction

Heat of reaction is the heat gained or lost in a chemical reaction. To calculate it for gasification, we consider an overall gasification reaction where 1 mol of biomass ($C_aH_bO_c$) is gasified in α mols of steam and β mols of oxygen. The overall equation is

$$\begin{aligned} C_aH_bO_c + \alpha H_2O + \beta O_2 = {} & A' \cdot C + B' \cdot CO_2 + C' \cdot CO + D' \cdot CH_4 \\ & + E' \cdot H_2O + F' \cdot H_2 + Q \end{aligned} \tag{6.17}$$

The equilibrium analysis of Section 5.5.2 gives the mole fraction A', B', C', D', E', and F' in the flue gas for given values of α and β. The chosen S/B ratio defines α, while the ER defines β. The heat of reaction, Q, for the overall gasification reaction (Eq. 6.22) may be found from the heat of formation of the products and reactants:

$$\begin{aligned} \text{Heat of reaction} = {} & \text{Heat of formation of product} \\ & - \text{heat of formation of reactant} \\ = {} & \text{heat of formation of } [A \cdot C + B \cdot CO_2 + C \cdot CO \\ & + D \cdot CH_4 + E \cdot H_2O + F \cdot H_2] \\ & - \text{heat of formation of } [\alpha \cdot H_2O \\ & + \beta \cdot O_2 + \text{biomass}] \end{aligned} \tag{6.18}$$

The heat of formation at 25 °C, or 298 K, is available in Table C.6 (Appendix C). The heat of formation at any other temperature, T K, can be found from the relation:

$$\Delta H_T^0 = \Delta H_{298}^0 + \sum\left(\int_{298}^{T} A' C_{p,i} dT\right)_{product} - \sum\left(\int_{298}^{T} \alpha C_{p,i} dT\right)_{reactants} \tag{6.19}$$

where $C_{p,i}$ is the specific heat of a substance i at temperature T K, and A', … , β are the stoichiometric coefficients of the products and reactants, respectively. The specific heat of gases as a function of temperature is given in Table C.4 (Appendix C).

The net heat, $Q_{gasification}$, to be supplied to the reactor is thus

$$Q_{gasification} = \Delta H_T \quad \text{kJ/mol} \tag{6.20}$$

This expression takes into account both exothermic combustion and endothermic gasification reactions. If the value of $Q_{gasification}$ works out to be negative, the overall process is exothermic, and so no net heat for the reactions is required.

Example 6.2

Find the heat of reaction for the following reaction at 1000 K:

$$C + H_2O\,(gas) = \frac{1}{2}CH_4 + \frac{1}{2}CO_2$$

Solution

Taking values at the reference temperature, 298 K, we have

$$\text{Heat of formation at 298 K for } C = 0;\ H_2O\,(g) = -241.8\ \text{kJ/mol};$$
$$\text{for } CH_4 = -74.8\ \text{kJ/mol};\ CO_2 = -393.5\ \text{kJ/mol}$$

$$\text{Total } \Delta H^0_{298} = \text{product} - \text{reactant} = \left[\frac{1}{2}(-74.8 - 393.5) - (-241.8)\right] = 7.65\ \text{kJ/mol}$$

Now, to find the value at 1100 K, we use Eq. (6.21):

$$\Delta H^0_{1100} = \Delta H^0_{298} + \sum\left(\int_{298}^{1100}(C_{p,CH_4} + C_{p,CO_2})\,dT\right)_{product} - \sum\left(\int_{298}^{1100} C_{p,H_2O}\,dT\right)_{reactants} \quad (6.21)$$

The specific heats of gases are taken from Table C.4 (Appendix C) as

$C_{p,CH_4} = 22.35 + 0.0481\ T$ kJ/kmol
$C_{p,CO_2} = 43.28 + 0.0114\ T - 818363/\ T^2$ kJ/kmol
$C_{p,H_2O} = 34.4 + 0.000628\ T + 0.0000056\ T^2$ kJ/kmol

Substituting these values and integrating the above expression, we get

$$\Delta H^{\circ}_{1100} = 7.65 + 104.58 - 33.578 = 78.65\ \text{kJ/mol}$$

Thus, this reaction is written as

$$C + H_2O\,(gas) \rightarrow \frac{1}{2}CH_4 + \frac{1}{2}CO_2 + 78.65\ \text{kJ/mol}$$

The reaction is endothermic.

Figure 6.22 shows the energy flow in and out of a gasifier. Biomass enters with its chemical energy and sensible heat. The gasifying agents enter with sensible heat at the reference temperature. External heat is added for heating the feeds to the gasification temperature, for meeting any shortfall in the reaction heat requirement, and for wall losses from the reactor. The product gas, with its chemical energy, leaves at the gasifier temperature. Unburnt char leaves with a potential energy in it. The unutilized steam and other gases also leave at the gasification temperature.

The overall energy balance may be written as

Energy input: Enthalpy of (biomass + steam + oxygen) at reference temperature + heating value of biomass + external heat

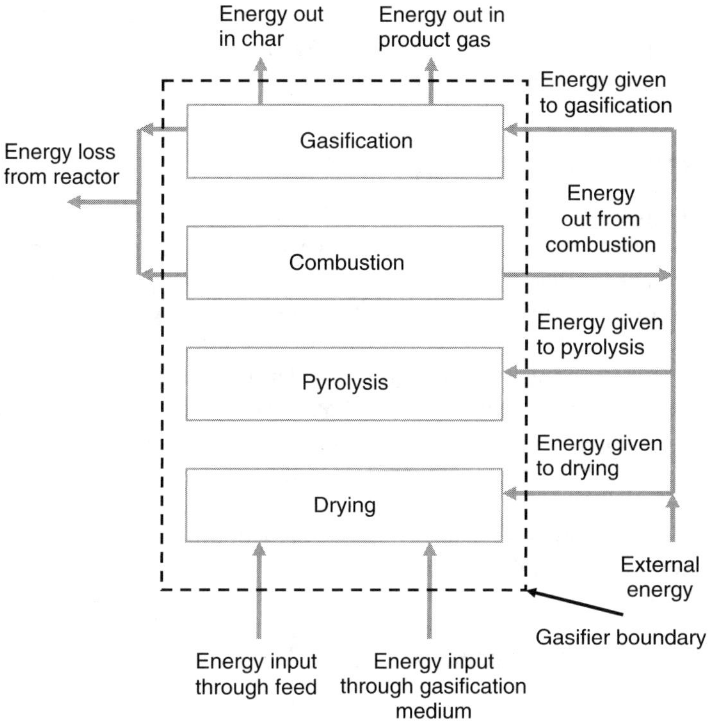

FIGURE 6.22 Energy flow in and out of a gasifier.

Energy output: Enthalpy of product gas at gasifier temperature + heating value of product gas + heat in unconverted char + heat loss from the reactor

If A is the amount of air needed and W is the total steam (from moisture or otherwise) needed to gasify F kg of fuel to produce 1 Nm^3 of product gas, we can write the energy balance of the gasifier taking 0 °C as the reference:

$$\begin{aligned} ACp_aT_0 + FCp_fT_0 + WH_0 + F \times HHV + Q_{ext} &= (C_{CO}V_{CO} + C_{CO_2}V_{CO_2} \\ &+ C_{CH_4}V_{CH_4} + C_{H_2}V_{H_2} + C_{O_2}V_{O_2} + C_{N_2}V_{N_2})T_g + (1-X_g)WH_g \\ &+ P_cq_c + Q_{gasification} + Q_{loss} + Q_{product} \end{aligned} \quad (6.22)$$

where H_0 and H_g are the enthalpies of steam at the reference temperature and the gasifier exit temperature; C_i and V_i are the volumetric specific heat and the volume of the gas species, i, at temperature T_g leaving the gasifier; $(1 - X_g)W$ is the net amount of steam remaining in the product gas of the gasification reaction; P_c is the amount of char produced; and q_c is the heating value of the char. Here, Q_{loss} is the total heat loss through the wall, radiation from the bed surface, ash drain, and entrained solids, corresponding to 1 Nm^3 of gas generation. This allows computation of external heat addition, Q_{ext} kJ/Nm^3 of product gas to the system. $Q_{product}$ is the amount of energy in the product gas, and $Q_{gasification}$ is the net heat of reaction.

6.7 PRODUCT GAS PREDICTION

A typical gasifier design starts with a desired composition of the product gas. Equilibrium and other calculations are carried out to check how closely that targeted composition is made through a choice of design parameters.

The product of combustion reactions is predominantly made up of carbon dioxide and steam, the percentages of which can be estimated with a fair degree of accuracy from simple stoichiometric calculations. For gasification reactions, this calculation is not straightforward; the fraction of the fuel gasified and the composition of the product gas needs to be estimated carefully. Unlike combustion reactions, gasification reactions do not always reach equilibrium, so only a rough estimate is possible through an equilibrium calculation. Still, this can be a reasonable start for the design until detailed kinetic modeling is carried out in the design optimization stage.

6.7.1 Equilibrium Approach

An equilibrium calculation ideally predicts the product of gasification if the reactants are allowed to react in a fully mixed condition for an infinite period of time. There are two types of equilibrium model. The first one is based on equilibrium constants (stoichiometric model). The specific chemical reactions used for the calculations have to be defined, so this model is not suitable for complex reactions where the chemical formulae of the compounds, the reaction path, or the reaction equations are not known. This requires the second model type, which involves minimization of the Gibbs free energy (nonstoichiometric model). This process is more complex but it is advantageous because the chemical reactions are not needed.

Stoichiometric Model

The stoichiometric model requires a selection of appropriate chemical reactions and information concerning the values of the equilibrium constants. Chapter 5 explains the calculation procedure, so it is not repeated here.

Nonstoichiometric Model

The nonstoichiometric model is based on the premise that at an equilibrium stage the total Gibbs free energy has to be minimized. It is described briefly in Chapter 5. Using Eq. (5.77) we can write the Gibbs free minimization equation for five gas species as follows:

$$\text{CH}_4: \frac{\left(\overline{\Delta G^o}_{CH_4}\right)}{RT} + \ln\left(\frac{n_{CH_4}}{n_{total}}\right) + \frac{1}{RT}\lambda_C + \frac{4}{RT}\lambda_H = 0 \quad (6.23)$$

$$\text{CO}_2: \frac{\left(\overline{\Delta G^o}_{CO_2}\right)}{RT} + \ln\left(\frac{n_{CO_2}}{n_{total}}\right) + \frac{1}{RT}\lambda_C + \frac{2}{RT}\lambda_O = 0 \quad (6.24)$$

$$\text{CO}: \frac{\left(\overline{\Delta G^o}_{CO}\right)}{RT} + \ln\left(\frac{n_{CO}}{n_{total}}\right) + \frac{1}{RT}\lambda_C + \frac{1}{RT}\lambda_O = 0 \quad (6.25)$$

$$\text{H}_2: \frac{\left(\overline{\Delta G^o}_{H_2}\right)}{RT} + \log\left(\frac{n_{H_2}}{n_{total}}\right) + \frac{2}{RT}\lambda_H = 0 \quad (6.26)$$

$$\text{H}_2\text{O}: \frac{\left(\overline{\Delta G^o}_{H_2O}\right)}{RT} + \ln\left(\frac{n_{H_2O}}{n_{total}}\right) + \frac{2}{RT}\lambda_H + \frac{1}{RT}\lambda_O = 0 \quad (6.27)$$

The five molar fractions of gases, such as (nCH_4/n_{total}), and the three Lagrangian constants, λ_H, λ_O, and λ_C, can be solved from the five equations and the three mass balance equations for C, H, and O derived from Eq. (5.74). Thus, for given feed and gasification medium and temperature, we can obtain the composition of the product gas.

Equilibrium models have limitations. The effect of tar is not considered here, even though tar can be a major problem in the gasification process and can affect plant operation. An equilibrium model may, for example, result in overestimation of the hydrogen produced. Kinetics, heat, and mass transfer determine the extent of chemicals participation in the chemical equilibrium in a given time or space domain (Florin and Harris, 2008). Furthermore, the equilibrium model assumes infinite speed of reaction and that all reactions will complete; these assumptions are not valid for most practical gasifiers. Nevertheless, equilibrium calculations give a good starting point, providing basic process parameters.

6.8 GASIFIER SIZING

The process design described in the previous section determines such operating parameters as gasification temperature, feed rates of fuel, and gasification medium. Now we can move to the next step, which involves the choice of gasifier configuration and type. Section 6.1.1 discusses the choice of gasifier. Table 6.5 compares the choices by their strength and weaknesses. By carefully examining these along with the type of plant to be designed, we can make a rational choice of gasifier type.

Once the gasifier type has been chosen, the designer can then proceed with the geometric design, where the basic sizes (the geometric dimensions of

TABLE 6.5 Comparison of Strength and Weaknesses of Different Gasifiers

Class	Types	Strength/Weakness	Power Production
Fixed bed	Downdraft	Low heating value, moderate dust, low tar	Small to medium scale
	Updraft	Higher heating value, moderate dust, high tar	
	Crossdraft	Low heating value, moderate dust	
Fluidized bed	Bubbling	Higher than fixed bed throughput, improved mass and heat transfer from fuel, higher heating value, higher efficiency	Medium scale

critical components) of the reactor are determined. At this stage, the designer decides on the geometric configuration of the reactor and its preliminary size. Both configuration and size depend on the reactor technology used.

6.8.1 Moving-Bed Gasifiers

A moving-bed gasifier may be designed on the basis of characteristic design parameters such as specific grate gasification rate, hearth load, and space velocity.

Specific grate gasification rate is the mass of fuel gasified per unit of cross-section area in unit time. The hearth load of a gasifier may be expressed in terms of the fuel gasified, the volume of gas that is produced, or the energy throughput.

$$\text{Hearth load}\left(\text{kg/s}\cdot\text{m}^2\right) = \frac{\text{Mass of fuel gasified}}{\text{Hearth cross-sectional area}}$$

$$\text{Hearth load}\left(\text{Nm}^3/\text{s}\cdot\text{m}^2\right) = \frac{\text{Volumetric gas production rate}}{\text{Hearth cross-sectional area}}$$

or

$$\text{Hearth load}\left(\text{MW/m}^2\right) = \frac{\text{Energy throughput in product gas}}{\text{Hearth cross-sectional area}} \tag{6.28}$$

The hearth load in volume flow rate of gas per unit of cross-section area is also known as *superficial gas velocity* or *space velocity*, as it has the unit of velocity (at reference temperature and pressure).

The following section discusses type-specific design considerations.

Updraft Gasifier

Updraft gasifiers are one of the simplest and most common types of gasifier for biomass. The maximum temperature increases when the feed of air or oxygen increases. Thus, the amount of oxygen feed for the combustion reaction is carefully controlled such that the temperature of the combustion zone does not reach the slagging temperature of the ash, causing operational problems. The gasification temperature may be controlled by mixing steam and/or flue gas with the gasification medium.

The hearth load of an updraft gasifier is generally limited to 2.8 MW/m^2 or 150 kg/m^2/h for biomass (Overend, 2004). For coal it might be higher. In an oxygen-based coal gasifier, for example, the hearth load of a moving bed can be greater than 10 MW/m^2. A higher hearth load increases the space velocity of gas through the hearth, fluidizing finer particles in the bed. Probstein and Hicks (2006) quote space velocities for coal on the order of 0.5 m/h for steam–air gasification and 5.0 m/h for steam–oxygen gasification. Excessive heat generation in such a tightly designed gasifier may cause slagging. Based on the characteristics of some commercial updraft coal gasifiers, Rao et al. (2004) suggest a specific grate gasification rate as 100 to 200 kg fuel/m^2h for RDF pellets, with the gas-to-fuel ratio in the range 2.5 to 3.0. Carlos (2005) obtained a rate of 745 to 916 kg/m^2h with air–steam and air preheat at temperatures of 350 and 830 °C, respectively.

For an updraft gasifier, the height of the moving bed is generally greater than its diameter. Usually, the height-to-diameter ratio is more than 3 : 1 (Chakraverty et al., 2003). If the diameter of a moving bed is too large, there may be a material flow problem, so it should be limited to 3 to 4 m in diameter (Overend, 2004).

Downdraft Gasifier

As we saw in Figures 6.4 and 6.6, the cross-sectional area of a downdraft gasifier may be nonuniform; it is narrowest at the throat. The hearth load is, therefore, based on the cross-sectional area of the throat for a throated gasifier, and for a throatless or stratified downdraft gasifier, it is based on the gasifier cross-sectional area. The actual velocity of gas is, however, significantly higher than the designed space velocity because much of the flow passage is occupied by fuel particles. The velocity is higher in the throat also because of the higher temperature there. Table 6.6 gives some characteristic values of these parameters.

In a downdraft gasifer, the gasification air is injected by a number of nozzles from the periphery (refer to Figure 6.6). The total nozzle area is typically 7 to 4% of the throat area. The number of nozzles should be an odd number so that the jet from one nozzle does not hit a jet from the opposite side, leaving a dead space in between. To ensure adequate penetration of nozzle air into the hearth, the diameter of a downdraft gasifier is generally limited to 1.5 m. This naturally restricts the size and capacity of a downdraft gasifier.

TABLE 6.6 Hearth Load for Downdraft Gasifiers Maximum Values Based on Throat Area

Plant	Gasifier Type	Medium	D_{throat} (m)	$D_{air\ entry}$ (m)	Superficial Velocity at Throat (m/s)	Hearth Load* (MW/m²)
Gengas	Imbert	Air	0.15	0.3	2.5	15
Biomass Corp.	Imbert	Air	0.3	0.61	0.95	5.7
SERI	Throatless	Air	0.15		0.28	1.67
Buck Rogers	Throatless	Air	0.61		0.23	1.35
Buck Rogers	Throatless	Air	0.61		0.13	0.788
Syngas	Throatless	Air	0.76		1.71	10.28
Syngas	Throatless	Oxygen	0.76		1.07	12.84
SERI	Throatless	Oxygen	0.15		0.24	1.42

*Based on throat area.
Source: Data compiled from Reed and Das, 1988, p. 36.

Table 6.7 lists typical sizes for the Imbert-type downdraft gasifier and shows the relation between throat size and air nozzle diameter.

6.8.2 Fluidized-Bed Gasifiers

No established design method for sizing a fluidized-bed gasifier is available in the literature because, though nearly a century old, this type is still evolving. This section presents a tentative method for determining size based on available information.

Cross-Sectional Area

The inside cross-sectional area of the fluidized-bed gasifier, A_b, is found by dividing the volumetric flow rate of the product gas flow, V_g, by the chosen superficial gas or fluidization velocity through it, U_g, at the operating temperature and pressure.

$$A_b = \frac{V_g}{U_g} \tag{6.29}$$

The volume of gas at the operating temperature and pressure, V_g, is estimated from the mass of air (or other medium), M_{fa}, required for gasification (Eq. 6.29) as well as for fluidization. Thus, V_g is necessarily the gas passing through the grate and the bed.

TABLE 6.7 Sizes of Imbert-Type Gasifiers

d_r/d_h (–)	d_h (mm)	d_r (mm)	$d_{r'}$ (mm)	h (mm)	H (mm)	R (mm)	A (no.)	d_m (mm)	Range of Gas Output (Nm^3/h)	Maximum Wood Consumption (kg/h)	Air Blast Velocity (m/s)
268/60	60	268	150	80	256	100	5	7.5	4–30	14	22.4
300/100	100	300	208	100	275	115	5	10.5	10–77	36	29.4
400/130	130	400	258	110	370	155	7	10.5	17–120	57	32.6
400/150	135	400	258	120	370	155	7	12	21–150	71	32.6
400/175	175	400	308	130	370	155	7	13.5	26–190	90	31.4
400/200	200	400	318	145	370	153	7	16	33–230	110	31.2

Variables not defined in the figure are defined as follows:
d_m = inner diameter of the tuyere
A = number of tuyeres
Source: Data compiled from Reed and Das, 1988.

In some designs, part of the gasifying medium is injected above the distributor grid. In that case, V_g is only the amount that passes through the grid. We can use the mass of gasification medium, M_{fa}, required for gasification for the computation of V_g:

$$V_g = \frac{M_{fa}}{\rho_g} \tag{6.30}$$

where ρ_g is the density of the medium at the gasifier's operating temperature and pressure.

Equation (6.29) requires choosing an appropriate value for the superficial gas (fluidizing) velocity, U_g, through the gasification zone. This is critical as it must be within acceptable limits for the selected particle size to ensure satisfactory fluidization and to avoid excessive entrainment.

Fluidization Velocity

The range of fluidizing velocity, U_g, in a bubbling bed depends on the mean particle size of the bed materials. The choice is made in the same way as for a fluidized-bed combustor. The range should be within the minimum fluidization and terminal velocities of the mean bed particles. The particle size may be within group B or group D of Geladart's powder classification (see Basu, 2006, Appendix I). The typical fluidization velocity for silica sand of about 1 mm mean diameter may, for example, vary between 1.0 and 2.0 m/s.

If the gasifier reactor is a circulating fluidized-bed type, the fluidization velocity in its riser (Figure 6.12) must be within the limits of fast fluidization, which favors groups A or group B particles. Typical fluidization velocity for particle size in the range 150 to 350 microns is 3.5 to 5.0 m/s in a CFB. This type of bed has another important operating condition to be satisfied for operation in the CFB regime. Solids, captured in the gas–solid separator at the gasifier exit, must be recycled back to the gasifier at a rate sufficiently high to create a "fast-fluidized" bed condition in the riser. Additional details about this are available in Basu (2006) or Kunii and Levenspiel (1991).

Gasifier Height

Since gasification involves only partial oxidation of the fuel, the heat released inside a gasifier is only a fraction of the fuel's heating value, and part of it is absorbed by the gasifier's endothermic reactions. Thus, it is undesirable to extract any further heat from the main gasifier column. For this reason, the height of a fluidized-bed gasifier is not determined by heat-transfer considerations as for fluidized-bed boilers. Instead, gas and solid residence times are major considerations.

The total height of the gasifier is made up of the height of the fluidized bed and that of the freeboard above it:

$$\text{Total gasifier height} = \text{bubbling bed height (depth)} + \text{freeboard height} \tag{6.31}$$

Fluidized-Bed Height

The bed height (or depth) of a bubbling fluidized-bed gasifier is an important design parameter. Gas–solid gasification reactions are slower than combustion reactions, so a bubbling-bed gasifier is necessarily deeper than a bubbling-bed combustor, which is typically 1.0- to 1.5-m deep for units larger than 1 m in diameter. Besides pilot plant data or design experience, there is presently no simple means of deciding the bed depth. A deeper bed allows longer gas residence time, but the depth should not be so great compared to its diameter as to cause slugging. The selection of bed height depends on economics. A higher bed height means a higher pressure drop and also a taller reactor. It also should provide a longer residence time for better carbon conversion.

The gasification agent, CO_2 or H_2O, entering the grid takes a finite time to react with char particles to produce the gas. The bulk of the gasifying agent travels up through the bubbles but very little reaction takes place in the bubble phase. Rather, the reaction takes place mostly in the emulsion phase. The extent to which oxygen or steam is converted into fuel gases thus depends on the gas exchange rate between the bubble and emulsion phases as well as on the char-gas reaction rate in the emulsion phase. This is best computed through a kinetic model of the gasifier as illustrated in Section 5.6.2. An alternative is to use an approach based on residence time, as described next.

Residence Time Design Approach A bubbling fluidized bed must be sufficiently deep to provide reactants the time to complete the gasification reactions. This is why residence time is an important consideration for determination of bed height. An approach based on residence time, developed primarily for coal gasification, can be used for biomass char gasification, which gives at least a first estimate of the bed height for a biomass-fueled bubbling fluidized-bed gasifier.

The residence time approach is based on the assumption that the conversion of char into gases is the slowest of all gasifier processes, so the reactor should provide adequate residence time for the char to complete its conversion to the desired level. Here is a simplified method.

Given the following assumption:

- The reactivity factor is f_o, = 1 (which lies between $0 < f_o \le 1$).
- The solid is in a perfectly mixed condition (i.e., continuous stirred-tank reactor).

Then, the volume of the fluidized bed, V, is calculated using the equation

$$V = \frac{W_{out}\theta}{\rho_b} \tag{6.32}$$

where W_{out} is the char moving out; kg/s = $(1-X)\ W_{in}$; X is the fraction of the char in the converted feed; ρ_b is the bed density, which can be estimated theoretically from fluidization hydrodynamics and regime (kg/m^3); and θ is the residence time of the char in the bed, or reaction time (s).

The residence time approach assumes that the water–gas reaction, ($C + H_2O \rightarrow CO + H_2$), as written in Eq. (6.33) is the main gasification reaction, where the char is consumed primarily by the steam gasification reaction for *n*th-order kinetics:

$$\frac{1}{m}\frac{dC}{dt} = k[H_2O]^n \tag{6.33}$$

where m is the initial mass of the biomass and C is the total amount of carbon gasified in time, t. Taking a logarithm of this,

$$\ln\left(\frac{1}{m}\frac{dC}{dt}\right) = \ln(k) + n\ln[H_2O] \tag{6.34}$$

experiments can be carried out taking a known weight of the biomass and measuring the change in carbon conversion at different time intervals for a given temperature, steam flow, and pressure. Using these data, graphs are plotted between $\ln\left(\frac{1}{m}\frac{dC}{dt}\right)$ and $\ln[H_2O]$. The y-intercept in this graph will give the value of k, and the slope will give the value of n. An example of such a plot is shown in Figure 6.23.

The experiment is carried out for different operating temperatures such as 700 °C, 800 °C, and 900 °C, so, for each temperature, one k value is obtained.

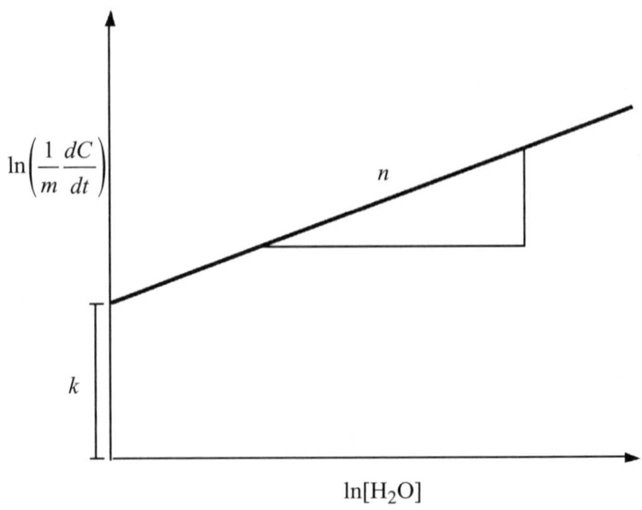

FIGURE 6.23 Plot of Eq. (6.34) for determination of residence time.

Now k can be expressed as

$$k = k_0 \exp\left(-\frac{E_a}{RT}\right)$$

$$\ln k = \ln k_0 - \frac{E_a}{RT} \tag{6.35}$$

This shows that if we plot a graph between ln k and $1/T$, the y-intercept will give the value of k_0 and the slope will give the value of $(-E_a/R)$.

The reaction rate for the steam gasification of biomass is given by

$$\frac{dC}{dt} = k_0 m \exp\left(-\frac{E_a}{RT}\right)[H_2O]^n \tag{6.36}$$

This gives the generalized reaction rate that shows the dependence of the gasification rate on temperature, mass of carbon or char, and concentration of steam/air/oxygen.

From a knowledge of the reaction rate, the residence time, θ, can be calculated as

$$\theta = C_0 \frac{X}{r} \tag{6.37}$$

where C_0 is the initial carbon in the biomass particle, kg; X is the required carbon conversion (–); and r is the steam gasification reaction rate (kg/s). We can avoid such experiments if there is a suitable expression for the rate of steam gasification of the designed biomass char (Sun et al., 2007).

From knowledge of the required solid residence time, θ, then, the bed volume, V_{bed}, is

$$V_{bed} = \frac{F[C]\theta}{(1-\varepsilon)\rho_s x_{char}} \tag{6.38}$$

where $F[C]$ is the char feed rate into the gasifier and ρ_s is the density of the bed solids. In a typical bubbling bed, the bed voidage is ~0.7. The bed generally contains 5 to 8% (by weight) of reacting char (x_{char}); the remaining solids are inert bed materials.

The bed height, H_{bed}, is known by dividing bed volumer by the bed area, A_b, which is known from chosen superficial velocity

$$H_{bed} = \frac{V_{bed}}{A_b} \tag{6.39}$$

Design charts for residence time, θ, of test coals for different feed conversions and S/C or O/C ratios are given in the *Coal Conversion Systems Technical Data Book* (U.S. DOE, 1978). The residence time may be adjusted for the reactivity of the char in question and for the reactivity of its partial gasification before it enters the gasifier.

Other Considerations

Although virgin biomass contains little or no sulfur, some waste biomass fuels do. For these, limestone is fed into the fluidized-bed gasifier for in-bed sulfur removal. The height of the gasifier (freeboard and bed) should be adequate to allow the residence time needed for the desired sulfur capture.

The tar produced should be thermally cracked inside the gasifier as far as possible. Therefore, the depth of the gasifier should be such that the gas residence time is adequate for the desired tar conversion/cracking.

The deeper the bed, the higher the pressure drop across it and the higher the pumping cost of air. Because bubble size increases with bed height, a deeper bed gives larger bubbles with reduced gas–solid mixing. Furthermore, if the bubble size becomes comparable to the smallest dimension of the bed cross-section, a highly undesirable slugging condition is reached. This imposes another limit on how deep the dense section of a fluidized bed can be.

Some biomass char, like that from wood, is fine and easily undergoes attrition in a fluidized bed. In such cases a deeper bed may not guarantee a longer residence time (Barea, 2009). Here, special attention must be paid to capturing the char and either combusting it in a separate chamber to provide heat required by the gasifier, or reinjecting it at an appropriate point in the bed where solids are descending.

A kinetic model (*n*th-order, shrinking particle, and shrinking core) may also be used to determine the residence time, the net solid holdup, and therefore the height of the dense bed.

Freeboard Height

Entrainment of unconverted fine char particles from the bubbling bed is a major source of carbon loss. The empty space above the bed, the *freeboard*, allows entrained particles to drop back into it. A bubbling, turbulent, or spouted fluidized bed must have such a freeboard section to help avoid excessive loss of bed materials through entrainment and to provide room for conversion of finer entrained char particles. The freeboard height must be sufficient to provide the required residence time for char conversion. It can be determined from experience or through kinetic modeling.

A larger cross-sectional area and a taller freeboard increase the residence time of gas/char and reduce entrainment. From an entrainment standpoint, the freeboard height need not exceed the transport disengaging height (TDH) of a bed because no further reduction in entrainment is achieved beyond this.

6.9 ENTRAINED-FLOW GASIFIER DESIGN

Because the gas residence time in an entrained-flow reactor is very short—on the order of a few seconds—to complete the reactions, the biomass particles must be ground to extremely fine sizes (less than 1 mm). The residence time requirement for the char is thus on the order of seconds. Section 6.9.1 describes some important considerations for entrained-flow gasifier design.

Although an entrained-flow gasifier is ideally a plug-flow reactor, in practice this is not necessarily so. The side-fed entrained-flow gasifier, for example, behaves more like a continuous stirred-tank reactor (CSTR). As we saw in Figure 6.15, at a certain distance from the entry point, fuel particles may have different residence times depending on the path they took to arrive at that section. Some may have traveled a longer path and so have a longer residence time. For this reason, a plug-flow assumption may not give a good estimate of the residence time of char.

6.9.1 Gasifier Chamber

Most commercial entrained-flow gasifiers operate under pressure and therefore are compact in size. Table 6.8 gives data on some of these operating in the United States and China.

A typical downflow entrained-flow gasifier is a cylindrical pressure vessel with an opening at the top for feed and another at the bottom for discharge of ash and product gas. The walls are generally lined with refractory and insulating materials, which serve three purposes: (1) they reduce heat loss through the wall, (2) they act as thermal storage to help ignition of fresh feed, and (3) they prevent the metal enclosure from corrosion.

TABLE 6.8 Characteristic Sizes of Some Entrained-Flow Gasifiers

Gasifier	Volume (m^3)	Reactor External Diameter (m)	Reactor Internal Diameter (m)	Reactor Height (m)
Tennessee Eastman	12.7	2.79	1.67	4.87
Cool water	17	3.17	2.13	3.73
Cool water	25.5	3.17	2.13	6
Cool water	12.7	2.79	1.67	4.62
Shandong fertilizer	12.7	2.79	1.67	4.87
Shanghai Chemical	12.7	2.79	1.67	4.87
Harbei fertilizer	12.7	2.79	1.67	4.87

Source: Data compiled from Zen, 2005.

The thickness of the refractory and insulation used is to be chosen with care. For example, biomass ash melts at a lower temperature and is more corrosive than most coal ash, so special care needs to be taken in designing the gasifier vessel for biomass feedstock.

The construction of a side-fed gasifier is more complex than that of a top-fed gasifier, as the reactor vessel is not entirely cylindrical and requires numerous openings. The bottom opening is for the ash drain, the top opening is for the product gas, and the side ports are for the feed. Additional openings may also be required depending on the design. Because of the complexity in the design of a pressure vessel operating at 30 to 70 bars and temperatures exceeding 1000 °C, any additional openings or added complexity in the reactor configuration must be weighed carefully against perceived benefits and manufacturing difficulties.

6.9.2 Auxiliary Items

The following subsections discuss the design of auxiliary systems in fluidized-bed gasifiers.

Position of Biomass Feeding Position

The feed points for the biomass should be such that entrainment of any particles in the product gas is avoided. This can happen when the feed points are located too close to the expanded bed surface of a bubbling fluidized bed. If they are in close proximity to the distributor plate, excessive combustion of the volatiles in the fluidizing air produced can occur. To avoid this, they should be some distance further above the grate.

Nascent tar is released close to the feed point, so tar cracking can be important for some designs. If tar is a major concern, the feed port should be close to the bottom of the gasifier so that the tar has adequate residence time to crack (Barea, 2009).

Distributor Plate

The distributor plate of a fluidized bed supports the bed materials. It is no different from that used for a fluidized combustor or boiler. The ratio of pressure drop across the bed and that across the distributor plate must be estimated to arrive at the plate design. More details are available in books on distributor plate design, including Basu (2006, Chapter 11). The typical open area in the air distributor grate is only a few percentage points.

Bed Materials

For the process design of a fluidized-bed gasifier, the choice of bed materials is crucial. These comprise mostly granular inorganic solids and some (<10%) fuel particles. For biomass, sand or other materials are used (as explained next);

coal gasification requires granular ash produced from the gasification process. Sometimes limestone is added with coal particles to remove sulfur. At different stages of calcination and sulfurization, the limestone can also form a part of the bed material.

Biomass has very little ash (less than 1% for wood), so silica sand is normally used as the inert bed material. This is a natural choice because silica is inexpensive and the most readily available granular solid. One major problem with silica sand is that it can react with the potassium and sodium components of the biomass to form eutectic mixtures having low melting points, thereby causing severe agglomeration. To avoid this, the following alternative materials can be used:

- Alumina (Al_2O_3)
- Magnesite ($MgCO_3$)
- Feldspar (a major component of Earth's crust)
- Dolomite ($CaCO_3 \cdot MgCO_3$)
- Ferric oxide (Fe_2O_3)
- Limestone ($CaCO_3$)

Magnesite (MgO) was successfully used in the first biomass-based IGCC plant in Värnamo, Sweden (Ståhl et al., 2001).

Tar is a mixture of higher-molecular-weight (higher than benzene) chemical compounds that condenses on downstream metal surfaces at lower temperatures. It can plug the passage and/or make the gas unsuitable for use. The bed materials, besides serving as a heat carrier, can catalyze the gasification reaction by increasing the gas yield and reducing the tar. Bed materials that act as a catalyst for tar reduction are an attractive option. Some are listed here (Pfeifer et al., 2005; Ross et al., 2005):

- Olivine
- Activated clay (commercial)
- Acidified bentonite
- Raw bentonite
- House brick clay

Common house brick clay can be effectively used in a CFB gasifier to reduce tar emission and enhance hydrogen production. The alkalis deposited on the bed materials from biomass may potentially behave as catalysts if their agglomerating effect can be managed (Ross et al., 2005).

Tar production can be reduced using olivine. The Fe content of olivine is catalytically active, and that helps with tar reforming (Hofbauer, 2002). Nickel-impregnated olivine gives even better tar reduction as nickel is active for steam tar reforming (Pfeifer et al., 2005).

Bingyan et al. (1994) reported using ash from the fuel itself (sawmill dust) as the bed material in a CFB gasifier. This riser is reportedly operated at a very low velocity of 1.4 m/s, which is 3.5 times the terminal velocity of the biomass particles. Chen et al. (2005) tried to operate a 1-MWe CFB gasifier with rice husk alone, but the system had difficulty with fluidization in the loop seal because of the low sphericity of the husk ash; however, the main riser reportedly operated in the fast bed regime without major difficulty.

6.10 DESIGN OPTIMIZATION

Design optimization generally starts after the preliminary design is complete and actual project execution is set to begin. It has two aspects: (1) process and (2) engineering.

Process optimization tells the designer if the preliminary design will give the best performance in terms of efficiency and gas yield, and how this is related to the operation and design parameters. Commercial simulation programs (mathematical models) or computational fluid dynamics codes are the most effective tools for this purpose. Engineering optimization involves optimizing the reactor configuration to enhance its operability, maintainability, and cost reduction.

6.10.1 Process Optimization

Process optimization enhances gasifier performance in terms of the following indicators:

- Cold- and hot-gas efficiency
- Unconverted carbon and tar concentration in the product gas
- Composition and heating value of the product gas

One can approach optimization either through experiments or through kinetic modeling.

Experiments are the best and most reliable means of optimizing process parameters, as they are based on the actual or prototype gasifier. However, they have several limitations, and are expensive. Furthermore, practical difficulties may not allow all operational parameters to be explored. An alternative is to conduct tests in a controlled laboratory-scale unit and to calibrate the resulting data to the full-scale unit. This allows the scale-up of data from the laboratory to the full-scale unit with a reasonable degree of confidence.

Optimization through Kinetic Modeling

With a kinetic model we can predict the performance of a gasifier already designed because it utilizes both configuration and dimensions of the reactor. Kinetic modeling can help optimize or fine-tune the operating parameters for best performance in a given situation. Section 5.6 described a kinetic model for gasifiers.

6.11 PERFORMANCE AND OPERATING ISSUES

Gasifier performance is measured in terms of both quality and quantity of gas produced. The amount of biomass converted into gas is expressed by gasification efficiency. The product quality is measured in terms of heating value as well as amount of desired product gas.

6.11.1 Gasification Efficiency

The efficiency of gasification is expressed as cold-gas efficiency, hot-gas efficiency, or net gasification efficiency. These are described in the following subsections.

Cold-Gas Efficiency

Cold-gas efficiency is the energy input over the potential energy output. If M_f kg of solid fuel is gasified to produce M_g kg of product gas with an LHV of Q_g, the efficiency is expressed as

$$\eta_{cg} = \frac{Q_g M_g}{LHV_f M_f} \tag{6.40}$$

where LHV_f is the lower heating value (LHV) of the solid fuel.

Example 6.3

Air–steam gasifier data include the mass composition of the feedstock:

C–66.5%
O–7%
H–5.5%
N–1%
Moisture–7.3%
Ash–12.7%
LHV–28.4 MJ/kg

and the volume composition of the product gas:

CO–27.5%
CO_2–3.5%
CH_4–2.5%,
H_2–15%
N_2–51.5%

The dry air supply rate is 2.76 kg/kg of feed; the steam supply rate is 0.117 kg/kg of feed; the moisture content is 0.01 kg of H_2O per kg of dry air; and the ambient temperature is 20 °C.

Find:

- The amount of gas produced per kg of feed
- The amount of moisture in the product gas

- The carbon conversion efficiency
- The cold-gas efficiency

Solution

Table C.3 (Appendix C) shows the mass fraction of N_2 and O_2 in air as 0.755 and 0.232, respectively. The nitrogen supply from air is

$$0.755 \times 2.76 = 2.08 \text{ kg } N_2/\text{kg feed}$$

The total nitrogen supplied by the feed air and the fuel feed, which carry 1% nitrogen, is

$$2.08 + 0.01 = 2.09 \text{ kg } N_2/\text{kg feed} = (2.09/28) = 0.0747 \text{ kmol } N_2/\text{kg feed}$$

noting that volume percent equals molar percent in a gas mixture.

Since the product gas contains 51.5% by volume of nitrogen, the amount of the product gas per kg of feed is

$$0.0747/0.515 = \mathbf{0.145\ kmol\ gas/kg\ feed}$$

Similarly, the oxygen from the air flow to the gasifier is

$$0.232 \times 2.76 = 0.640 \text{ kg/kg feed}$$

The steam supplied per kg of fuel is 0.117 kg, so the oxygen associated with the steam supply is

$$0.117 \times (8/9) = 0.104 \text{ kg/kg feed}$$

Oxygen also enters through the 7.3% moisture in the fuel and the 1% moisture in the air feed. The total oxygen from moisture is

$$0.073 \times (8/9) + 0.01 \times 2.76 \times (8/9) = 0.065 + 0.0245 = 0.0895 \text{ kg/kg feed}$$

The total oxygen flow to the gasifier, including the 7% that comes with the fuel, is

$$0.640 + 0.104 + 0.0895 + 0.07 = 0.9035 \text{ kg } O_2/\text{kg feed}$$

Hydrogen Balance

The total hydrogen inflow to the gasifier with fuel, steam, and moisture in the fuel and moisture in the air is

$$0.055 + 0.117/9 + 0.073/9 + 0.0276/9 = \mathbf{0.0792\ kg/kg\ feed}$$

The hydrogen leaving with H_2 and CH_4 in dry gas, noting that 1 mole of CH_4 contributes 2 mols of H_2, is

$$(0.15 + 2 \times 0.025) \times 0.145 = 0.029 \text{ kmol/kg feed} = 0.029 \times 2$$
$$= \mathbf{0.058\ kg\ hydrogen/kg\ feed}$$

To find the moisture in the product gas, we deduct the hydrogen in the dry gas from the total hydrogen inflow obtained earlier, using the hydrogen balance:

$$\text{Hydrogen inflow} - \text{hydrogen out through dry product gas}$$
$$= 0.0792 - 0.058 = 0.0212 \text{ kg/kg feed}$$

The steam or moisture associated with this hydrogen in the gas is

$$0.0212 \times (18/2) = \mathbf{0.1908\ kg/kg\ feed}$$

Carbon Balance

The carbon-bearing gases—CO, CO_2, and CH_4—in the dry gas each contain 1 mol of carbon. So the total carbon in 0.145 kmol/kg of fuel product gas is

$$(0.275 + 0.035 + 0.025) \times 0.145 = 0.0485\ \text{kg mol/kg feed} = 0.0485 \times 12$$
$$= 0.583\ \text{kg/kg feed}$$

The carbon input, as found from the composition of the feed, is 0.665 kg/kg feed. The carbon conversion efficiency is found by dividing the carbon in the product gas by that in the fuel:

$$= (0.583/0.665) \times 100 = \mathbf{87.6\%}$$

Energy Balance

The heats of combustion for different gas constituents are taken from Table C.2 (Appendix C). They are:

CO–12.63 MJ/nm^3
Hydrogen–12.74 MJ/nm^3
Methane–39.82 MJ/nm^3

We note that 1 kg of feed produces 0.145 kmol of gas, the volumetric composition of which is

CO–27.5%
CO_2–3.5%
CH_4–2.5%
H_2–15%
N_2–51.5%

By multiplying the heating value of the appropriate constituents of the product gas, we can find the total heating value of the product gas (the volume of 1 kmol of any gas is 22.4 nm^3):

$$(12.63 \times 0.275 + 12.74 \times 0.15 + 39.82 \times 0.025)\text{MJ/nm}^3 \times 0.145\ \text{kmol/kg feed}$$
$$\times 22.4\ \text{nm}^3/\text{kmol} = \mathbf{20.6\ MJ/kg\ feed}$$

The total energy input is equal to the heating value of the feed, which is 28.4 MJ/kg.

From Eq. (6.40), the cold-gas efficiency is

$$(20.6/28.4) \times 100 = \mathbf{72.5\%}$$

Hot-Gas Efficiency

Sometimes gas is burned in a furnace or boiler without being cooled, creating a greater utilization of the energy. Therefore, by taking the sensible heat of the hot gas into account, the hot-gas efficiency, η_{hg}, can be defined as

$$\eta_{hg} = \frac{Q_g M_g + M_g C_p (T_f - T_0)}{LHV_f M_f} \tag{6.41}$$

where T_f is the gas temperature at the gasifier exit or at the burner's entrance, and T_0 is the temperature of the fuel entering the gasifier. The hot-gas efficiency assumes the heating of the unconverted char to be a loss.

Example 6.4

The gas produced by the gasifier in Example 6.3 is supplied directly to a burner at the gasifier exit temperature, 900 °C, to be burnt for co-firing in a boiler. Find the hot-gas efficiency of the gasifier.

Solution

The product gas enters the burner at 900 °C (1173 K). To find the enthalpy of the product gas, we add the enthalpies of its different components. Specific heats of individual components are calculated using the relations from Table C.4 (Appendix C). For example, the specific heat of CO at 1173 K is

$$27.62 + 0.005 \times 1173 = 33.48 \text{ kJ/kmol.K}$$

From Example 6.3, the amount of product gas is 0.145 kmol/kg fuel. The enthalpy of CO in the product gas that contains 27.5% CO above the ambient temperature, 25 °C or 298 K, is

$$(0.145 \times 0.275)\text{kmol/kg feed} \times 33.48 \text{ kJ/kmol.K} \times (1173 - 298)\text{K} \times 10^{-3} \text{ MJ/kJ} = 1.168 \text{ MJ/kg feed}$$

Similarly enthalpy of other products,

CO_2: (0.145 × 0.035) × 56.06 × (1173–298) × 10^{-3} = 0.249 MJ/kg feed
H_2: (0.145 × 0.15) × 31.69 × (1173–298) × 10^{-3} = 0.0603 MJ/kg feed
N_2: (0.145 × 0.515) × 32.13 × (1173–298) × 10^{-3} = 0.21 MJ/kg feed
CH_4: (0.145 × 0.025) × 78.65 × (1173–298) × 10^{-3} = 0.249 MJ/kg feed

The amount of steam in the flue gas was calculated as 0.1908 kg/kg of feed. To find the enthalpy of this steam above 298 K, we take values of the steam enthalpy at 1 bar of pressure at 1173 K and 298 K. The values are 4398.05 and 104.93 kJ/kg, respectively, so the enthalpy in water is

$$H_2O: 0.1908\text{x}\,(4398.05 - 104.93) \times 10^{-3} = 0.819 \text{ MJ/kg feed}$$

Adding these we get the total enthalpy of the product gas at 900 C.

$$1.168 + 0.249 + 0.060 + 0.21 + 0.249 + 0.819 = 2.76 \text{ MJ/kg feed}$$

The total thermal energy is

$$\text{Heating value} + \text{enthalpy} = 20.6 + 2.76 = 23.34 \text{ MJ/kg coal}$$

The total gasifier efficiency is

$$\text{Total thermal energy/heat in feedstock} = \frac{23.34}{28.4} \times 100\% = \mathbf{82.2\%}$$

Net Gasification Efficiency

The enthalpy or energy content of the gasification medium can be substantial, and so, for a rigorous analysis, these inputs should be taken into consideration. At the same time, part of the input energy is returned (energy credit) by the tar or oil produced as well as by any recovery of the heat of vaporization in the product gas. A more rigorous energy balance may thus be written as

- Total gross energy input = fuel energy content + heat in gasifying mediums
- Net energy input = total energy input – energy recovered through burning tar, oil, and condensation of steam in the gas

The net gasification efficiency can be written as

$$\eta_{net} = \frac{\text{Net energy in the product gas}}{(\text{Total energy input to the gasifer} - \text{credits})} \tag{6.42}$$

Example 6.5

In most steam-fed gasifiers, a large amount of steam remains unutilized. For the given problem, find the amount of unutilized steam. Also find the cold-gas and net gasification efficiency of a fixed-bed gasifier that uses steam and oxygen to gasify grape wastes (HHV = 21,800 kJ/kg). The product gas composition (mass basis) is

CO–31.8%
H_2–3.1%
CO_2–38.2%
CH_4–1.2%
C_3H_8–0.9%
N_2–1%
H_2O–44.8%

The HHV of the product gas is 8.78 MJ/kg.

The ultimate and proximate analyses of the biomass are as given in Table 6.9. The total fuel feed rate is 25 kg/s; the oxygen feed rate is 5.3 kg/s. The steam is fed into the gasifier at a rate of 27 kg/s at 180 °C and 5 bars of pressure. The product contains dry gas, condensable moisture, and tar. The tar production rate is 1.3 kg/s and is analyzed to contain 85% carbon and 15% hydrogen by weight. The heating value of the tar is 42,000 kJ/kg. The oxygen is produced from air using an oxygen-separation unit (OSU) that consumes 4000 kJ of energy/kg of the oxygen produced (assume full conversion of char).

Find the amount of product gas produced and the fraction of steam that remains unutilized.

Solution

Hydrocarbon hydrogen from the ultimate analysis is 5.83 × (1 – 0.04) = 5.6%. Additional hydrogen also in the moisture is 0.04 × (2/18) = 0.44%. Thus, the total

TABLE 6.9 Analyses of Ultimate and Proximate Biomass

	Proximate Analysis in Mass (%)		Ultimate Analysis (dry basis) in Mass (%)
Ash	4.2	Carbon	55.59
Volatile matter	70.4	Hydrogen	5.83
Fixed carbon	21.4	Nitrogen	2.09
Moisture	4.0	Sulfur	0.21
		Oxygen	32.08
Total	100.0	Ash	4.2

hydrogen, on an as-received basis, is 5.6 + 0.40 = 6.0%. The feed rate of the total hydrogen through the fuel is 25 × 6.0/100 = 1.5 kg/s.

A mass balance between input and output helps determine the production rate of the gas. Output equals input, so

Product + ash + tar and oil = fuel + oxygen + steam
Product + (25 × 0.042) + 1.3 = 25 + 5.3 + 27
Product = 54.95 kg/s

The product contains gas, the composition of which was given previously (M_{gas}), as well as the condensate, M_{cond}. To find the gas we carry out a carbon balance from its measured composition.

Part (a) Carbon Balance

The total carbon in the gas (%) is

$$\text{Molecular weight of carbon} \times (\text{\% mass of compound/molecular weight of the compound}) = 12\,(31.8/28 + 38.2/44 + 1.2/16 + 0.9/44) = 33.29\%$$

The carbon balance gives

$$\text{Carbon in gas} + \text{carbon in tar and oil} = \text{carbon in fuel}$$

$$M_{gas} \times (33.29.9/100) + 1.3 \times 0.85 = 25 \times 0.5559$$

$$M_{gas} = 38.4\ \text{kg/s}$$

$$\text{Total product} = 54.95 = 38.4 + M_{cond}$$

$$M_{cond} = 16.55\ \text{kg/s}$$

Part (b) Water Balance

Water enters the gasifier through the steam as well as through the moisture in the fuel, so

$$\text{Water in steam} + \text{water in fuel} = \text{water used in gasification} + \text{water leaving as waste steam/water}$$

The water used in gasification is

$$27 + 25 \times 0.04 - 38.4 \times 0.448 = 9.8 \text{ kg/s}$$

Therefore, the percent of steam not utilized is

$$1 - 9.8/27 = 63.7\%$$

6.11.2 Operational Considerations

A large number of operational issues confront a biomass gasifier. Universal to all gasifier types are problems related to biomass handling and feeding. Bridging of biomass over the exit of a hopper is common for plants that use low-shape-factor (flaky) biomass such as leaves and rice husk. This problem is discussed in more detail in Chapter 8.

Fixed-Bed Gasifier

Charcoal particles become porous and finer during their time in the gasification zone. Thus, in a downdraft gasifier, when fine charcoal drops into the ash pit, the product gas can easily carry the particles as dust. Escaping particles can be a source of carbon loss, and they often plug downstream equipment.

The movement of solids in any layer of a moving-bed gasifier should be equal to the feed rate of the fuel at the top. Even with that balance, if the fuel is dry, the pyrolysis zone may, in an updraft gasifier, travel upward faster, thus consuming the layer of fresh fuel above and leading to premature pyrolysis. The gas lost in this way may result in lower gasification efficiency.

On the other hand, if the fuel is moist, its pyrolysis may be delayed. This may move the pyrolysis zone downward. In the extreme case, the cooler pyrolysis zone may sink sufficiently to extinguish the gasification and combustion reaction. Clearly, a proper balance of rates of fuel flow and air flow is required for stabilization of each of these zones in respective places.

Fluidized-Bed Gasifier

The startup of a fluidized-bed gasifier is similar to the startup of a fluidized bed combustor. The inert bed materials are preheated either by an overbed burner or by burning gas in the bed. Once the bed reaches the ignition temperature of the fuel, the feed is started. Combustion is allowed to raise the temperature. After that, the air/oxidizer-to-fuel ratio is slowly adjusted to switch to gasification mode.

One major problem with fluidized-bed gasifiers is the entrainment (escape) of fine char with the product gas. The superficial velocity in a fluidized bed is often sufficiently high to transport small and light char particles, contributing to major carbon loss. A tall freeboard can reduce the problem, but that has a cost penalty. Instead, most fluidized-bed gasifiers use a cyclone and a recycle system to return the entrained char particles back to the gasifier.

Entrained-Flow Gasifier

The startup procedure for an entrained-flow gasifier takes a long time because a startup burner must heat up the reactor vessel wall. During this time, the reactor vessel is not pressurized. Once oil or gas flame heats up the thick refractory wall to ~1100 °C, the startup burner is withdrawn and the fuel is injected along with the oxidizer (Weigner et al., 2002). The hot reactor wall serves as an igniter for the fuel, which once ignited continues to burn in the combustion zone, consuming the oxygen. For this reason, the fuel injector in an entrained-flow reactor is also called the burner. The reactor is pressurized slowly once the main fuel is ignited.

The gasifying medium is rarely premixed with the fuel. The fuel and the medium are often injected coaxially, as in a pulverized-coal (PC) burner in a boiler or furnace. They immediately mix on entering the reactor. The operation of a gasifier "burner" is similar to that of conventional burners, so design methods for PC or oil burners can be used for a rough and an initial sizing. The use of a separate startup burner involves replacing it with a fuel injector. This is especially difficult for water-cooled walls because their lower thermal inertia cannot hold the wall temperature long enough. Integration of the startup burner in the existing fuel injector is the best option.

Tar Cracking

Several options for tar control and destruction are available; these were discussed in Chapter 4. In fixed-bed gasifiers, thermal cracking or burning has been used with success. In one such design, as shown in Figure 6.24, the air entering the gasifier passes through an aspirator that entrains the tar vapor. The mixture is then burnt in the combustion zone. The aspirator can be outside or inside the gasifier.

Symbols and Nomenclature

A_b = cross-sectional area of the fluidized bed (m^2)
ASH = fractional of ash in the fuel in dry basis (–)
C = fractional of carbon in the fuel in dry basis (–)
C_i = volumetric specific heat of gas i (kJ/nm^3.K)
C_o = initial carbon in the biomass (kg)
C_p = specific heat of the gas (kJ/kg.C)
E_a = activation energy (kJ/mol)
EA = excess air coefficient (–)
ER = equivalence ratio (–)
F = amount of dry fuel required to obtain 1 Nm^3 of product gas (kg/nm^3)
$F[C]$ = char feed rate into the gasifier (kg/s)

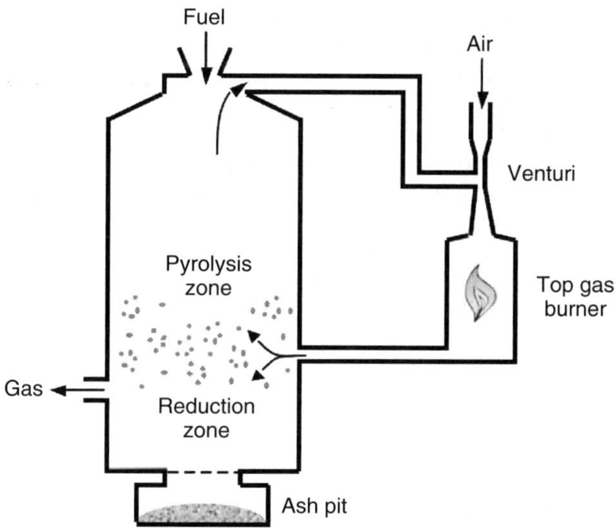

FIGURE 6.24 Gasifier with an aspirator for cracking tar. Fresh air picks up the tar from the gasifier and injects it into the high-temperature combustion zone.

H = fractional of hydrogen in the fuel in dry basis (–)
HHV = higher heating value (kJ/kg)
HHV_d = higher heating value of biomass on dry basis (MJ/kg)
HHV_{daf} = higher heating value of biomass on dry ash–free basis (MJ/kg)
H_{bed} = height of the bed (m)
H_g = enthalpy of steam at gasification temperature (kJ/kg)
H_{in} = heat of the input gas (kJ)
$[H_2O]$ = concentration of steam (–)
k = rate constant (s^{-1})
k_0 = pre-exponential constant in the Arrhenius equation (s^{-1})
LHV_{bm} = lower heating value of the biomass (MJ/kg)
LHV_{daf} = lower heating value of biomass on dry ash-free basis (MJ/kg)
LHV_f = lower heating value of the solid fuel (MJ/Nm^3)
LHV_g = lower heating value of the produced gas (MJ/Nm^3)
m = mass-flow rate of carbon or char (kg/s)
m_{th} = theoretical air requirement for complete combustion of a unit of biomass (kg/kg)
M_a = amount of air required for gasification of unit mass of biomass (kg/kg)
M = fractional of moisture in the fuel (–)
M_{daf} = moisture based on dry ash-free basis
M_f = fuel flow rate (kg/s)
M_{fh} = quantity of steam (kg/s)
M_g = gas produced (kg/s)
n = order of reaction (–)
n_i = number of moles of species i (–)
N = fractional of nitrogen in the fuel in dry basis (–)
n_{total} = total number of moles

O = fractional of oxygen in the fuel in dry basis (–)
P_c = amount of char produced per nm^3 of product gas (kg/nm^3)
q_c = heating value of char (kJ/kg)
Q = power output of the gasifier (MWth)
Q_{ext} = external heat addition to the system (kJ/Nm^3)
Q_g = Lower heating value of the product gas from gasification (MJ/Nm^3)
$Q_{gasification}$ = heat supplied to gasify 1 mol of biomass (kJ/mol)
Q_{loss} = heat loss from the gasifier (kJ/Nm^3)
r = steam gasification reaction rate (kg/s)
R = universal gas constant (0.008314 kJ/mol.K)
S = fractional of sulfur in the fuel in dry basis (–)
SC = steam to carbon molar ratio (–)
t = time (s)
T = temperature (K)
T_f = gas temperature at the exit (°C)
T_g = gas temperature (°C)
T_0 = gas temperature at the entrance (°C)
U_g = fluidizing velocity (m/s)
V = volume of the fluidized bed (m^3)
V_{bed} = volume of the bed (m^3)
V_{daf} = volatile based on dry mass-free basis
V_g = gas generation rate (m^3/s)
V_g = volumetric flow rate of product gas (Nm^3/s)
V_i = volumetric fraction of gas species i (–)
W = total steam needed in Eq. 6.22 (kg/s)
W_{in} = rate of the char moving in (kg/s)
W_{out} = rate of the char moving out (kg/s)
x_{char} = weight of the reacting char (kg)
X = fraction of char in the feed converted (–)
X_c = fixed carbon fraction in the fuel (kg carbon/kg dry fuel)
X_{char} = char fraction in bed (–)
X_g = fraction of steam used up in gasification
ε = voidage of the bed (–)
λ_I = Lagrangian multiplier for species i (–)
ρ_g = density of air at the opening temperature and pressure of the gasifier (kg/m^3)
θ = residence time of char in bed or reactor (s)
ρ_b = bed density (kg/m^3)
ρ_s = density of bed solids (kg/m^3)
η_{gef} = gasifier efficiency (–)
η_{ceff} = cold gas efficiency (–)
η_{cg} = cold gas efficiency of the gasifier (–)
η_{hg} = hot gas efficiency of the gasifier (–)
η_{net} = net gasification efficiency of the gasifier (–)
ΔH_T = heat of formation at temperature T (kJ/mol)

Chapter 7

Hydrothermal Gasification of Biomass

7.1 INTRODUCTION

In the mid-1970s Sanjay Amin, a graduate student working at the Massachusetts Institute of Technology (MIT), was studying the decomposition of organic compounds in hot water (steam reforming):

$$C_6H_{10}O_5 + 7H_2O \rightarrow 6CO_2 + 12H_2 \tag{7.1}$$

While conducting an experiment in subcritical water, he observed that in addition to producing hydrogen and carbon dioxide, the reaction was producing much char and tars. Herguido et al. (1992) also made similar observations in the steam gasification of biomass at atmospheric pressure.

Sanjay interestingly noted that when he raised the water above its "critical state," the tar that formed in the subcritical state disappeared entirely (Amin et al., 1975). This important finding kick-started research and development on supercritical water oxidation (SCWO) for disposal of organic waste materials (Tester et al., 1993), which has now become a commercial option for disposal of highly contaminated organic wastes (Shaw and Dahmen, 2000).

Biomass in general contains substantially more moisture than do fossil fuels like coal. Some aquatic species, such as water hyacinth, or waste products, such as raw sewage, can have water contents exceeding 90%. Thermal gasification, where air, oxygen, or subcritical steam is the gasification medium, is very effective for dry biomass, but it becomes very inefficient for a high-moisture biomass because the moisture must be substantially driven away before thermal gasification can begin; in addition, a large amount of the extra energy (~2260 kJ/kg moisture) is consumed in its evaporation. For example, Yoshida et al. (2003) saw the efficiency of their thermal gasification system reduce from 61 to 27% while the water content of the feed increased from 5 to 75%. So, for gasification of very wet biomass, some other means such as anaerobic digestion (see Section 2.2) and hydrothermal gasification in high-pressure hot water are preferable because the water in these processes is not a

Biomass Gasification and Pyrolysis. DOI: 10.1016/B978-0-12-374988-8.00007-6

liability as it is in thermal gasification. Instead it serves as a reaction medium and a reactant.

The efficiencies of these processes do not decrease with moisture content. For anaerobic digestion and supercritical gasification, Yoshida et al. (2003) found the gasification efficiency to remain nearly unchanged, at 31% and 51%, respectively, even when the moisture in the biomass increased from 5 to 75%.

A major limitation of anaerobic digestion is that it is very slow, with a relatively low efficiency and, most important, it produces methane only, no hydrogen. If hydrogen is the desired product, as is often the case, an additional step of steam reforming the methane ($CH_4 + H_2O = CO + 3H_2$) must be added to the anaerobic digestion process.

Hydrothermal gasification involves gasification in an aqueous medium at very a high temperature and pressure exceeding or close to its critical value. While subcritical water has been used effectively for hydrothermal reaction, supercritical water has attracted more attention owing to its unique features. Supercritical water offers rapid hydrolysis of biomass, high solubility of intermediate reaction products, including gases, and a high ion product near (but below) the critical point that helps ionic reaction. These features make supercritical water an excellent reaction medium for gasification, oxidation, and synthesis.

This chapter deals primarily with hydrothermal gasification of biomass in supercritical water. It explains the properties of supercritical water and the biomass conversion process in it. The effects of different parameters on SCW gasification and design considerations for the SCW gasification plants are also presented.

7.2 SUPERCRITICAL WATER

Water above its critical temperature (374.29 °C) and pressure (22.089 MPa) is called *supercritical* (Figure 7.1). Water or steam below this pressure and temperature is called *subcritical*. The term *water* in a conventional sense may not be applicable to SCW except for its chemical formula, H_2O, because above the critical temperature SCW is neither water nor steam. It has a waterlike density but a steam like diffusivity. Table 7.1 compares the properties of subcritical water and steam with those of SCW, indicating that SCW's properties are intermediate between the liquid and gaseous states of water in subcritical pressure; descriptions of each follow the table.

Figure 7.1 shows that the higher the temperature, the higher the pressure required for water to be in its liquid phase. Above a critical point the line separating the two phases disappears, suggesting that the division between the liquid and vapor phases disappears. Temperature and pressure at this point are known as *critical temperature*, and *critical pressure*, above which water attains supercritical state and hence is called *supercritical* (SCW).

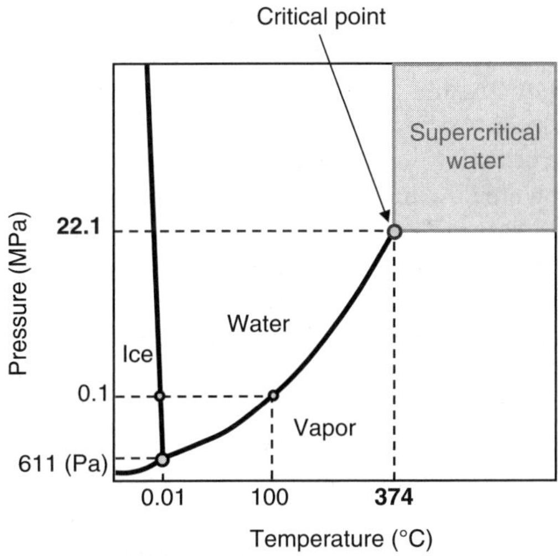

FIGURE 7.1 Phase diagram of water showing the supercritical region.

TABLE 7.1 Properties of Supercritical and Subcritical Water

Property	Subcritical Water	Supercritical Water	Supercritical CO_2	Subcritical Steam
Temperature (°C)	25	400	55	150
Pressure (MPa)	0.1	30	28	0.1
Density, kg/m^3	997*	358*	835	0.52*
Dynamic viscosity, μ (kg/m.s)	890.8×10^{-6}	43.83×10^{-6}*	0.702×10^{-6}	14.19×10^{-6}*
Diffusivity of small particles (m^2/s)	$\sim 1.0 \times 10^{-9}$**	$\sim 1.0 \times 10^{-8}$**		$\sim 1.0 \times 10^{-5}$**
Dielectric constant***	78.46	5.91		1.0
Thermal conductivity, λ (w/m.k)	607×10^{-3}*	330×10^{-3}*		28.8×10^{-3}*
Prandtl number, $C_p\mu/\lambda$	6.13	3.33		0.97

*Haar et al., 1984; **Serani et al., 2008; ***Uematsu and Franck, 1980.

Subcritical water ($T < T_{sat}$; $P < P_c$). When the pressure is below its critical value, P_c, and the temperature is below its critical value, T_c, the fluid is called *subcritical*. If the temperature is below its saturation value, the fluid is known as *subcritical water*, as shown in the lower left block of Figure 7.1.

Subcritical steam ($T > T_{sat}$; $P < P_c$. Note: T may be above T_c). When water (below critical pressure) is heated, it experiences a drop in density and an increase in enthalpy; this change is very sharp when the temperature of the water just exceeds it saturation value, T_{sat}. Above the saturation temperature, but below the critical value, the fluid (H_2O) is called *subcritical steam*. This regime is shown below the saturation line in Figure 7.1.

Supercritical water ($T > T_c$; $P > P_c$). When heated above its critical pressure, P_c, water experiences a continuous transition from a liquidlike state to a vapor like state. The vaporlike, supercritical, state is shown in the upper right block in Figure 7.1. Unlike in the subcritical stage, no heat of vaporization is needed for the transition from liquidlike to vaporlike. Above the critical pressure, there is no saturation temperature separating the liquid and vapor states. However, there is a temperature, called *pseudo-critical temperature*, that corresponds to each pressure ($>P_c$) above which the transition from liquidlike to vaporlike takes place. The pseudo-critical temperature is characterized by a sharp rise in the specific heat of the fluid.

The pseudo-critical temperature depends on the pressure of the water. It can be estimated within 1% accuracy by the following empirical equation (Malhotra, 2006):

$$T^* = (P^*)^F$$
$$F = 0.1248 + 0.01424P^* - 0.0026(P^*)^2 \qquad (7.2)$$
$$T^* = \frac{T_{sc}}{T_c}; \; P^* = \frac{P}{P_c}$$

where T_{sat} is the saturation temperature at pressure P; P_{sat} is the saturation pressure at temperature T; P_c is the critical pressure of water, 22.089 MPa; T_c is the critical temperature of water, 374.29 °C; and T_{sc} is the pseudo-critical temperature at pressure P ($P > P_c$).

7.2.1 Properties of Supercritical Water

The critical point marks a significant change in the thermophysical properties of water (Figure 7.2). There is a sharp rise in the specific heat near the critical temperature followed by a similar drop. The thermal conductivity of water drops from 0.330 W/m.K at 400 °C to 0.176 W/m.K at 425 °C. The drop in molecular viscosity is also significant, although the viscosity starts rising with temperature above the critical value. Above this critical point, water experiences a dramatic change in its solvent nature primarily because of its loss of hydrogen bonding. The dielectric constant of the water drops from a value of about 80 in the ambient condition to about 10 at the critical point. This changes the water from a highly polar solvent at an ambient condition to a nonpolar solvent, like benzene, in a supercritical condition.

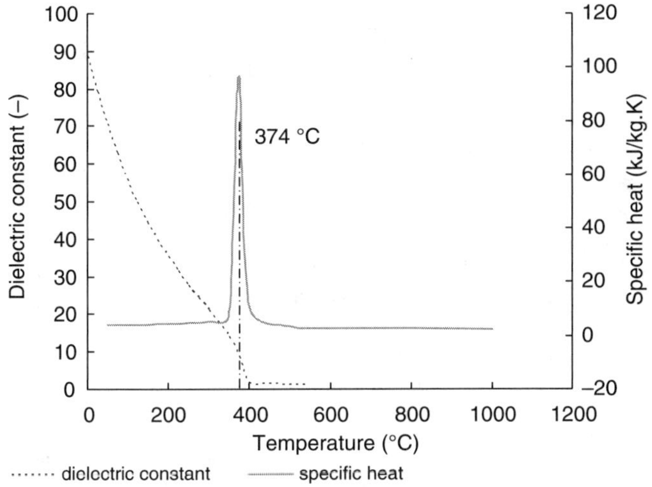

FIGURE 7.2 Specific heat of water above its critical pressure shows a peak at its pseudo-critical temperature. Dielectric constant at 22.1 MPa, also plotted on this graph, shows rapid decline closer to the critical temperature.

The change in density in supercritical water across its pseudo-critical temperature is much more modest, however. For example, at 25 MPa it can drop from about 1000 to 200 kg/m^3 while the water moves from a liquidlike to a vaporlike state. At subcritical pressure, however, there is an order of magnitude drop in density when the water goes past its saturation temperature. For example, at 0.1 MPa or 1 atm of pressure, the density reduces from 1000 to 0.52 kg/m^3 as the temperature increases from 25 to 150 °C (refer to Table 7.1).

The most important feature of supercritical water is that we can "manipulate" and control its properties around its critical point simply by adjusting the temperature and pressure. Supercritical water possesses a number of special properties that distinguish it from ordinary water. Some of those properties relevant to gasification are as follows:

- The solvent property of water can be changed very strongly near or above its critical point as a function of temperature and pressure.
- Subcritical water is polar, but supercritical water is nonpolar because of its low dielectric constant. This makes it a good solvent for nonpolar organic compounds but a poor one for strongly polar inorganic salts. SCW can be a solvent for gases, lignin, and carbohydrates, which show low solubility in ordinary (subcritical) water. Good miscibility of intermediate solid organic compounds as well as gaseous products in liquid SCW allows single-phase chemical reactions during gasification, removing the interphase barrier of mass transfer.

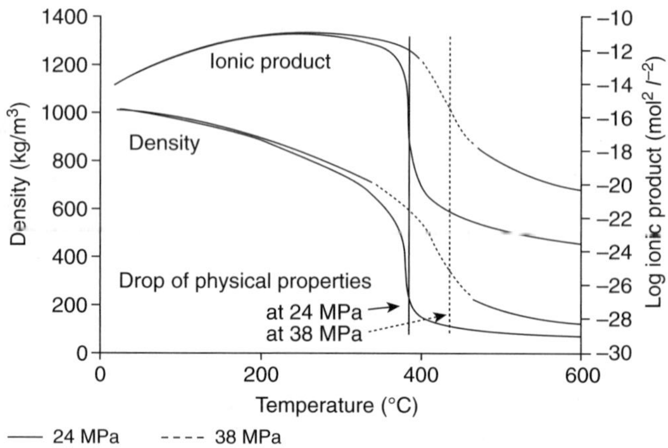

FIGURE 7.3 At a given temperature, both the density and the ion product of water increase with pressure. Data plotted for 24 MPa show that these values reduce quickly at the critical temperature, but are slower at 38 MPa. The ion product is plotted as negative log.kw. *(Source: Adapted from Kritz, 2004; used with permission.)*

- SCW has a high density compared to subcritical steam at the same temperature. This feature favors the forward reaction between cellulose and water to produce hydrogen.
- Near its critical point, water has higher ion products ($[H^+][OH^-]\sim10^{-11}$ $(mol/l)^2$) than it has in its subcritical state at ambient conditions ($\sim10^{-14}$ $(mol/l)^2$) (Figure 7.3). Owing to this high $[H^+]$ and $[OH^-]$ ion, the water can be an effective medium for acid- or base-catalyzed organic reactions (Serani et al., 2008). Above the critical point, however, the ion product drops rapidly ($\sim10^{-24}$ $(mol/l)^2$ at 24 MPa), and the water becomes a poor medium for ionic reactions.
- Most ionic substances, such as inorganic salts, are soluble in subcritical water but nearly insoluble under typical conditions of SCW gasifiers. As the temperature rises past the critical point, the density as well as the ionic product decreases (Figure 7.3). Thus, highly soluble common salt (NaCl) becomes insoluble at higher temperatures above the critical point. This tunable solubility property of SCW makes it relatively easy to separate the salts as well as the gases from the product mixture in an SCW gasifier.
- Gases, such as oxygen and carbon dioxide, are highly miscible in SCW, allowing homogeneous reactions with organic molecules either for oxidation or for gasification. This feature makes SCW an ideal medium for destruction of hazardous chemical waste through SCWO.
- SCW possesses excellent transport properties. Its density is lower than that of subcritical water but much higher than that of subcritical steam. This, along with other properties like low viscosity, low surface tension (surface tension of water reduces from 7.2×10^{-2} at 25 °C to 0.07 at 373 °C), and high diffusivity greatly contribute to the SCW's good transport property,

which allows it to easily enter the pores of biomass for effective and fast reactions.

- Reduced hydrogen bonding is another important feature of SCW. The high temperature and pressure break the hydrogen-bonded network of water molecules.

Table 7.1 compares some of these water properties under subcritical and supercritical conditions.

7.2.2 Application of Supercritical Water in Chemical Reactions

Chemical reactions involve the mixing of reactants. If the mixing is incomplete, the reaction will be incomplete, even if the right amounts of reactant and the right temperature are available. The mixing is better when all reactants are either in the gas phase or in the liquid phase compared to that when one reactant is in the solid phase and the other is in the gas or liquid phase. The absence of interphase resistance in a monophase reaction medium greatly improves the mixing. The conventional thermal gasification of solid biomass in air or steam involves heterogeneous mixing, and therefore the gas–solid interphase resistance limits the conversion reactions.

Supercritical water allows reactions to take place in a single phase, as most organic compounds and gases are completely miscible in it. It is thus a superior reaction medium. Because the absence of interphase mass transfer resistance facilitates better mixing and therefore higher conversion, SCW can be an excellent medium for the following three types of reactions:

Hydrothermal gasification of biomass. SCW is an ideal medium for gasification of very wet biomass, such as aquatic species and raw sewage, which ordinarily have to be dried before they can be gasified economically. SCW gasification produces gas at high pressure and thus obviates the need for an expensive product gas compression step for transport or use in combustion.

Synthesis reactions. A variety of organic reactions like hydrolysis and molecular rearrangement can be effectively carried out in SCW, which serves as a solvent, a reactant, and sometimes a catalyst. There is no need for acid or base solvents, the disposal of which is often a problem.

Supercritical water oxidation. Complete miscibility of oxygen in SCW helps harmful organic compounds to be easily oxidized and degraded. Thus, SCW is an attractive means of turning pollutants into harmless oxides.

7.2.3 Advantages of SCW Gasification over Conventional Thermal Gasification

The following are two broad routes for the production of energy or chemical feedstock from biomass:

Biological: Direct biophotolysis, indirect biophotolysis, biological reactions, photofermentation, and dark fermentation are the five major biological processes.

Thermochemical: Combustion, pyrolysis, liquefaction, and gasification are the four main thermochemical processes.

Thermal conversion processes are relatively fast, taking minutes or seconds to complete, while biological processes, which rely on enzymatic reactions, take much longer, on the order of hours or even days. Thus, for commercial use, thermochemical conversion is preferred.

Gasification may be carried out in air, oxygen, subcritical steam, or water near or above its critical point. This chapter concerns hydrothermal gasification of biomass above or very close to the water's critical point to produce energy and/or chemicals.

Conventional thermal gasification faces major problems from the formation of undesired tar and char. The tar can condense on downstream equipment, causing serious operational problems, or it may polymerize to a more complex structure, which is undesirable for hydrogen production. Char residues contribute to energy loss and operational difficulties. Furthermore, very wet biomass can be a major challenge to conventional thermal gasification because it is difficult to economically convert if it contains more than 70% moisture. The energy used in evaporating fuel moisture (2257 kJ/kg), which effectively remains unrecovered, consumes a large part of the energy in the product gas.

Gasification in supercritical water (SCWG) can largely overcome these shortcomings, especially for very wet biomass or organic waste. For example, the efficiency of thermal gasification of a biomass containing 80% water in conventional steam reforming is only 10%, while that of hydrothermal gasification in SCW can be as high as 70% (Dinjus and Kruse, 2004). Gasification in near or supercritical water therefore offers the following benefits:

- Tar production is low. The tar precursors, such as phenol molecules, are completely soluble in SCW and so can be efficiently reformed in SCW gasification.
- SCWG achieves higher thermal efficiency for very wet biomass.
- SCWG can produce in one step a hydrogen-rich gas with low CO, obviating the need for an additional shift reactor downstream.
- Hydrogen is produced at high pressure, making it ready for downstream commercial use.
- Carbon dioxide can be easily separated because of its much higher solubility in high-pressure water.
- Char formation is low in SCWG.
- Heteroatoms like S, N, and halogens leave the process with aqueous effluent, avoiding expensive gas cleaning. Inorganic impurities, being insoluble in SCW, are also removed easily.

- The product gas of SCWG automatically separates from the liquid containing tarry materials and char if any.

7.3 BIOMASS CONVERSION IN SCW

There are three major routes for SCW-based conversion of biomass into energy as follows:

Liquefaction: Formation of liquid fuels above critical pressure (22.1 MPa) but near critical temperature (300–400 °C).
Gasification to CH_4: Conversion in SCW in a low-temperature range (350–500 °C) in the presence of a catalyst.
Gasification to H_2: Conversion in SCW with or without catalysts at higher (>600 °C) temperatures.

Here we discuss only the last two gasification options.

7.3.1 Gasification

Supercritical biomass gasification takes place typically at around 500 to 750 °C in the absence of catalysts, and at an even lower (350–500 °C) temperature with catalysts. The biomass decomposes into char, tar, gas, or other intermediate compounds, which are reformed into gases like CO, CO_2, CH_4, and H_2. The process is schematically shown in Figure 7.4. If the biomass is represented by the general formula $C_6H_{12}O_6$, the gasification process may be described by the following overall reaction:

$$mC_6H_{12}O_6 + nH_2O \rightarrow wH_2 + xCH_4 + yCO + zCO_2 \quad (7.3)$$

Gasification in SCW involves, among other reactions, hydrolysis and oxidation reactions. A brief description of these reactions follows.

7.3.2 Hydrolysis

Hydrolysis (meaning "splitting with water") is the reaction of an organic compound with water. Here, a bond of an organic molecule is broken, and the water molecule is also broken into $[H^+]$ and $[OH^-]$. The organic molecules are cleaved into two parts by the water molecule: One part gains the $[H^+]$ ion; the other

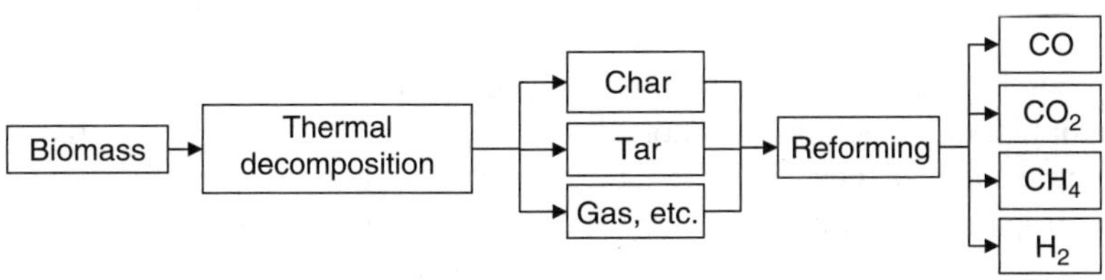

FIGURE 7.4 Biomass gasification process.

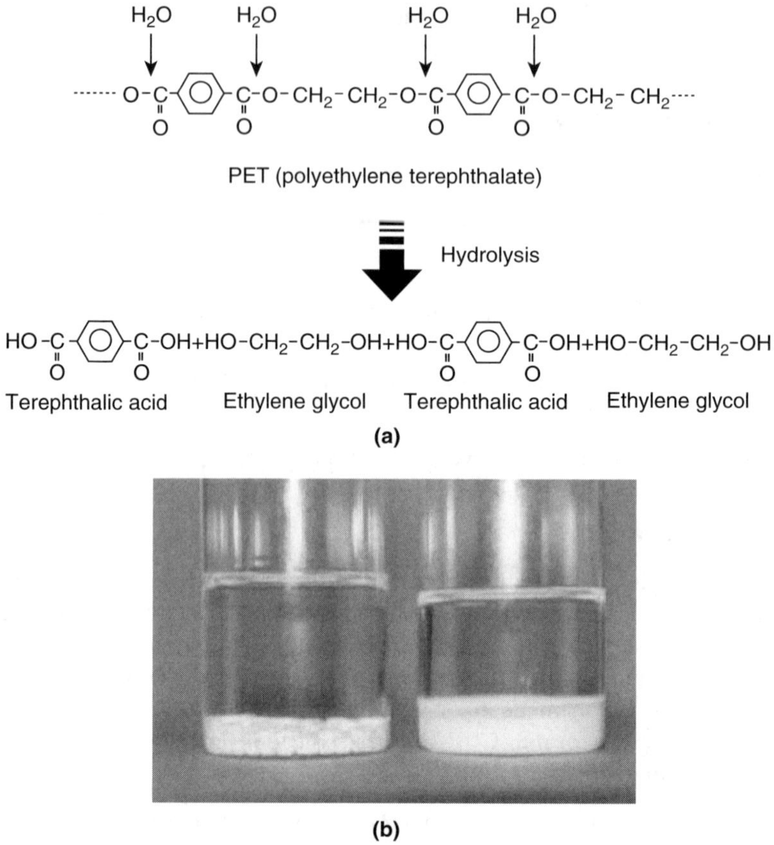

FIGURE 7.5 (a) Hydrolysis of PET in supercritical water; photograph of PET pellets in subcritical water and (b) that of its product in SCW after hydrolysis. *(Source: Adapted from Kobe Steel.)*

part, the [OH^-] ion. Hydrolysis reactions are generally catalyzed by acid or base catalysts. Water near its critical point (at high temperature and pressure) has a high ion product, so the hydrolysis reaction is catalyzed by the water itself.

A simplified representation of the reaction scheme is shown in Figure 7.5(a) with polyethylene terephthalate (PET) as an example. The hydrolysis of PET into terephthalic acid and ethylene glycol in SCW is a better option than other reactions (e.g., methanolysis or glycololysis) because it does not require solvents and catalysts like others. Here, water near its critical point is used to accomplish this reaction in a shorter time. Additionally, SCW avoids the need to recover and dispose of external solvents or catalysts. Figure 7.5(b) is a photograph of polyethylene terephthalate in ordinary water before and after hydrolysis in SCW into fine particles of teraphthalic acid in ethylene glycol solution.

7.3.3 Supercritical Water Oxidation

Supercritical water that exhibits complete miscibility with oxygen is a homogeneous reaction medium for the oxidation of organic molecules. This feature of SCW allows oxidation of harmful or toxic substances at low temperature in a process known as supercritical water oxidation (SCWO) or cold combustion. In a typical SCWO unit, the entire mixture (water, oxygen, and waste) remains as a single fluid phase with no interphase transport limitations. This allows very rapid and complete (>99.9%) oxidation of the organic wastes to harmless lower-molecular-weight compounds like H_2O, N_2, and CO_2. Unlike thermal incineration, SCWO produces toxic by-products such as dioxin. This method of waste treatment is especially attractive for highly dilute toxic wastes in water.

One important shortcoming of this process is the production of highly corrosive liquid effluents because chlorine, sulfur, and phosphorous, if present in the waste, are converted into their corresponding acids (Serani et al., 2008). The destruction of polychlorinated biphenyls (PCBs) in supercritical water, producing carbon dioxide and hydrochloric acid, may be represented by the following simple reaction:

$$C_{12}H_{10-m}Cl_m\ (PCB)+(19+m)/2\ O_2+(5-m)\ H_2O=12CO_2+m HCl \quad (7.4)$$

Conventional thermal incineration uses very high temperature to destroy by-products like dioxin, which results in the production of another pollutant, NOx. This is not the case with SCWO owing to its low-temperature operation (450–600 °C).

7.3.4 Scheme of an SCWG Plant

A typical SCWG plant includes the following key components:

- Feedstock pumping system
- Feed preheater
- Gasifier/reactor
- Heat-recovery (product-cooling) exchanger
- Gas–liquid separator
- Optional product-upgrading equipment

The feed preheating system is very elaborate and accounts for the majority (~60%) of the capital investment in an SCW gasification plant.

Figure 7.6 describes the SCWG process using the example of an SCWG plant for gasifying sewage sludge. Biomass is made into a slurry for feeding. It is then pumped to the required supercritical pressure. Alternatively, water may be pressurized separately and the biomass fed into it. In any case, the feedstock needs to be heated to the designed inlet temperature for the gasifier, which must be above the critical temperature and well above the designed gasification temperature because the enthalpy of the water provides the energy

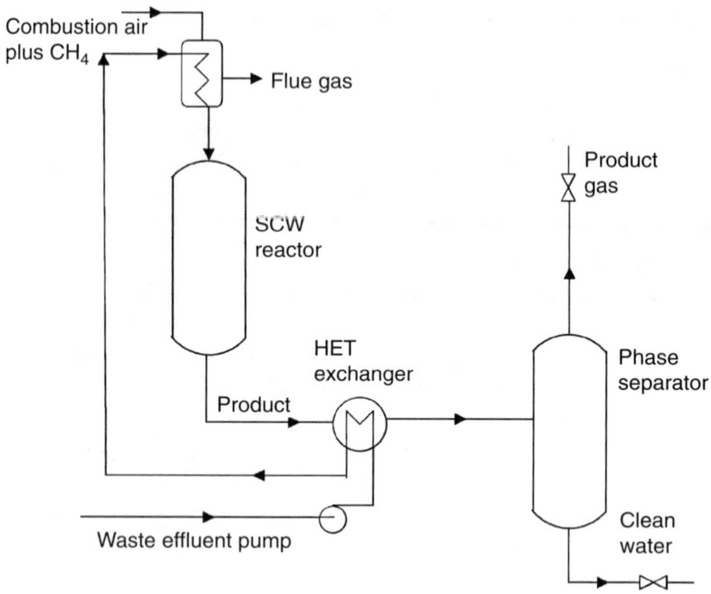

FIGURE 7.6 Schematic of a pilot plant for supercritical water gasification of biomass.

required for the endothermic gasification reactions. This temperature is a critical design parameter.

The sensible heat of the product of gasification may be partially recovered in a waste heat-recovery exchanger and used for partial preheating of the feed (Figure 7.6). For complete preheating, additional heat may be obtained from one of the following:

- Externally fired heater (Figure 7.6)
- Burning of a part of the fuel gas produced to supplement the external fuel
- Controlled burning of unconverted char in the reactor system (refer to Figure 7.12 later in chapter)

After gasification, the product is first cooled in the waste heat-recovery unit. Thereafter, it cools to room temperature in a separate heat exchanger by giving off heat to an external coolant.

The next step involves separation of the reaction products. The solubility of hydrogen and methane in water at low temperature but high pressure is considerably low, so they are separated from the water after cooling while the carbon dioxide, because of its high solubility in water, remains in the liquid phase. For complete separation of CO_2, the gas may be scrubbed with additional water (refer to Figure 7.14 later in chapter). The gaseous hydrogen is separated from the methane in a pressure swing adsorber. The CO_2-rich liquid is depressurized to the atmospheric pressure, separating the carbon dioxide from the water and unconverted salts.

7.4 EFFECT OF OPERATING PARAMETERS ON SCW GASIFICATION

The product of gasification is defined by its yield and composition, which are influenced by a number of gasifier design and operating parameters. For proper design and operation of an SCW gasifier, a good understanding of the influence of the following parameters is important:

- Reactor temperature
- Catalyst use
- Residence time in the reactor
- Solid concentration in the feed
- Heating rate
- Feed particle size
- Reactor pressure
- Reactor type

7.4.1 Reactor Temperature

Temperature has an important effect on the conversion, the product distribution, and the energy efficiency of an SCW gasifier, which typically operates at a maximum temperature of nearly 600 °C. The overall carbon conversion increases with temperature; at higher temperatures hydrogen yield is higher while methane yield is lower. Figure 7.7 shows the temperature dependence of gasification efficiency and product distribution in a reactor operated at 28 MPa (30-s residence, 0.6-M glucose) (Lee et al., 2002). We see that the hydrogen yield increases exponentially above 600 °C, while the CO yield, which rises gently with temperature, begins to drop above 600 °C owing to the start of the shift reaction (Eq. 5.52).

Gasification efficiency is measured in terms of hydrogen or carbon in the gaseous phase as a fraction of that in the original biomass. Carbon conversion efficiency increases continually with temperature, reaching close to 100% above 700 °C. Hydrogen conversion efficiency (the fraction of hydrogen in glucose converted into gas) also increases with temperature. It appears strange that at 740 °C, the hydrogen conversion efficiency exceeds 100%, reaching 158%. This clearly demonstrates that the extra hydrogen comes from the water, confirming that water is indeed a reactant in the SCWG process as well as a reaction medium.

Hydrothermal gasification of biomass has been divided into three broad temperature categories: high, medium, and low with their desired products (Peterson et al., 2008). Table 7.2 shows that the first group targets production of hydrogen at a relatively high temperature (>500 °C); the second targets production of methane at just above the critical temperature (~374.29 °C) but below 500 °C; and the third gasifies at subcritical temperature, using only simple organic compounds as its feedstock. The last two groups, because of their low-temperature operation, need catalysts for reactions.

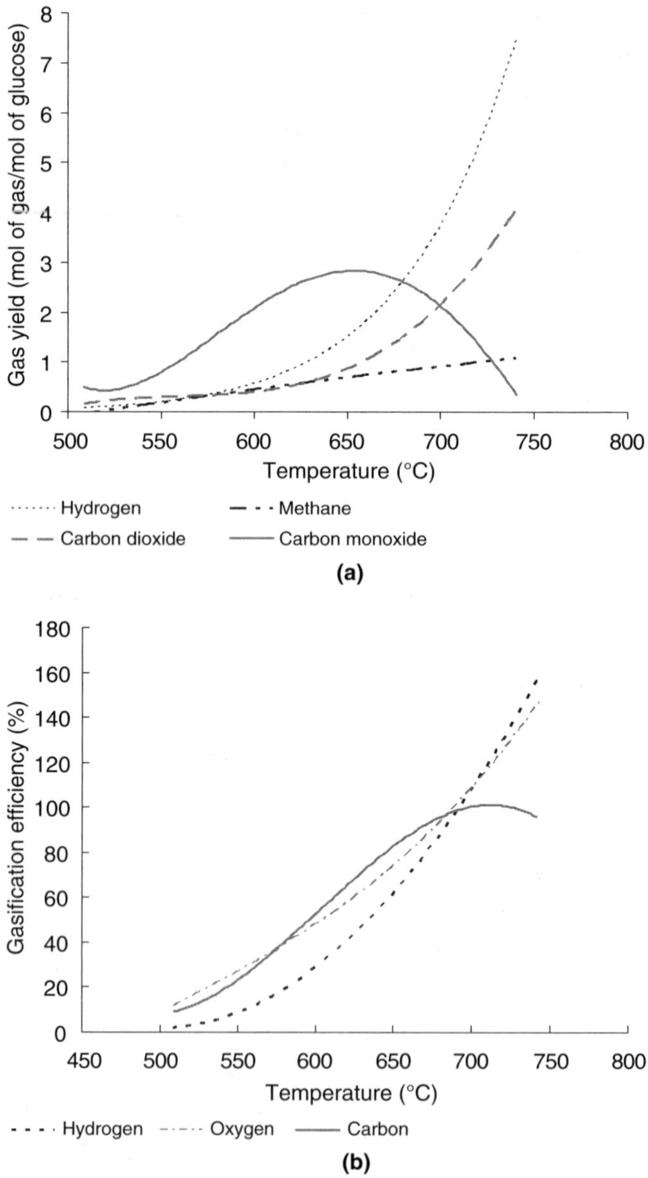

FIGURE 7.7 Effect of temperature on gas yield (a), and effect of temperature on gasification efficiency (b). *(Source: Adapted from Lee et al., 2002.)*

TABLE 7.2 Hydrothermal Gasification Temperature Categories Based on Target Product (T_c ~374.29 °C)

Temperature (°C)	Catalyst	Target Product
High (>500)	Not needed	Hydrogen-rich gas
Medium (T_c – 500)	Needed	Methane-rich gas
Low (<T_c)	Essential	Other gases of smaller organic molecules

7.4.2 Catalysts

An effective degradation of biomass and the gasification of intermediate products of thermal degradation into lower-molecular-weight gases like hydrogen require the SCW reactor to operate in the high-temperature range (>600 °C). The higher the temperature, the better the conversion, especially for production of hydrogen, but the lower the SCW's energy efficiency. A lower gasification temperature is therefore desirable for higher thermodynamic efficiency of the process.

Catalysts help gasify the biomass at lower temperatures, thereby retaining, at the same time, high conversion and high thermal efficiency. Additionally, some catalysts also help gasification of difficult items like the lignin in biomass. Watanabe et al. (2003) noted that the hydrogen yield from lignin at 400 °C and 30 MPa is doubled when a metal oxide (ZrO_2) catalyst is used in the SCW. The yield increases four times with a base catalyst (NaOH) compared to gasification without a catalyst. The three principal types of catalyst used so far for SCW gasification are: (1) alkali, (2) metal, and (3) carbon-based.

An important positive effect of catalysts in SCWG is the reduction in required gasification temperature for a given yield. Minowa et al. (1998) noted a significant reduction in unconverted char while gasifying cellulose with an Na_2CO_3 catalyst at 380 °C. Base catalysts (e.g., NaOH, KOH) offer better performance, but they are difficult to recover from the effluent. Some alkalis (e.g., NaOH, KOH, Na_2CO_3, K_2CO_3, and $Ca(OH)_2$) are also used. They, too, are difficult to recover.

The special advantage of metal oxide catalysts is that they can be recovered, regenerated, and reused. Commercially available nickel-based catalysts are effective in SCW biomass gasification. Among them, Ni/MgO (nickel supported on an MgO catalyst) shows high catalytic activity, especially for biomass (Minowa et al., 1998).

Metal catalysts have a severe corrosion effect at the temperatures needed to secure high yields of hydrogen. To overcome this problem, Antal et al. (2000) used carbon (e.g., coal-activated and coconut shell–activated carbon and macadamia shell and spruce wood charcoal). The carbon catalysts resulted in high yields of gas without tar formation.

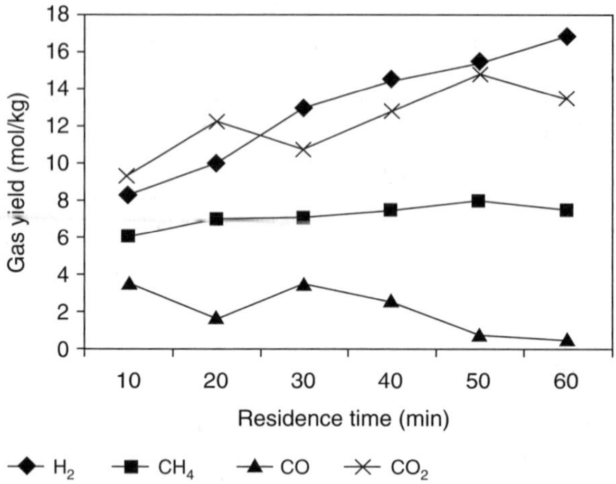

FIGURE 7.8 Effect of residence time on the gasification of 2% rice husk in supercritical water at 650 °C and 30 MPa in a batch reactor.

7.4.3 Residence Time

A longer residence for the reactants in the reactor gives a better yield. Lu et al. (2006) experimented with 2% (by weight) sawdust and 2% carboxymethyl cellulose (CMC) in a flow reactor at 650 °C and 25 MPa. Mettanant et al. (2009) experimented with 2% rice husk in a batch reactor under the same conditions. Both found a steady increase in hydrogen and a moderate increase in methane (Figure 7.8) when the residence time was increased by three times and six times, respectively. Total organic carbon in the liquid product decreases with residence time, whereas carbon and hydrocarbon gasification efficiencies increase. This implies that a longer residence time is favorable for SCW biomass gasification. The optimum residence time, beyond which no further improvement in conversion efficiency is possible, depends on several factors. At a higher temperature, the residence time required for a given conversion is shorter.

7.4.4 Solid Concentration in Feedstock

Unlike in other gasification methods, solids in the feed have an important effect on the gasification in supercritical water. Thermodynamic calculations suggest that the conversion of carbon to gases in SCW declines rapidly when the solid content in a liquid feed exceeds 50% (Prins et al., 2005), but experimental results show this to occur for a much lower concentration. Experimental data (Mettanant et al., 2009; Schmieder et al., 2000) indicate that gasification efficiency starts to decline when the solid concentration exceeds a value as low as 2%.

Table 7.3 presents data (Mozaffarian et al., 2004) that show the effect of solid content in feed. Although experimental conditions and feedstock vary, we can broadly classify these results into groups of low, medium, and high solid feedstock. For a lower feed concentration (<2%), carbon conversion efficiency is in the range 100 to 92% and reduces to 60 to 90% for an intermediate concentration (2–10%) and to 68 to 80% for a >10% concentration. An SCW gasifier thus needs a very low solid concentration in the feed for high carbon conversion efficiency. This requires higher pumping costs and liquid effluent disposal, which may be a major impediment in commercialization of SCW gasification.

The reactor type also influences how solid concentration affects gasification efficiency. For example, Kruse et al. (2003) noted that a stirred reactor shows opposite results—that is, higher gasification efficiency at higher solid content (1.8 to 5.4%) in feed. This contrasts with data from Schmieder et al. (2000) from tumbling and tubular reactors that indicate a decrease in gasification efficiency with solid content (0.2–0.6 M). In stirred reactors, reactants are very well mixed, resulting in a heating rate that is faster than achieved in other reactor types. This may be the explanation for the higher gasification efficiency where there is a higher solid content. The exact reason for this decrease is not clear and is a major issue in the development of commercial SCW gasifiers. Catalysts, high gasification temperatures, and high heating rates can avoid the drop in conversion of a high-solid-content feedstock (Lu et al., 2006).

7.4.5 Heating Rate

Limited data obtained by Sinag et al. (2004) suggest that at a higher heating rate the yield of hydrogen, methane, and carbon dioxide increases while that of carbon monoxide decreases. Further investigation is needed to elucidate this point.

7.4.6 Feed Particle Size

The effect of biomass particle size is not well researched. With limited data, Lu et al. (2006) showed that smaller particles result in a slightly improved hydrogen yield and higher gasification efficiency. However, Mettanant et al. (2009) did not observe any effect when they varied the size of rice husk particles in the range of 1.25 to 0.5 mm. Even if the size effect is confirmed with further data, it remains to be seen if the extra energy required for grinding is worth the improvement.

TABLE 7.3 Effect of Solid Content in Feed and Other Operating Parameters on Gasification

	$C < 2$ wt.%			$2 < C < 10$ wt.%			$C > 10$ wt.%	
Investigators	**Holgate, 1995**	**Yu, 1993**	**Kruse, 1999**	**Hao, 2003**	**Xu, 1996**	**Kruse, 2003**	**Yu, 1993**	**Xu, 1996**
Feedstock	Glucose	Glucose	Wood	Glucose	Formic acid	Baby food	Glucose	Glucose
Feed concentration in SCW (%/weight)	0.01	1.8	1	7.2	2.8	5.4	14.4	22
Pressure (bar)	246	345	350	250	345	300	345	345
Temperature (°C)	600	600	450	650	600	500	600	600
Reactor type	Flow reactor	Tubular flow reactor	Autoclave	Tubular flow reactor (9 mm)	Tubular flow reactor	SCTR	Tubular flow reactor	Tubular flow reactor
Residence time (s)	6	34	7200	210	34	300	34	34
Carbon conversion efficiency (%)	100	90	91.8	89.6	93	60	68	80
Gas composition:								
H_2	61.3	61.6	28.9	21.5	49.2	44	25	11
CO_2	36.8	29	48.4	35.5	48.1	41	16.6	5.7
CO	—	2	3.3	18.3	1.7	0.4	41.6	62.3
CH_4	1.8	7.2	19	15.8	1	14.6	16.7	16.5
$C_{2,3}$	-		-	5.3	-	-	-	4.5

C = concentration of solid in feed
Source: Compiled from Mozaffarian et al., 2004.

7.4.7 Pressure

Experiments by Van Swaaij et al. (2003) in their microreactor over the range of 19 to 54 MPa, those by Kruse et al. (2003) in a stirred tank (30–50 MPa, 500 °C), and those by Lu et al. (2006) in a plug-flow reactor (18–30 MPa, 625 °C) showed no major effect of pressure on carbon conversion or product distribution. Nor did Mettanant et al. (2009) see much effect in their temperature and pressure range, although they noted a clear positive effect of pressure at 700 °C. This issue needs further exploration.

7.4.8 Reactor Type

The reactors used so far for SCWG research have been either batch or continuous (flow). Depending on their type of mixing, they can be further divided as follows:

- Autoclave
- Tubular steel
- Stirred tank
- Quartz capillary tube
- Fluidized bed

A batch reactor is simple, does not require a high-pressure pump and can be used for almost all biomass feedstock. However, its reaction processes are not isothermal and it needs time to heat up and cool down. During heat-up many reactions occur that cause transformation of the feedstock; this does not happen in a continuous-flow reactor.

Reactor type has an important effect on the influence of feed concentration. The drop in gasification efficiency with feed concentration, noted in tubular reactors, was not found in the stirred-tank reactor studied by Matsumura et al. (2005). However, the reactor used was exceptionally small (1.0 mm in diameter), so validation of this finding in a reasonably large reactor (Matsumura and Minowa, 2004) is necessary. The process development of SCW gasifiers is lagging laboratory research because of engineering difficulties and the high cost of pilot plant construction.

7.5 APPLICATION OF BIOMASS CONVERSION IN SCWG

Three major areas of application for biomass SCWG are: (1) energy conversion, (2) waste remediation, and (3) chemical production.

7.5.1 Energy Conversion

All three of the followng important feedstocks for the energy industry can be produced by biomass conversion in supercritical water:

- *Bio-oil:* Potential use in the transport sector
- *Methanol:* Though a chemical feedstock, may be used for combustion
- *Hydrogen:* Potential use in fuel cells

The overall efficiency of an energy conversion system depends on the technology route, on the wetness of the biomass, and on many other factors. Yoshida et al. (2003) compared the effect of moisture content on the net efficiency of seven options for electricity generation, including an SCWG combined cycle. Interestingly, the SCWG-based system shows a total efficiency independent of moisture content, while for all other systems, total efficiency decreases with increasing moisture. Total electricity generation efficiency is even higher than that for conventional combustion-based systems. Integrated gasification combined cycle (IGCC) efficiency is higher than that of SCWG for biomass containing less than 40% moisture. Above 40%, its efficiency drops below that of SCWG (Figure 7.9).

Yoshida et al. (2003) also compared the total heat utilization efficiency of seven energy conversion processes:

- Direct combustion of biomass
- Combustion of biomass-oil produced by liquefaction or pyrolysis

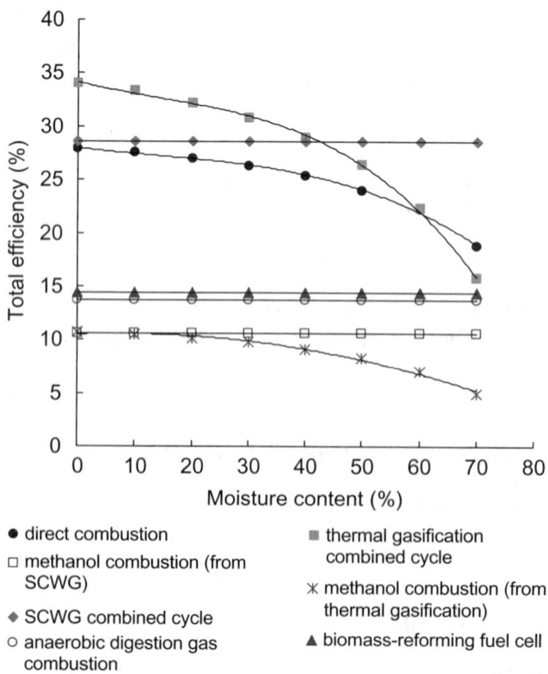

FIGURE 7.9 Dependence of net electricity generation efficiency of different biomass-based processes on biomass moisture content. *(Source: Adapted from Yoshida et al., 2003.)*

- Combustion of methanol produced by thermal gasification
- Combustion of methanol produced by SCWG
- Combustion of biogas produced by thermal gasification
- Combustion of methanol produced by supercritical water gasification
- Anaerobic digestion

Supercritical water gasification has the distinction of easily separating CO_2 from the product gas. This makes it an optimal technology for generation of electricity and heat from biomass when CO_2 emission limits become binding.

Fuel cells have the highest energy conversion efficiency for electricity generation, but they need hydrogen as their fuel. For hydrogen production, from very wet biomass, SCW gasification could be an attractive route. However, the capital costs of a fuel cell and that of a gasification plant have an important bearing on the economic viability of this generation option.

7.5.2 Waste Remediation

Waste treatment is another SCWG application. As explained in Section 7.3.3, in supercritical water even highly toxic wastes can be oxidized to harmless disposable residues. The agricultural industry produces large volumes of non-toxic but unhealthy products such as animal extracts and farm wastes that need to be disposed of productively. Many of these contain so much moisture that economical combustion or thermal gasification is not possible. Anaerobic digestion is a widely used alternative, especially in developing countries for production of useful gas (mostly methane) from animal extracts. Along with methane, anaerobic digestion produces fermentation sludge, which can be used as fertilizer.

Nevertheless, anaerobic gasification is orders of magnitude slower than thermal and other gasification processes, even with the use of catalysts. As a result, this makes large-scale commercial operation of anaerobic digesters difficult. Furthermore, the attractiveness of this method depends on the price of fertilizer, which can vary as a result of over- or undersupply in the market (Matsumura, 2002).

SCWG or SCWO is an alternative suitable for waste treatment because it does not depend on the production of sludge and is much faster than anaerobic digestion. Matsumura (2002) noted that supercritical water gasification has better energy efficiency, cheaper gas production, and faster CO_2 payback time (64.8%, 3.05 yen/MJ, and 4.19 years, respectively) in comparison with biomethanation (49.3%, 3.74 yen/MJ, and 5.05 years, respectively).

7.5.3 Chemical Production

Solvents are an important component of many chemical reactions. SCW acts as a solvent, but can also be a reactant and/or a catalyst. Ordinary subcritical water is popular as a solvent for reactions, especially because it is inexpensive and easily disposed of. Many organics, however, do not react efficiently in it. For these reactions, acid or base solvents are needed, which are good for synthesis reactions but, unless they can be efficiently recovered, are expensive and hazardous to dispose of.

Owing to its unique properties, SCW can act as a solvent for some reactions. Based on their studies of the following reactions Krammer et al. (1999) noted that many hydration, dehydration, as well as hydrolysis reactions can take place in supercritical water with good selectivity and high space/time yield, with no acids or bases as support materials.

- Dehydration of 1,4-butandiol and glycerine
- Hydrolysis of ether acetate, acetonitrile, and acetamide
- Reaction of acetone cayano hydrine

Production of useful chemicals from biomass is another use for SCW gasification. During its degradation in SCW, biomass produces phenols. Phenol production increases with feed concentration (Kruse et al., 2003). Because phenol is an important feedstock for the green resin, wood composite, and laminate industries, SCW provides an effective medium for green chemistry.

7.6 REACTION KINETICS

Limited information is available on the global kinetics of SCW gasification. Lee et al. (2002) studied the kinetics of glucose (used as the model biomass) in SCWG with a plug-flow reactor.

$$C_6H_{12}O_6 + 6H_2O = 6CO_2 + 12H_2 \tag{7.5}$$

We define the reaction rate, r, as the depletion of the biomass carbon fraction, C, with time. Assuming pseudo-first-order kinetics, we can write

$$r = -\frac{dC}{d\tau} = k_g C \tag{7.6}$$

where k_g is the reaction rate constant.

The fraction of carbon converted into gas, X_c, may be related to the current carbon fraction, C, and the initial carbon fraction, C_0, in the fuel:

$$X_c = 1 - \frac{C}{C_0} \tag{7.7}$$

Now replacing the carbon fraction in Eq. (7.6) and integrating we get:

$$k_g = -\frac{\ln(1 - X_c)}{\tau} \tag{7.8}$$

Table 7.4 presents some data on the global kinetics for SCWG of model compounds. The rates measured by Metanant et al. (2009), Lee et al. (2002), and

TABLE 7.4 Global Kinetic Gasification Rate for Model Compound in Supercritical Water

	Blasi et al., 2007	Metanant et al., 2009	Lee et al., 2002	Kabeymela et al., 1997
Feed	Wastewater from wood gasifier/TOC	Rice husk	Glucose/COD	Glucose
Reactor	Plug	Batch	Plug	Plug
Temperature (K)	723–821	673–873	740–1023	573–673
Residence time (s)	60–120	3600	16–50	0.02–2
Solid content	7.0–1.0 gm/l	9.4 mol/L	0.6 mol/l	0.007 mol/l
Pre-exponential factor, A (s^{-1})	1018 ± 494	184	897 ± 29	
Activation energy, E (kJ/mol)	75.7 ± 22	77.4	71 ± 3.9	96
k_g (s^{-1})		0.0002–0.006	0.01–0.55	0.15–9.9

Kabyemela et al. (1997) show how the reaction rate decreases with increasing solid carbon in the feed.

7.7 REACTOR DESIGN

Because SCWG is likely to enter the market once its major development barriers are removed this section discusses important considerations for design of an SCWG reactor. The discussion is based on limited information available in laboratory units and on the design of thermal gasifiers (see Chapter 6).

The major design parameters for an SCWG reactor are temperature, residence time, pressure, catalysts, and feed concentration. Important design considerations for auxiliary or support equipment are (1) waste heat-recovery exchanger and feed-preheating system, (2) the biomass feed system, and (3) product separation. The following subsections present a brief discussion of some of the design parameters.

7.7.1 Reactor Temperature

The temperature and pressure of an SCWG must be above the critical value of 374.21 °C and 22.089 MPa, respectively. As explained in Section 7.4.7, pressure has a minor effect on biomass conversion, but the effect of gasification temperature is a major one (see also Section 7.4.1).

Because feedstock (biomass and water) must be heated to the reaction temperature using energy from an external source, the lower the designed reactor temperature, the lower the energy required for feed preheat and the more efficient the process. The gasification temperature should be above 600 °C for a reasonable hydrogen yield, but it can be lower if catalysts are used.

For synthetic natural gas (SNG) production, high methane and low hydrogen are required; therefore, we can choose a reaction temperature of 350 to 500 °C, but catalysts are necessary for a reasonable yield. With catalysts, methane-rich gas may be produced even just below the critical temperature (~350 °C) (Mozaffarian et al., 2004).

7.7.2 Catalyst Selection

The choice of catalyst influences reactor temperature, product distribution, and plugging potential. Section 7.4.2 discussed the catalysts used in SCW gasification. They are selected on the basis of the desired product. Catalyst deactivation is an issue assigned to most catalyzed reactions because the deactivated catalysts must be regenerated. If they are deactivated because of carbon deposits, as happens in a fluid catalytic cracker (FCC), they can be combusted by adding oxygen, preferably in a separate chamber. The combustion reaction reactivates the catalysts and can additionally provide enough heat for preheating the feed.

7.7.3 Reactor Size

Consider a simple reactor receiving W_f of feed while producing W_p of product per unit of time. The product comprises a number of hydrocarbon components represented by species i. The total carbon in the product gas is its total in the individual gaseous hydrocarbons:

$$\text{Total carbon production in the product gas} = \sum W_p C_i \alpha_i \quad \text{kmol/s} \tag{7.9}$$

where $\alpha_{i\tau}$ is the number of carbon atoms in component i in the gas product; C_i is mole fraction of i in the gas product; and W_p is the product gas flow rate (kmol/s). The amount of carbon in the feed is known from the feed rate, W_f (kg/s), and its carbon fraction, F_c. The carbon gasification yield, Y, is defined as the ratio of gasified carbon to the carbon in the feed:

$$Y = \frac{\sum_i 12 W_p \alpha_i C_i}{W_f F_c} \tag{7.10}$$

where 12 is the carbon's molecular weight (kg/kmol).

From Eq. (7.8) the reaction rate is given in terms of conversion as

$$k_g = -\frac{\ln(1 - X_c)}{\tau} \tag{7.11}$$

where τ is the residence time in a reactor of volume V.

For a continuous stirred-tank reactor,

$$\tau = \frac{V}{\text{Volume flow rate of feed at reactor condition}} \quad \text{s} \tag{7.12}$$

Thus, for a known reaction rate, k_g, and a desired conversion, X_c, we can estimate the reactor volume required for gasification.

7.7.4 Heat-Recovery Heat-Exchanger Design

A feedstock preheater is the second most important part of an SCW gasifier system. The heat required to preheat the feedstock (water and biomass) is a significant fraction of the potential heating value of the product gas. Without efficient recovery of heat from the product gas, the external energy needed for gasification may exceed the energy produced, making the gasifier a net energy consumer. The feedstock should therefore obtain as much of its enthalpy as possible from the sensible heat of the product. This is one of the most important aspects of SCW plant design.

Figure 7.10 compares the capital costs of different components of an SCWG plant. We can see that the heat-recovery exchanger represents 50 to 60% of the total capital cost of the plant, which makes it a critical component.

Efficient heat exchange between the feed and the product is the primary goal of an SCWG heat-recovery system. However, for supercritical water intended for hazardous waste reduction (SCWO) or synthesis reaction (SCWS), it may not be all that important since the primary purpose of these systems is the production of chemicals, not energy as in a supercritical gasifier.

The heat-exchange efficiency, η, defines how much of the available heat in the product stream can be picked up by the feed stream.

$$\eta = \frac{H_{product\text{-}out} - H_{product\text{-}in}}{H_{feed\text{-}in} - H_{product\text{-}in}} \tag{7.13}$$

where H is the enthalpy, and the subscripts define the liquid it refers to.

Theoretically, the heat-exchange efficiency can be 100% if no heat of vaporization is required to heat the feed and an infinite heat-exchange surface area is available. Of course, these conditions are not possible. Figure 7.11 shows variations in heat-exchange efficiency with changes in tube surface area and water pressure.

The specific heat of water rises sharply close to its critical point and then drops equally sharply as the temperature increases (Figure 7.2). Thus, around

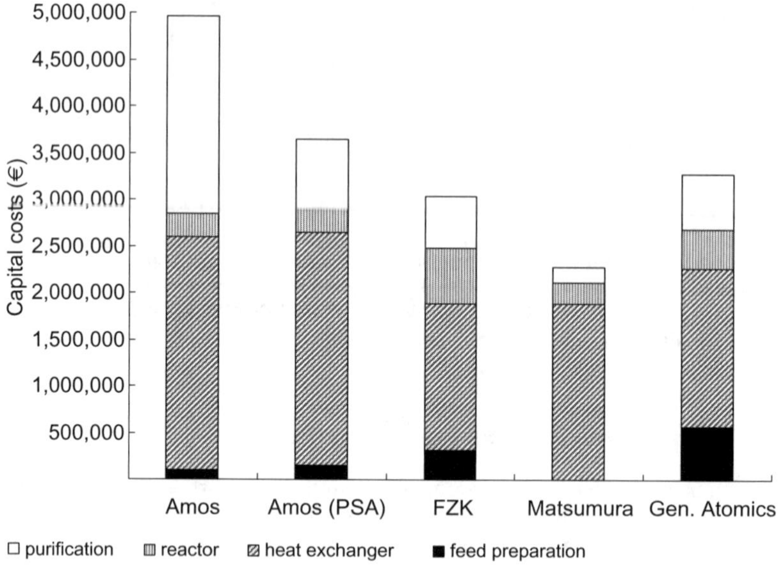

FIGURE 7.10 Investment cost of different SCW plant designs based on a throughput of 5000 kg/h of sewage sludge. *(Source: Adapted from Gasafi et al., 2008.)*

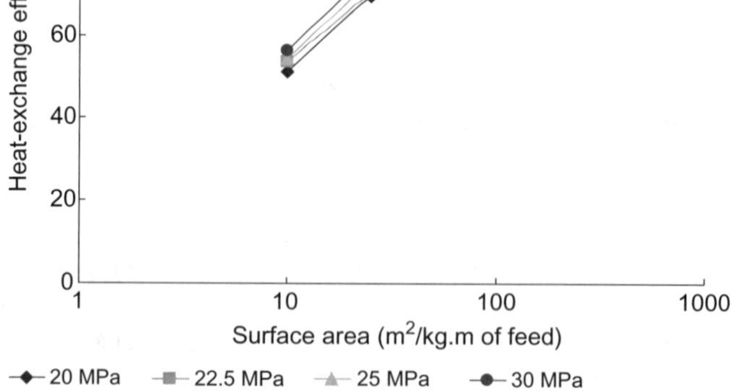

FIGURE 7.11 Calculated heat-exchange efficiency for water–water countercurrent heat exchanger at different pressures when feed and product are entered at 24 °C and 600 °C, respectively. *(Source: Data taken from Knoef, 2005.)*

TABLE 7.5 Sample Data for VERENA Pilot Plant Product-to-Feed Heat Exchanger

Flow Rate (kg/h) (Methanol %)	Product in (°C)	Product Out (°C)	Feed In (°C)	Feed Out (°C)	Reactor Temperature (°C)
100 (10%)	561	168	26	405	582
90 (20%)	524	155	22	388	537

Note: Heat-exchanger surface area: 1.1 m^2; heat-transfer coefficient: 920 W/m^2C (Boukis et al., 2005).

the critical point we may expect a modest temperature rise along the heat-exchanger length.

Thermal conductivity in SCW is lower than that in subcritical water because SCW's intermolecular space is greater than that in liquid. A slight increase in conductivity is noticed as the fluid approaches the critical point. This increase is due to an increase in the agitation of molecules when the change from a liquidlike to a gaslike state (SCW) takes place. Above the critical point, thermal conductivity decreases rapidly with temperature.

The heat-transfer coefficient varies with temperature near its pseudo-critical value (see Section 7.2) because of variations in the thermophysical properties of water. As the temperature approaches the pseudo-critical value, conductivity and viscosity decrease but specific heat increases. The drop in viscosity and the peak of specific heat at the pseudo-critical temperature overcome the effect of decreased thermal conductivity so as to increase the overall heat-transfer rate.

As the temperature further increases, beyond the pseudo-critical point, the specific heat decreases sharply; the drop in thermal conductivity continues as well, and therefore the heat-transfer coefficient reduces. For a given heat flux, the wall temperature rises for the drop in heat-transfer coefficient. Generally, for high heat flux and low mass flux, the heat transfer deteriorates, leading to hot spots in the tube.

Heat Transfer in Supercritical Water

Table 7.5 illustrates the operation of a typical heat-recovery exchanger for supercritical water gasification. The data are taken from a large operating near-supercritical gasification plant. The fluid-to-wall heat-transfer coefficient in clean supercritical water in the tube may be calculated by the correlation of Yamagata et al. (1972):

$$\mathrm{Nu} = 0.0135\,\mathrm{Re}_b^{\,0.85}\,\mathrm{Pr}_b^{\,0.8} \tag{7.14}$$

Based on Yamagata's experiments with isobutane, Hsu (1979) found that the Sieder-Tate equation, as follows, is in better agreement:

$$\mathrm{Nu} = 0.027\,\mathrm{Re}^{0.8}\,\mathrm{Pr}^{0.33}(\mu_b/\mu_b)^{0.14} \tag{7.15}$$

Heat transfer in SCWG may vary because of solids in the fluid. Thus, applicability of these equations to SCWG is uncertain. Information on this aspect of heat transfer is presently unavailable.

7.7.5 Carbon Combustion System

Because gasification and pyrolysis reactions are endothermic, heat from some external source is required for operation of the reactor. In thermal gasification systems, the reaction temperature is very high (800–1000 °C), so a large amount of energy is required for production of fuel gases from biomass or other feedstock. This heat is generally provided by allowing part of the hydrocarbon or carbon in the feed to combust in the gasifier, but then a part of the energy in the feedstock is lost.

A SCW gasifier operates at a much lower (450–650 °C) temperature and thus requires a much lower but finite amount of heat. Thermodynamically, the heat recovered from the gasification product is inadequate to raise the feed to the gasification temperature (450–600 °C) and provide the required reaction heat. This shortfall is made up either by an external source or by combustion of part of the product gas in a heater.

Both options are expensive. For example, a study of an SCWG design for gasification of 120 t/day (5000 kg/h) of sewage sludge with 80% water showed that 122 kg/h of natural gas is required to provide the gasification heat. This, along with an electricity consumption of 541 kW, constitutes 23% of the total revenue requirement for the plant (Gasafi et al., 2008). A better alternative would be controlled combustion of the unconverted char upstream of the gasifier, which would make SCWG energy self-sufficient.

Although SCWG is known for its low char and tar production, in practice we expect some char formation. Furthermore, as shown previously in Figure 7.7, gasification efficiency is low at lower temperatures. A low gasification temperature is thermodynamically more efficient, but raises the char yield. If this char can be combusted in SCW, it can provide the extra heat needed for preheating the feed, thereby improving the efficiency of the overall system.

Combustion of char offers an additional benefit for an SCWG that sometimes uses solid catalysts, which are deactivated after being coated with unconverted char in the gasifier. A combustor can burn the deposited carbon and regenerate the catalyst. The generated heat is carried to the gasifier by both solid catalysts and the gasifying medium (SCW and CO_2).

Recycling of solid catalysts is an issue for plug-flow reactors. Special devices such as fluidized beds may be used for these, as shown in Figure 7.12. Here, the catalysts or their supports are granular solids, which are separated from the product fluid leaving the reactor in a hydrocyclone operating in an SCW state. The separated solids drop into a bubbling fluidized-bed combustor,

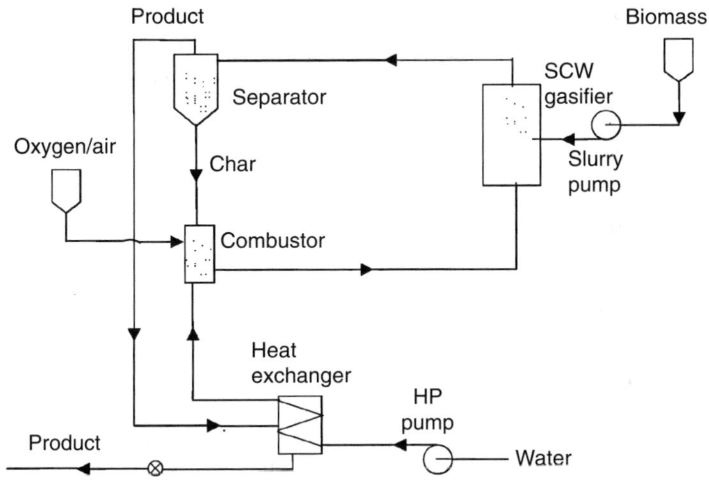

FIGURE 7.12 Conceptual system for combustion of residual carbon deposited on solid catalysts to provide heat for SCW biomass gasification.

where oxygen or air is injected to facilitate burning of the deposited carbon. The bed is fluidized by pressurized water already heated above its critical temperature in a heat-recovery heat exchanger.

Under supercritical conditions, oxidation or combustion reactions occur in a homogeneous phase where carbon is converted to carbon dioxide.

$$C + O_2 = CO_2 - 393.8\ \text{kJ/mol} \tag{7.16}$$

Because these reactions are exothermic, the process can become thermally self-sustaining with the appropriate concentration of oxygen. Heated water from the combustor carries the regenerated catalysts to the gasification reactor, into which the biomass is fed directly.

Under supercritical conditions, water acts as a nonpolar solvent. As a result, the supercritical water fully dissolves oxygen gas. The mass transfer barrier that is between dissolved oxygen and solid char may be lower than that between gas and char. This, along with its high-density feature, may allow the SCW to conduct the combustion reaction quickly and efficiently. Another advantage of low-temperature combustion is that it avoids formation of toxic by-products.

7.7.6 Design of Gas–Liquid Separator System

In an SCWG system, the product gas mixture is separated from water in two stages. In the first stage, initial separation takes place in a high-pressure but low-temperature separator. In the second stage, final separation occurs under low pressure and low temperature.

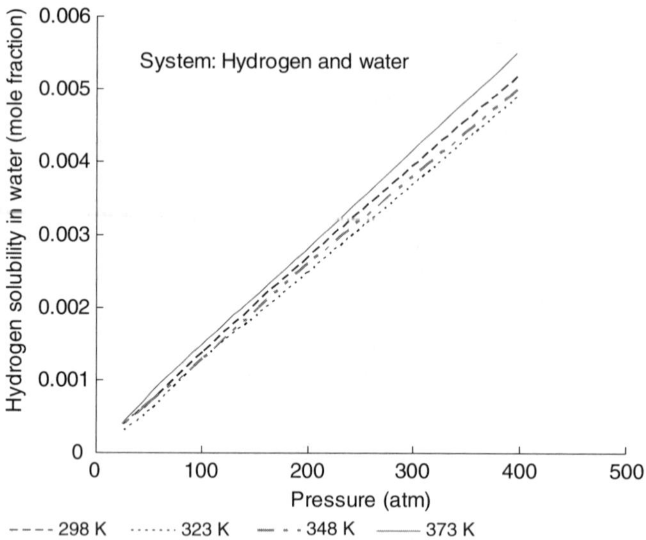

FIGURE 7.13 Hydrogen solubility in water. *(Source: Adapted from Ji et al., 2006.)*

At low temperatures (25–100 °C), hydrogen or methane has very low solubility (0.001–0.006) in water, even at high pressure (Figure 7.13). So the bulk of the hydrogen is separated from the water when cooled. Figure 7.14 shows one such scheme where S1 is the hydrogen separator. Other gases like CO_2 are also separated from the water but to a limited extent. As we can see from Figure 7.15, the solubility of carbon dioxide is an order of magnitude higher (0.01–0.03) that of hydrogen at this low temperature and high pressure.

This feature can be exploited to separate the hydrogen from the carbon dioxide, but the CO_2's equilibrium concentration may not be sufficient to dissolve it entirely in the high-pressure water. Additional water may be necessary to dissolve all of these gases except hydrogen so that the hydrogen alone remains in the gas phase (S1, Figure 7.14). The equilibrium concentration of these gases in water can be calculated from the equation of state, such as Peng Robinson or SAFT.

The liquid mixture is next depressurized through a pressure regulator before it enters the second separator (S2, Figure 7.14). The solubility of most gases reduces with a decrease in pressure, so the second unit separates the rest of the CO_2 from the gas.

Feng et al. (2004a,b) calculated the phase equilibrium of different gases in water for a plant using different relations. Values calculated using SAFT equilibrium showed the best agreement with experimental results. These results are shown in Table 7.6 to illustrate the process. It is apparent that at 25 °C the solubility of CO_2 is orders of magnitude higher than that of methane and hydrogen. The solubility of methane and hydrogen is similar at nearly all pressures. For

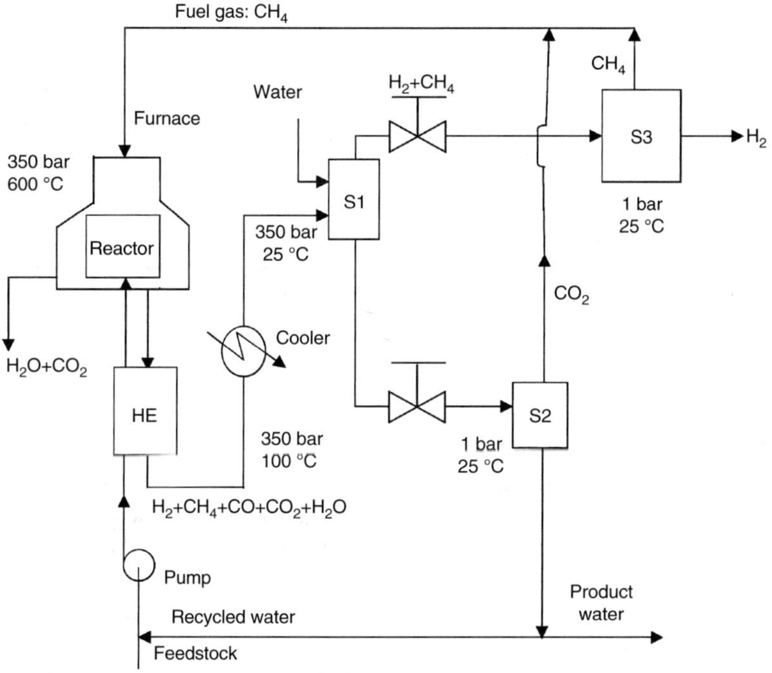

FIGURE 7.14 Gas–liquid separation scheme for an SCW gasifier plant. HE is the waste heat-recovery heat exchanger; S1 is the hydrogen separator; S2 is the carbon dioxide separator; S3 is the pressure swing adsorber for separation of methane from hydrogen.

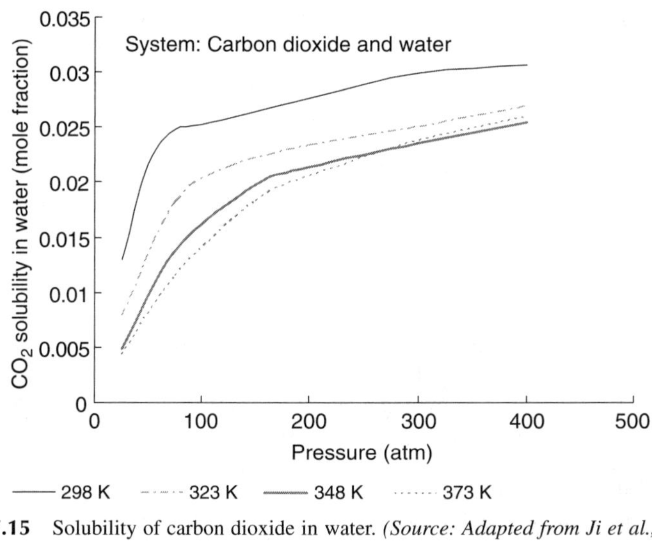

FIGURE 7.15 Solubility of carbon dioxide in water. *(Source: Adapted from Ji et al., 2006.)*

TABLE 7.6 Solubility of Three Gases in Water at 25 °C and Various Pressures

Pressure (bar)	60	90	120	140	200	300	400	600	1000
CH_4 ($cm^3/g\ H_2O$)	1.8	2.34	2.9	3.3					
H_2($cm^3/g\ H_2O$)	1.0	1.5	2.0	2.1	3.0	4.5	7.9	9.0	15
CO_2($cm^3/g\ H_2O$)	27		32		33		39		

Source: Collected from experimental and calculated values of Feng et al. (2004a,b).

their separation, then, it is necessary to use a system such as a pressure swing adsorber (S3), as shown in Figure 7.14.

An important consideration is the additional water required to keep the carbon dioxide dissolved while the hydrogen is being separated. The amount, which may be considerable, can be expressed as the ratio of water to gaseous product (R) on a weight basis. When pressure and R increase, the purification of hydrogen increases but the amount of hydrogen in the gas phase decreases. Therefore, we can recover more hydrogen with less purity or less hydrogen with more purity. This depends on an adjustment of the pressure and R. Example 7.1 illustrates the computation.

Example 7.1

Design a separator to produce 79% pure hydrogen from an SCWG operating at 250 bars of pressure. Assume the following overall gasification equation, which produces hydrogen, methane, carbon dioxide, and carbon monoxide.

$$C_6H_{10}O_5 + 4.5H_2O = 4.5CO_2 + 7.5H_2 + CH_4 + 0.5CO$$

Solution

We use the carbon dioxide solubility curve in Figure 7.15 to design the separator. Here, at 250 bars of pressure and 25 °C, we find the solubility of CO_2 to be 0.028 mole fraction. This implies that 1 mol of water is needed to dissolve 0.028 mol of carbon dioxide.

To separate gaseous hydrogen from liquid water, we reduce the ambient temperature to 25 °C. From Figure 7.13 we find that the hydrogen solubility is only 0.0031 at 250 bars and 25 °C, so (1 – 0.0031) or 0.9969 fraction of hydrogen produced will be in the gas phase here. The gas may, however, contain other gases, so to ensure that the hydrogen is 79% pure, we need to add water to the separator. If we know the operating temperature, pressure, and weight ratio of the water to the gas mixture, the amount of product in the liquid and vapor phases can be calculated according to an equation of state. Here we use Figure 7.16 computed by the Peng-Robinson equation. For 250 bars of pressure and a mole

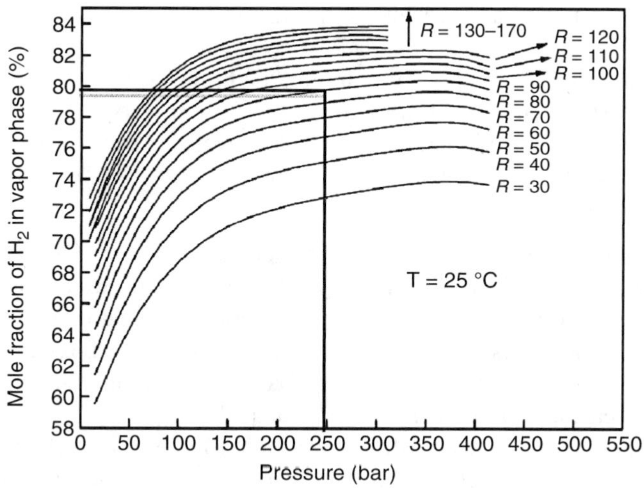

FIGURE 7.16 Equilibrium mole fraction of hydrogen in the gas phase for different water-to-product gas ratios. *(Source: Adapted from Ji et al., 2006.)*

fraction of 79% in the gas phase, we get R = 80. Thus, the amount of water required is 80 × (the mass of product gas).

From the overall gas equation, the mass of product gas is 4.5 × 44 + 7.5 × 2 + 1 × 16 + 0.5 × 28 = 243 g/mol of biomass. The mass of water is 80 × 243 = 19,440 g = 19.4 kg.

From the property table of water, we get the density of water at 25 °C and 250 bars, which is 1008.5 kg/m³. The volume of water is 19.4kg/1008.5 kg/m³ = 0.0192 m³.

$$\begin{aligned}\text{Volume of product gas} &= \Sigma(nRT/P)\\ &= (4.5+7.5+1+0.5)\times 10^{-3}\ \text{kmol}\times\\ &\quad(8.314\ \text{kPa.m}^3\text{.kmol}^{-1}\text{.K}^{-1}\times 298\ \text{K})/(250\times 10^2\ \text{kPa})\\ &= 0.00134\ \text{m}^3\end{aligned}$$

Therefore, the total volume of biomass that is gasified is 0.0192 + 0.00134 = 0.0205 m³/mol.

7.7.7 Biomass Feed System

The feeding of biomass into a high-pressure (>22 MPa) reactor is a formidable challenge for an SCW gasifier. If the feed is a dilute stream of organics, the problem is not so severe, as pumps can handle light slurries. However, if it is fibrous solid granular biomass that needs to be pumped against high pressure, the problem is especially difficult for the reasons that follow:

- The irregular size and the low shape factor of biomass makes it difficult to flow.
- Pulverization is necessary for pumping the biomass, but it is very difficult to pulverize. Pretreatment of the feedstock is necessary.
- Fibrous by nature, biomass does not flow well through an augur or gear pump, and it is difficult to make a uniform slurry for pumping through impellers.

Most of the research work on SCWG generally used model water-soluble biomass such as glucose, digested sewage sludge, and wastewater (Blasi, 2007), which are easy to pump. For other types of biomass, Antal et al. (2000) used additives or emulsifiers such as corn starch gel, sodium carboxymethyl cellulose (CMC), and xanthan to make pumpable slurries. In an industrial application, large-scale use of emulsifiers is impractical.

A sludge pump was successfully used in a 100-kg/h pilot plant; however, the solids had to be ground to less than 1-mm particles and pretreated before pumping. Even then grass and fibrous materials clogged the membrane pump's vents (Boukis et al., 2006). Cement pumps have been suggested but, to date, have not been tried for pumping biomass in an SCW gasifier (Knoef, 2005).

Another important problem is plugging of the feed line during the preheating stage, in which the feed being heated can start breaking down. Char and other intermediate products can deposit on the tube walls, blocking the passage and thereby creating a dangerous situation.

Carbon buildup on the reactor wall has an adverse effect. It reduces the gas yield when the reactor is made of metals that have catalytic effects, although it is not associated with the feed system. Lu et al. (2006) showed that gas yields, gasification efficiency, and carbon efficiency are reduced by 3.25 mol/kg, 20.35%, and 17.39%, respectively, when carbon builds up on the reactor wall compared to when the reactor is clean. Similar results were found by Antal et al. (2000).

7.8 CORROSION

In an SCWG or SCWO, where the temperature can go as high as 600 °C and the pressure can be in excess of 22.089 MPa, water becomes highly corrosive. SCWG and SCWO plants work with organic compounds, which react with oxygen in supercritical water oxidation to produce mostly CO_2 and H_2O, or hydrolyze in SCWG. Halogen, sulfur, and phosphorous in the feed are converted into mineral acids such as HCl, H_2SO_4, or H_3PO_4. High-temperature water containing these acids along with oxygen is extremely corrosive to stainless steels and nickel-chromium alloys (Friedrich et al., 1999).

After oxidation of neutral or acidic feeds, the pH of SCWO solutions is low, making it as corrosive as hydrochloric acid (Boukis et al., 2001). Chlorine is especially corrosive in SCW. Interestingly, a supercritical steam boiler, which is one of the most common uses of supercritical water, is relatively free from corrosion because the water used in the boiler is well treated and contains no corrosive species such as salts and oxygen or only very low concentrations.

The following sections briefly describe the mechanism and the prevention of corrosion in biomass SCWG plants. More details are available in reviews presented by Kritz (2004) and Marrone and Hong (2008).

7.8.1 Mechanism of Corrosion

Metal surfaces are generally protected by a oxide layer that forms on them and guards against further attack from corrosive elements. This protective layer can be destroyed through chemical or electrochemical dissolution.

In chemical dissolution, the protective layer is removed by a chemical process using either an acidic or an alkaline solution depending on the pH valuc in the local region. In electrochemical dissolution, depending on the electrochemical potential, the metal can undergo either transpassive or active dissolution. All forms of electrochemical corrosion require the presence of aggressive ionic species (as reactants, products, or both), which in turn requires the existence of an aqueous environment capable of stabilizing them.

Stainless and nickel-chromium alloys experience high corrosion rates at supercritical pressure but subcritical temperatures because of transpassive dissolution (Friedrich et al., 1999), where the nickel or iron cannot form a stable insoluble oxide that protects the alloy. Under supercritical conditions, the acids are not dissociated and ionic corrosion products cannot be dissolved by the solution because of the solvent's low polarity. Consequently, corrosion drops down to low values.

Electrochemical corrosion requires the presence of ionic species like halides, nickel-based alloys, and compounds. These show high corrosion rates, which decrease at higher temperatures. High-pressure water in an SCW reactor provides favorable conditions for this, but once the water enters the supercritical domain the solubility and concentration of ionic species in it decrease, although the reaction rate continues to be higher because of higher temperatures. The total corrosion reduces because of decreased concentration of the reacting species. Thus, corrosion in a plant increases with temperature, reaching a peak just below the critical temperature, and then reduces when the temperature is supercritical. The corrosion rate increases downstream, where the temperature drops into the subcritical region.

At a relatively low supercritical pressure (e.g., 25 MPa), the salt NaCl is not soluble. Thus, in an SCW a reaction that produces NaCl, the salt can precipitate on the reactor wall. Sometimes water and brine trapped between the salt deposit and the metal can create a local condition substantially different from conditions in the rest of the reactor in terms of corrosion. This is known as *underdeposit corrosion*.

In general, a reaction environment that is characterized by high density, high temperature, and high ion concentration (e.g., acidic) is most conducive to corrosion in an SCW reactor. Rather than the severity of corrosion in terms of whether the flow is supercritical or subcritical, the density of the water should be the major concern.

7.8.2 Prevention of Corrosion

According to Marrone and Hong (2008), corrosion prevention in a supercritical water unit is broadly classified in these four ways: (1) contact avoidance, (2) corrosion-resistant barriers, (3) process adjustments, and (4) corrosion-resistant materials.

Contact Avoidance

The following are some innovative options that may be used to reduce contact between corroding species and the reactor wall:

- A transpiring wall on which water constantly washes down, preventing any corroding material's contact with the wall surface.
- A centrifugal motion created in the reactor to keep lighter reacting fluids away from the wall.
- In a fluidized bed, neutralizing or retaining of the corrosive species by the fluidized particles.

Corrosion-Resistant Barriers

Corrosion-resistant liners are used inside the reactor to protect the vessel wall. These are required to withstand the reactor's high temperature but not its high pressure. Titanium is corrosion resistant, but in large quantities, such as required for the reactor shell, it is not recommended because of the risk of fire if it comes in contact with high concentrations of oxidant, particularly when pure oxygen is used in an SCWO. In much smaller quantities, titanium can be as a liner; alternatively, some type of sacrificial liner can be used.

Process Adjustments

Changes in process conditions may reduce or even avoid corrosion in some cases, but they may not be practical in many situations. For example, if the corrosion is as a result of acidic reaction, the addition of a base to the feed may preneutralize the reactant. Since most of the corrosion occurs just below critical temperature, the water without the feed may be preheated to a sufficiently high temperature such that on mixing with the cold feed the reaction zone quickly reaches the design reactor's temperature; then the biomass may be fed directly into the reactor to reduce the corrosion in the feed preheat section.

Corrosion-Resistant Materials

If corrosion cannot be avoided altogether, it can be reduced by the use of highly corrosion-resistant materials. Choosing one of these as the primary construction material in an SCWO system is the simplest and most basic means of corrosion control. The following materials have been tried in supercritical environments. Of course, no single material can meet all design requirements, so some optimization is required. The materials listed are arranged in the order of least-to-most corrosion resistant.

- Stainless steel
- Nickel-based alloys
- Titanium
- Tantalum
- Niobium
- Ceramics

7.9 ENERGY CONVERSION EFFICIENCY

Matsumura (2002) estimated the energy required for SCW gasification of water hyacinth. His analysis came up with a high overall efficiency. Gasafi et al. (2008) carried out a similar analysis for sewage sludge that came up with a much lower efficiency. The energy consumption of these two biomass types is compared in Table 7.7. We note that the energy required to pump and preheat the feed is a substantial fraction of the energy produced in a supercritical water plant.

Overall efficiency may depend on the type of feedstock used. Yoshida et al. (2003) studied options for electricity generation from biomass, including

TABLE 7.7 Energy Consumption in Gasification of Water Hyacinth and Sewage Sludge

	Matsumura et al. (2002)	Gasafi et al. (2008)
Feedstock	Water hyacinth	Sewage sludge
Potential energy in feed	4.44 MW	1.44 MW
Energy in product gas	3.32MW	1.38 MW
Electricity consumption in pumping and other equipment	0.54 MW	0.05 MW
External energy for feed preheating (MW)	1.69	0.33
Net energy production (MW)	1.09	0.99
Overall efficiency (%)	24.5	68.6

SCWG combined cycle, thermal gasification, and direct combustion. They concluded that the SCWG combined cycle offers the highest efficiency for high-moisture biomass, but it does not for low-moisture fuels.

7.10 MAJOR CHALLENGES

Commercialization of SCW biomass gasification must overcome the following major challenges:

- Supercritical water gasification requires a large heat input for its endothermic reactions and for maintenance of its moderately high reaction temperature. This heat requirement greatly reduces energy conversion efficiency unless most of the heat is recovered from the sensible heat of the reaction product. For this reason, the efficiency of the heat exchanger and its capital cost greatly affect the viability of supercritical water gasification.
- The feeding of wet solid biomass, which is fibrous and widely varying in composition, is another major challenge. A slurry pump has been used to feed solid slurry into high-pressure reactors, but it has not been tested for feeding biomass slurry into a supercritical reactor with ultra-high pressure.
- The drop in gasification efficiency and gas yield with an increase in dry solids in the feed may be a major obstacle to commercial SCW gasification. Efforts are being made to improve this ratio using different catalysts, but a cost-effective method has yet to be discovered.
- Separation of carbon dioxide from other gases may require the addition of large amounts of water at high pressure (see Section 7.7.6). This can greatly increase the system's cost and reduce its overall energy efficiency.
- The heating of biomass slurry in the heat exchanger and reactor is likely to cause fouling or plugging because of the tar and char produced during the preheating stage. Further research is required to address this important challenge. A final problem that might inhibit commercialization of SCW gasification is the corrosion of the reactor wall.

Symbols and Nomenclature

A = cross-sectional area (m^2)
C_i = mole fraction of the component i in the gas product
F_c = carbon fraction in feed
k_g = reaction rate (s^{-1})
H = enthalpy of products for product-out, product-in, and feed-in (kJ)
L = length of gasifier reactor (m)
Q' = volume flow rate through reactor (m^3/s)
V = volume of reactor (m^3)
W_p = product gas flow rate (kmol/s)
W_f = feed rate (kg/s)

X_c = carbon conversion fraction
Y = gasification yield
μ = viscosity (N.s/m^2)
μ_i = number of carbon atoms of component i in the gas product
η = heat exchange efficiency
τ = residence time (s)

Chapter 8

Biomass Handling

8.1 INTRODUCTION

Liquids and gases are relatively easy to handle because they continuously deform under shear stress—they easily take the shape of any vessel they are kept in and flow easily under gravity if they are heavier than air. For these reasons, storage, handling, and feeding of gases or liquids do not generally pose a major problem. On the other hand, solids can support shear stress and do not flow freely. This problem is most evident when they are stored in conical bins and are withdrawn from the bottom. Because they do not deform under shear stress, solids can form a bridge over the cone and cease to flow.

Biomass is particularly notorious in this respect, because of its fibrous nature and nonspherical shape. The exceptionally poor flow characteristic of biomass poses a formidable challenge to both designers and operators of biomass plants. The cause of many shutdowns in these plants incidents can be traced to the failure of some parts of the biomass-handling system.

This chapter describes the design and operating issues involved in the flow of biomass through the system. It discusses options for the handling and feeding of biomass in a gasification plant.

8.2 DESIGN OF A BIOMASS ENERGY SYSTEM

A typical biomass energy system comprises farming, collection, transportation, preparation, storage, feeding, and conversion. This is followed by transmission of the energy produced to the point of use. The concern here is with the handling of biomass upstream of the conversion system—that is, a gasifier in the present context. Biomass farming is a subject by itself and is beyond the scope of this chapter.

Biomass fuel can be procured from the following sources:

- Energy crop or forestry
- Ligno-cellulose wastes that are from forestry, agriculture, wood, or other industries
- Carbohydrates such as fat, oil, and other wastes

Biomass Gasification and Pyrolysis. DOI: 10.1016/B978-0-12-374988-8.00008-8

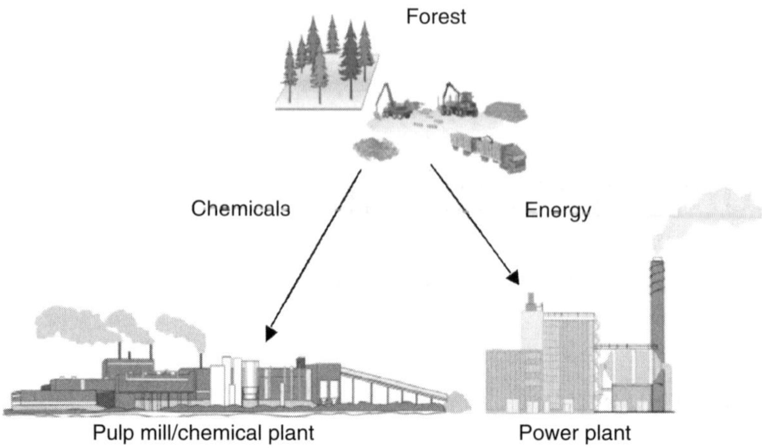

FIGURE 8.1 Biomass is used for the production of energy or for commercial products such as paper or chemicals.

Biomass has two major (Figure 8.1) applications: (1) energy production through gasification or combustion, and (2) production of chemicals and fiber-based items (e.g., paper).

The collection methods for biomass vary depending on its type and source. Forest residues are a typical ligno-cellulose biomass used in gasification plants. They are collected by various pieces of equipment and transported to the gasification plant by special trucks (or rail cars in some cases). There, the biomass is received, temporarily stored, and pretreated as needed. Sometimes the plant owner purchases prepared biomass to avoid the cost of onsite pretreatment. The treated biomass is placed in storage bins located in line with the feeder, which feeds it into the gasifier at the required rate.

Biomass typically contains only a small amount of ash, but it is often mixed with undesirable foreign materials. These materials require an elaborate system for separation. If the plant uses oxygen for gasification, it needs an air-separation unit for oxygen production. If it uses steam, a steam generator is necessary. Thus, a biomass plant could involve several auxiliary units. The capacity of each of these units and the selection of equipment depend on a large number of factors. These are beyond the scope of this chapter.

Forestry and agriculture are two major sources of biomass. In forestry, large trees are cut, logged, and transported to the market. The logging process involves delimbing, and taking out the large-diameter tree trunks as logs. The processes involved in biomass harvesting, such as delimbing, deburking, and chipping, produce a large amount of woody residue, all of which constitutes a major part of the forest residue. The entire operation involves chopping the tree into chips and then using those chips to make fuels or feedstock for pulp industries.

8.3 BIOMASS-HANDLING SYSTEM

A typical biomass gasification plant comprises a large number of process units, of which the biomass-handling unit is the most important. Unlike coal-fired boiler plants, an ash-handling system is not a major component of a biomass gasification plant because biomass contains a relatively small amount of ash. Normally it does not produce a large volume of spent catalysts or sorbents. Transportation, feed preparation, and feeding are more important for biomass than they are for coal- or oil–gas-fired units.

The biomass-handling system can be broadly classified into the following components:

- Receiving
- Storage and screening
- Feed preparation
- Conveying
- Feeding

The design of the handling system is very similar to that of a biomass-fired steam plant. Figure 8.2 illustrates the layout of a typical plant showing receiving, screening, storage, and conveying.

Major considerations for the design of feeding and handling systems are transportation, sealing, and injection. The feed should be transported smoothly

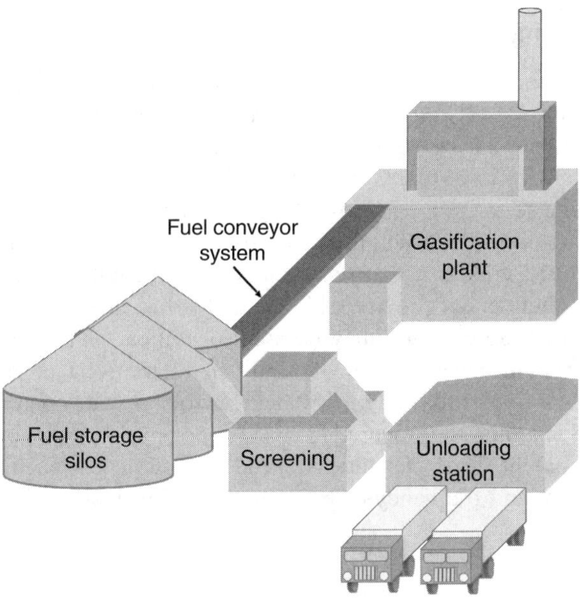

FIGURE 8.2 Plant layout for biomass gasification. The fuel, delivered by truck (or rail car), is cleaned of foreign materials before it is stored in silos.

FIGURE 8.3 Biomass delivery truck tilted to unload at the gasification plant. *(Source: Photograph by the author.)*

from the temporary storage to the feed system, which must be sealed against the gasifier's pressure and temperature. The fuel is then injected into the gasifier. Metering or measurement of the fuel feed rate is an important aspect of the feed system, as it controls the entire process.

The following subsections discuss the individual components of a solids-handling system for biomass. They assume the biomass to be solid, although some biomass, such as sewage sludge, is in slurry or semisolid form.

8.3.1 Receiving

Biomass is brought to the plant typically by truck or, sometimes, by rail car. For large biomass plants, unloading from the truck or rail car is a major task. Manual unloading can be strenuous and uneconomical except in very small plants. This is why large plants use truck hoisters, wagon tipplers, or bottom-discharge wagons. Figure 8.3 shows a typical system where a truck hoister unloads the biomass. The truck drives onto the hoist platform and is clamped down. The hoister tilts to a sharp angle, allowing the entire load to drop into the receiving chute under gravity. This method is fast and economical.

A bottom-discharge wagon may be used for rail cars. The wagon drops its load into a large bin located below the rail. An alternative is a standard open-top wagon and a tippler to rotate it 180 degrees to empty its contents into a bin underneath. Such units are procured from the suppliers of various bulk material-handling equipment. Their capacities depend on a number of factors, including plant throughput and frequency of truck and/or rail arrival.

8.3.2 Storage

The primary purpose of storage is to retain the biomass in a good condition and in a position convenient for easy transfer to the next stage of operation, such as drying or feeding into the gasifier. The stored biomass should be protected from rain, snow, and infiltration of groundwater.

Once unloaded, the biomass is moved by belt conveyers to the storage yard, where it is stored in piles according to usage patterns. If the biomass is from several sources and is to be mixed before use, the piles are arranged in such a way that they can be mixed conveniently into the desired proportions. Because of the large volume of biomass, indoor storage may not be economical. Open-air storage is most common, though it can cause absorption of additional moisture from rain or snow and produce dust pollution. Storage can be of two types: above ground, for large-volume biomass, or enclosure in a silo or bunker.

Figure 8.2 showed the general arrangement of the solids-handling system in a typical biomass-fired plant. A truck-receiving station unloads into an underground hopper from which a belt conveyor takes the biomass to a screening station. After removal of foreign materials, the biomass is crushed and screened to the desired size range and then transported into silos for covered storage. From there, it is taken to the plant as required. Figure 8.4 is a photograph of receiving, size-screening, and above-ground outdoor storage.

Underground bunker storage is very convenient and cost effective from a fuel delivery standpoint, as it protects the biomass from rain and snow. However, because it needs good ventilation and drainage for safety and environmental protection, its capital cost is higher than that of above-ground storage. The hygroscopic nature of biomass is a major issue, as it causes the prepared biomass to absorb moisture even if stored indoors. Moreover, long-term storage can cause physical and chemical changes in the biomass that might adversely

FIGURE 8.4 Biomass is conveyed to the storage pile; the scrapper collects it when needed and transfers it to conveyors that take it to the fuel preparation plant. *(Source: Photograph by the author.)*

affect its flow and gasification properties. For these reasons, it is desirable to occasionally turn the biomass. A simple and practical way of doing this is to draw it at a rate higher than that required and return the excess to the top of the pile.

Moving or retrieving the biomass from the storage piles to the gasifier plant requires careful design, because interruptions or delays can have a major effect on the operation of the plant. Generally, it is desirable to withdraw biomass from the bottom of the pile such that the *first in–first out* principle is followed to allow a relatively uniform shelf life.

The properties of the biomass determine the ease with which it is retrieved or handled. Oversized materials, frozen chunks (in cold countries), and compaction can lead to poor or interrupted fuel flow. If the fuel bin is not filled uniformly, erratic operation can result, creating problems for hydraulic scrapers and bridging over the unloaders. Sticks, wires, and gloves, for example, can jam augers. Mobile loaders normally achieve uniformity in above-ground storage buildings or in live-bottom unloaders and augers in bins and silos. For large plants, a scraper connected to a conveyor, as shown in Figure 8.4, is more efficient for reclamation.

The following are some common methods for retrieval of biomass from storage:

- Simple gravity feed or chute
- Screw-type auger feed
- Conveyor belt
- Pneumatic blower
- Pumped flow
- Bucket conveyor
- Frontloader
- Bucket grab

Walking beams are sometimes used on the floors of large bunkers or storage buildings to facilitate the movement of biomass to the discharge end of the storage.

Above-Ground Outdoor Storage

In large-scale plants, above-ground outdoor storage is the only option (Figure 8.4). Indoor storage is usually too expensive. Biomass needs to be piled in patterns that allow maximum flexibility in retrieval as well as in delivery. Furthermore, it is necessary to ensure the *first in–first out* principle. In some cases, an emergency or strategic reserve is kept separate from the regular flow of biomass. This is a special consideration for long-term storage.

Good ventilation is important in storage design. Biomass absorbs moisture. Ventilation prevents condensation of moisture and the formation of moulds that can pose serious health hazards. It also prevents composting (formation of methane), which not only reduces the energy content of the biomass, but also may run the risk of fire. Because tall storage piles are difficult to ventilate, the maximum height of a wood chip storage pile should not exceed 8 to 10 m (Biomass Energy Centre, 2009). For an indoor facility, water or moisture accumulation may occur inadvertently. Unless moved periodically, the biomass may form fungi and cause a health hazard. Drainage is an important issue, especially for outdoor storage.

Silos and Bins for Storage of Biomass

Improper storage not only makes retrieval difficult, it also can adversely affect the quality of the biomass. Retrieval or reclamation from storage is equally important, if not more so. It represents one of the most trouble-prone areas of biomass plant operation. The handling system and its individual components must be designed to ensure uninterrupted flow to the gasifier at a measured rate.

Bunkers, silos, and bins provide temporary storage in a protective environment. Bunkers are a type of large-scale storage. Although the term *bunker* is generally associated with underground shelter, here it refers to the indoor storage of fuel in power or process plants that is not necessarily underground. Silos could be fairly large in diameter (4–10 m) and are very tall, which is good for storing grain-type biomass. For example, Figure 8.5 shows a tower silo for cattle feed. Bins are for smaller-capacity temporary storage.

FIGURE 8.5 Typical grain silo for storing cattle feed. *(Source: Photograph by the author.)*

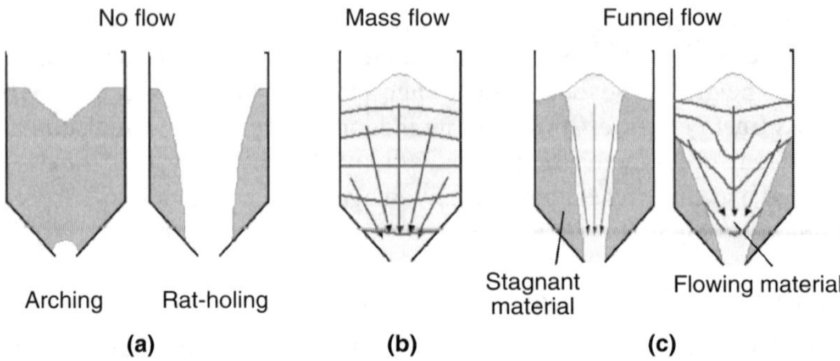

FIGURE 8.6 Schematic of three types of solids flow through a hopper: (a) no flow, (b) mass flow, and (c) funnel flow.

Hopper Design

Hoppers or chutes facilitate withdrawal of biomass or other solids from temporary storage such as a bunker. Major issues in their design include (1) mode of solids flow, (2) slope angle of discharge, and (3) size of discharge end.

Funnel flow is characterized by an annular zone of stationary solids and a moving core of solids at the center. In this case, the solids flow primarily through the core of the hopper. Solids in the periphery either remain stationary (Figure 8.6c, *left*) or move very slowly (Figure 8.6c, *right*). Fine particles tend to move through the core while coarser particles stay preferentially in the annulus. The particles from the top surface can flow into the funnel, thus violating the doctrine of first in–first out. If that does not happen, a stationary annulus is formed and the discharge stops, causing a rat hole to form through the hopper that becomes void and stops the flow. The rest of the solids in the hopper stay in the annulus (Figure 8.6a, *right*), which prevents the hopper from emptying completely. The only positive thing about a funnel-flow hopper is that it requires a lower height.

Mass flow (Figure 8.6b) is the preferred flow mode because the solids flow across the entire hopper cross-section. Though there may be some difference in velocity, this allows an uninterrupted and consistent flow with very little radial size segregation, which permits the hopper to effectively follow the first in–first out norm. However, because of the solids' plug-flow behavior, there can be more wear on the hopper walls with abrasive solids. Therefore, the required height of a mass-flow hopper must be greater than that of a funnel-flow hopper. The steeper the cone angle of a hopper, the higher the probability of a mass flow of solids through it. Some common operating problems with hoppers are

- Rat holing
- Funnel flow
- Arching

- Flushing
- Insufficient flow and incomplete emptying
- Caking

Two of the most common problems experienced in an improperly designed silo or bin (hereafter referred to as silo) are no flow and erratic flow. No flow from a silo can be due to either arching or rat holing (Figure 8.6a).

Rat holing (Figure 8.6a, *right*) most often happens in the flow of biomass with particles that are cohesive and rough. This is a serious problem in hoppers. To facilitate solids flow, the rat hole must be collapsed by proper aeration in the hopper or by vibrations on the hopper wall.

Arching occurs when cohesive particles form an obstruction over the exit (Figure 8.6a, *left*), usually in the shape of an arch or a bridge above the hopper outlet that prevents further discharge. The arch can be interlocking, with the particles mechanically locking to form the obstruction, or it can be cohesive. Coarse particles can also form an arch while competing for an exit, as a traffic jam results from a large number of vehicles trying to pass through a narrow road in an unregulated manner. By making the outlet size at least 8 to 12 times the size of the largest particle, this type of arching can be avoided (Jacob, 2000).

Flushing results in the uncontrolled flow of fine solids—Geldart's group A or group C particles (Basu, 2006, p. 443)—through the exit hole. It it is uncommon in relatively coarse biomass, but it can happen if the hopper is improperly aerated in an attempt to collapse a rat hole.

Another problem influenced by hopper design is inadequate emptying. This can happen if the sloped base of the hopper is improperly inclined, causing some solids to remain on the floor that cannot flow by themselves.

Erratic flow from an inappropriately designed hopper often results from alternating between an arch and a rat hole. A rat hole may collapse because of an external force, such as vibrations created by a plant pulverizer (mill), a passing train, or a flow-aid device such as an air cannon or vibrator. Some biomass discharges as the rat hole collapses, but the falling material can compact over the outlet and form an arch. The arch may break because of a similar external force, and the material flow will resume until the flow channel is emptied and a rat hole is once again formed (Hossfeld and Barnum, 2007).

Material discharge problems can also occur if the biomass stays in the bunker for a very long time, forming cakes because of humidity, pressure, and temperature. This easily results in arching or rat holes. To avoid this, renewal of solids in the hopper is necessary.

There are some special problems in fuel-handling systems. For example, spontaneous ignition of coal can occur if fine coal particles stay stagnant in a bunker for too long. Even in an operating silo, a stagnant region can be a problem for fuels like coal, which are prone to spontaneous combustion. Fine dust in the silo may lead to dust explosion. If the fuel flows through a channel in the silo, the fuel outside of the channel remains stagnant for a long time. The residence time of such fuels in the silo should be limited by emptying the silo frequently or by using a first in–first out mass-flow pattern (Figure 8.6b), where all of the material is in motion whenever the fuel is discharged. Biomass is relatively free of this problem as most of it is not prone to spontaneous ignition.

Achieving Mass Flow

To achieve mass flow, the following conditions are to be met:

- The hopper wall must be sufficiently smooth for mass flow.
- The hopper angle should be adequately steep to force solids to flow at the walls.
- The hopper outlet must be large enough to prevent arching.
- The hopper outlet must be adequately large to achieve the maximum discharge rate.

The required smoothness and sloping angle for mass flow in a hopper depends on the friction between the particles and the hopper surface. This friction can be measured in a laboratory using a standard test (ASTM, 2000).

Several factors can affect wall friction for a given fuel:

- Wall material
- Surface texture or roughness of the wall
- Moisture content and variations in solids composition and particle size
- Length of time solids remain unmoved
- Corrosion of wall material due to reaction with solids
- Scratching of wall material caused by abrasive materials

To enhance the smoothness of the surface, sometimes the hopper is coated or a smooth lining is applied. Lining materials that can be used include polyurethane sheets, TIVAR-88, ultra-high-molecular-weight polyethylene plastic, and krypton polyurethane.

Mass flow can be adversely affected by the narrowness of the hopper outlet. A too-narrow outlet (compared to particle size) permits the particles to interlock when exiting and form an arch over the outlet. The probability of this happening increases when

- The particles are large compared to the outlet width.
- There is high moisture in the solids.
- The particles are of a low shape factor and have a rough surface texture.
- The particles are cohesive.

Wedge-shaped hoppers require a smaller width than conical hoppers do in order to prevent bridging. Slotted outlets must be at least three times as long as they are wide.

Hopper Design for Mass Flow

The design of the hopper outlet significantly affects the flow of solids. When solids flow through the hopper, air or gas enters, dilating the particles. It is essential for powder solids to flow freely through the outlet. Air drag, which is proportional to surface area, must be balanced by gravitational force that is equal to the weight of the particle. Fine particles have a lower ratio of weight to surface area compared to coarser particles. So, for fine particles, this force balance becomes an important consideration. For such particles, the following expression is used (Carleton, 1972):

$$\frac{4V_0^2 \sin\theta}{B} + 15\frac{\left(\rho_a \mu^2 V_0^4\right)^{1/3}}{\rho_p d_p^{5/3}} = g \text{ For } d_p < 500\ \mu\text{m} \tag{8.1}$$

where

V_0 is the average solid velocity through the outlet, m/s
ρ_a, ρ_p is the density of the air and solids, respectively, kg/m^3
d_p is the particle size, m
μ is the viscosity of the air, kg/m.s
θ is the semi-included angle of the hopper
g is the acceleration due to gravity, 9.81 m/s^2
B is the parameter

The mass-flow rate, m, is given in terms of the bulk solid density, ρ_b, and the outlet area, A:

$$m = \rho_b A V_0 \tag{8.2}$$

For coarse particles (>500 μm), an alternative relation is used (Johanson, 1965):

$$m = \rho_b A \sqrt{\frac{Bg}{2(1+C)\tan\theta}} \text{ kg/s for } d_p > 500\ \mu\text{m} \tag{8.3}$$

Values of the parameters A, B, and C are given as

Parameter	Conical outlet	Symmetric slot
B	Outlet diameter, D	Slot width, W
A	$\frac{\pi}{4}D^2$	Width x breadth
C	1.0	0

Design Steps

Hopper design involves determining particle properties, such as interparticle friction, particle-to-wall friction, and particle compressibility or permeability. With these properties known, the outlet size, hopper angle, and discharge rate are found.

Dedicated experiments like shear tests are carried out to determine the interparticle friction. A parameter, such as angle of repose, has little value in hopper

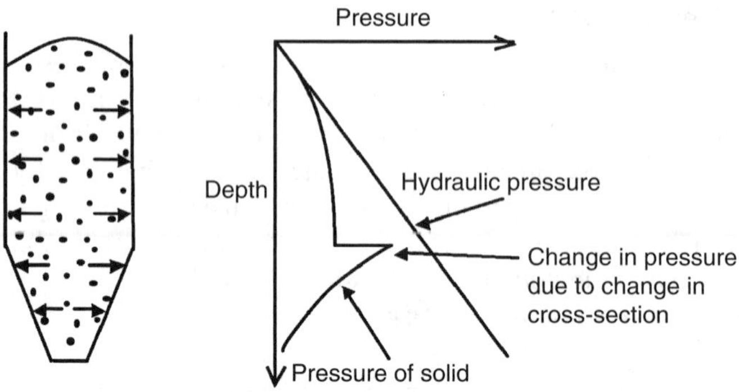

FIGURE 8.7 Wall pressure distribution along the height of a hopper filled with solids. It is noted that the pressure profile changes in the inclined section, which is not the case for the hydraulic pressure that would be the wall pressure if the hopper was filled with water.

design, as it simply gives the heap angle when solids are poured in. Particle–wall friction should also be measured by purpose-designed experiments.

The stress distribution on the silo wall is important, especially for a tall unit. Figure 8.7 compares the wall pressure in a biomass-filled silo with that of a liquid-filled silo. As we can see, the wall pressure in a solid-filled silo does not vary linearly with height, but it does in a liquid-filled silo. In the former case, the pressure increases with depth, reaching an asymptotic value that depends on the diameter of the hopper rather than on the height. Because there is no further increase in wall stress with height, large silos are smaller in diameter but taller.

To find the stress in the barrel, or the vertical wall section, of a hopper, we consider the equilibrium of forces on a differential element, dh, in a straight-sided silo (Figure 8.8):

- Vertical force due to pressure acting from above: $P_v A$
- Weight of material in element: $\rho A g\, dh$
- Vertical force due to pressure acting from below: $(P_v + dP_v) A$
- Solid friction on the wall acting upward: $\tau \pi D\, dh$

The force balance on the elemental solid cross-section gives

$$(P_v + dP_v)A + \tau \pi D\, dh = P_v A + \rho A g\, dh \tag{8.4}$$

The wall friction is equal to the particle–wall friction coefficient, k_f, times the normal pressure on wall, P_w:

$$\tau = k_f P_w \tag{8.5}$$

Janssen (1895) assumed the lateral pressure to be proportional to the vertical pressure, as shown in the following equation:

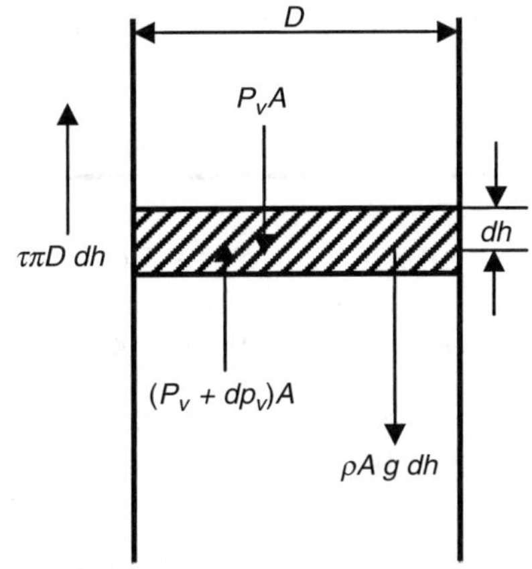

FIGURE 8.8 Force balance on an element of a storage silo.

$$P_w = KP_v \tag{8.6}$$

where K is the Janssen coefficient. For liquids, the pressure is uniform in all directions, so K is 1.0. This relation is not strictly valid for all solids, but for engineering approximations we can start with this assumption.

Substituting Eqs. (8.5) and (8.6) in Eq. (8.4), we get

$$AdPv = \rho Ag\, dh - k_f KP_v \pi D\, dh \tag{8.7}$$

Boundary conditions for this equation are $h = 0$, $P_v = 0$; $h = H$, $P_y = P_0$. With this, Eq. (8.7) is integrated from $h = 0$ to $h = H$ to get the pressure at the base of the silo's vertical section, P_0. Substituting

$$A = \frac{\pi D^2}{4}$$

we get

$$P_0 = \frac{\rho Dg}{4k_f K}\left(1 - \exp\left(-\frac{4Hk_f K}{D}\right)\right) \tag{8.8}$$

This is known as the Janssen equation.

Figure 8.7 illustrated the pressure distribution along the height of a silo. The straight line shows the pressure we expect if the stored substance is a liquid; the discontinuous exponential curve is the one predicted for solids. There is a sharp increase in pressure at the beginning of the inclined wall. The pressure decreases with height (Figure 8.7).

The stress on the inclined section is different from that calculated from the preceding. To calculate this, we use the Jenike equation, which states that the radial pressure is proportional to the distance of the element from the hopper apex, which is the point where inclined surfaces would meet if they were extended (Jenike, 1964). It can be seen that the magnitude of stress at the hopper exit is the lowest, although this is the lowest point in the hopper.

Example 8.1

Find the wall stress at the bottom of a large silo, 4.0 m in diameter and 20 m in height, that uses a flat bottom for its discharge. Compare the stress when the silo is filled with wood chips (bulk density 300 kg/m^3) with that when it is filled with water.

Given that the wall-to-wood chip friction coefficient, k_f, is 0.37, assume the Janssen coefficient, K, to be 0.4.

Solution

We use Eq. (8.8) to calculate the vertical pressure, P_0, in the silo. Data given are as follows:

- The bulk density of the wood chips, ρ, is 300 kg/m^3.
- The wall–solid friction coefficient, k_f, is 0.37.
- The diameter, D, is 4.0 m.
- The height, H, is 20 m.
- The Janssen coefficient, K, is 0.4.

$$P_0 = \frac{\rho D g}{4k_f K}\left(1-\exp\left(-\frac{4Hk_f K}{D}\right)\right)$$
$$= \frac{300\times 4\times 9.81}{4\times 0.37\times 0.4}\left[1-\exp\left(-\frac{4\times 20\times 0.37\times 0.4}{4}\right)\right] = 18{,}854\ \text{Pa}$$

Since the lateral pressure, P_w, is proportional to the vertical pressure, P_v,

$$P_w = KP_0 = 0.4\times 18{,}854 = 7542\ \text{Pa}$$

For water, the vertical pressure is the weight of the liquid column:

$$P_0 = \rho g H$$

Because the lateral and vertical pressures are the same (i.e., K = 1.0), we can write

$$P_w = P_0 = 1000\times 9.81\times 20 = 196{,}200\ \text{Pa}$$

The lateral pressure for water is therefore (196,200/7542) or 26 times greater than that for wood chips.

Chute Design

In a silo, the solids are withdrawn through chutes at the bottom. Previous discussions examined solids flow through the silo. Now, we will look at the flow out of the silo through the chute, which connects the silo to the feeder. A proper chute design ensures uninterrupted flow from storage to feeder. Improper

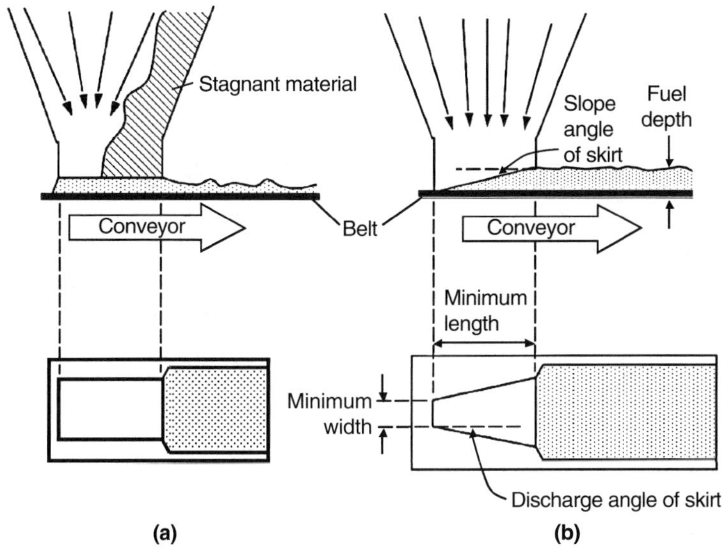

FIGURE 8.9 Two feed chutes between a hopper and a belt conveyor: (a) a simple design that causes partial flow; (b) a design that provides complete flow.

design results in nonuniform flow. Figure 8.9 illustrates the problem, showing partial solids flow with a uniform-area chute and full solids flow with a properly designed chute. As the solids accumulate on the belt, their uniform flow through the hopper prevents them from accumulating at the chute's downstream section. The chute's expanded and lifted opening helps the solids spread well, allowing uniform withdrawal. For this reason, the modified design of Figure 8.9(b) shows the skirt on the chute to be lifted and expanded (in plan view) to facilitate uniform solid discharge from the hopper. These angles (slope and discharge) should be in the range of 3 to 5 degrees.

Figure 8.10 is another illustration of this phenomenon, this time with a rotary feeder. Here the design on the left (Figure 8.10a) is without the short vertical section like that on the right (Figure 8.10b). Solids are compressed in the direction of rotation and pushed up through the hopper. The design on the right uses a short vertical chute that limits this backflow only to the chute height, giving a relatively steady flow.

The two key requirements for chute design are: (1) the entire cross-section of the outlet must be active, permitting the flow of solids; and (2) the maximum discharge rate of the chute must be higher than the maximum handling rate of the feeder to which it is connected.

A restricted outlet, caused by a partially open slide gate, results in funnel flow with a small active flow channel regardless of hopper design. A rectangular outlet ensures that feeder capacity increases in the direction of the flow. With a belt feeder, the increase in capacity is achieved by a tapered interface. The

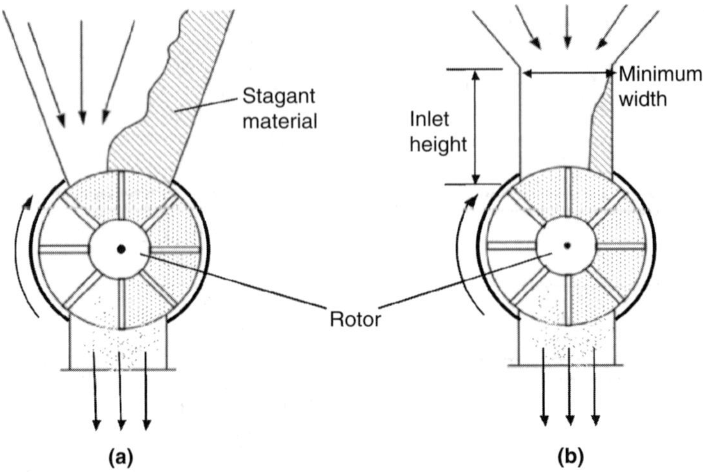

FIGURE 8.10 (a) Feeder without vertical rise and (b) feeder with a vertical section.

capacity increase along the feeder length is achieved by the increase in height and width of the interface above the belt.

Poor feeder design is a common cause of flow problems, as it prevents smooth withdrawal of solids. If the discharge rate of the chute is lower than the maximum designed feeding rate of the feeder, the feeder can be starved of solids and its flow control will be affected.

8.3.3 Feed Preparation

Biomass received from its source cannot be fed directly into the gasifier for the following reasons:

- Presence of foreign materials (e.g., rocks and metals)
- Unacceptable level of moisture
- Too large (or uneven in size)

Such undesirable conditions not only affect the flow of solids through the feeder, but they also affect operation of the gasifier. It is thus necessary to eliminate them and prepare the collected biomass appropriately for feeding. Foreign materials pose a grave problem in biomass-fired plants. They jam feeders, form arches in silos, and affect the gasifier operation, so it is vitally important to remove them as much as possible. The three main foreign materials are: (1) stones, (2) ferrous metals (e.g., iron), and (3) nonferrous metals (e.g., aluminum).

Some of the equipment used to remove foreign materials from the collected biomass are as follows:

De-stoner. The basic purpose of a de-stoner is the separation of heavier-than-biomass materials such as glass, stones, and metals. Typical de-stoners

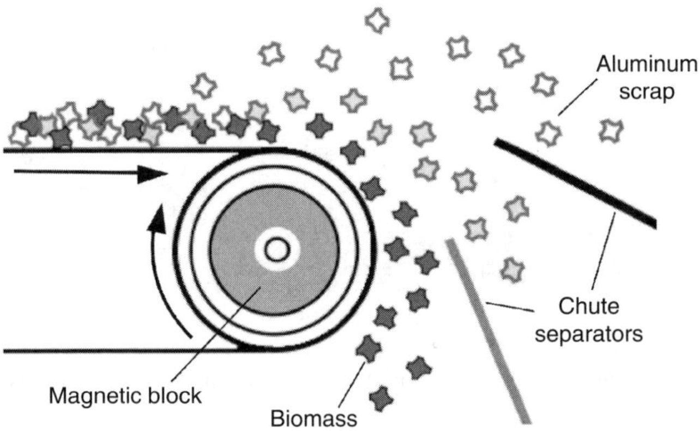

FIGURE 8.11 Separation of nonferrous metals from biomass using eddy current separation.

use vibration in tandem with suitable air flow to stratify heavy materials according to their specific gravity.

Nonferrous metal separators. Separation of nonferrous metals like aluminum has always been a challenge. One solution is an eddy current separator—essentially a rotor with magnet blocks—which, depending on the application, is made of either standard ferrite ceramic or a more powerful rare-earth magnet. The rotors are spun at high revolutions (more than 3000 rpm) to produce an "eddy current," which reacts differently with different metals according to their specific mass and resistivity to create a repelling force on the charged particle. If a metal is light yet conductive, such as aluminum, it is easily levitated and ejected from the normal flow of the product stream, making separation possible (Figure 8.11). Separation of stainless steel is also possible depending on its grade. Particles from material flows can be sorted down to a minimum size of 3/32 in. (2 mm) in diameter. Eddy current separators are crucial in the recycling industry because of their ability to separate nonmagnetic materials.

Magnetic metal separation. The use of powerful magnets to separate iron and other magnetic materials from the feed is a standard procedure in many plants. Magnets are located at several places along the feed stream. They are generally suspended above the belt to attract magnetic materials, which are then discharged away.

Size Reducers

Biomass comes from different sources, so the presence of oversized solids or trash is very common in the fuel delivered. Woody biomass may be sized and classified at the source or at the plant. The list following the figure contains some of the equipment used for its preliminary sizing along with the typical sizes produced (Van Loo and Koppejan, 2008, p. 64):

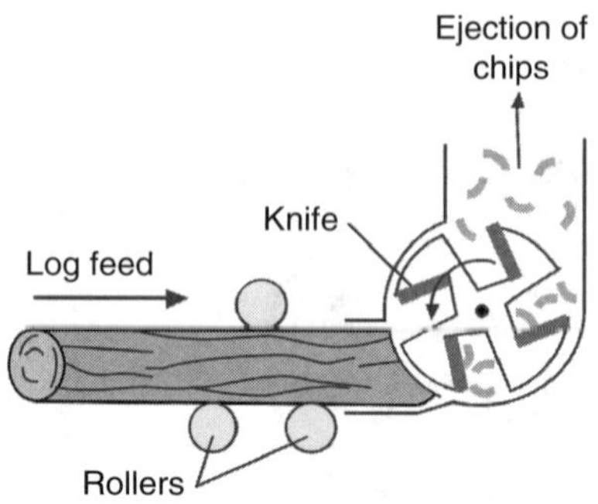

FIGURE 8.12 A drum chipper uses a number of knife edges mounted in a drum as shown.

- Chunker: 250 to 50 mm
- Chipper: 50 to 5 mm
- Grinder: <80 mm
- Pulverizer: <100 μm (dust)

Different types of equipment are necessary for sizing biomass. One example is a chunker with multiple blades. Another is a spiral chunker with a helical cutter mounted on a shaft; as the wood is fed into the machine, the cutter draws it in and slices it into chunks. The power consumption of a chunker is relatively low.

Chippers are used to break wood into small pieces. Disc and drum chippers are two common types. In a disc chipper, the wood is fed from the side, meeting a large disc with several rotating knives. In a drum chipper, several knives are embedded in grooves (Figure 8.12); as the drum rotates, wood fed at one end is chipped. The chips, which are now uniform in size, are carried away and thrown to the other end by the grooves.

Size-reducing machines consume energy in proportion to the reduction in size. Chipping typically consumes energy equivalent to 1 to 3% of the energy content of the wood (Van Loo and Koppejan, 2008, p. 65).

Grinding is needed when a finer size (<80 mm) of biomass particle is needed. Hammer mills may be used for this purpose. The wood is thrown to the wall of the mill and crushed by hammers. A conventional biomass combustor or gasifier does not require biomass to be ground to such a fine size. However, direct co-firing in pulverized-coal fired-boilers and the use of entrained beds for gasification require the biomass to be ground extremely fine so that the particles can be conveyed like pulverized coal.

Size Classification

Oversized materials often cause major problems in a biomass plant. They jam belts, bunkers, and other components. Sometimes trummels are used in the fuel yard to separate the oversize pieces before feeding to the plant. A trummel (Figure 8.13) is a rotating drum, with holes of various sizes, that separates the

FIGURE 8.13 Portable trummel used in the fuel yard for size classification. *(Source: Photograph by the author).*

smaller and larger feed. Several belt runs in the feed-processing stream divert oversized feed from the gasifier.

Drying

Freshly cut biomass can contain up to 40 to 60% surface moisture when harvested, but thermal gasification typically requires a moisture content of less than 10 to 15%—this moisture is inherent in the biomass. Furthermore, biomass is hygroscopic, so even after dried it can still absorb moisture from the atmosphere; only after torrefaction does the biomass stop absorbing moisture (see Section 3.8). This could happen even when the dried biomass is stored in a shed. Because biomass is bulky, with low energy density, a very large storage space is necessary for the typical fuel inventories required in an energy conversion (boiler or gasifier) plant. For this reason, the biomass is often stored outdoors, though it could absorbs additional moisture from rain and snow. Leaving freshly harvested biomass outdoors can at times have some positive effect. For example, straw is sometimes left in the field for a few days or weeks to lose moisture and then put in bales. Leaving wood logs outside over the summer can reduce moisture by as much as 20% (Van Loo and Koppejan, 2008, p. 70).

The moisture in biomass must be reduced before use because it represents a large drain on a plant's deliverable energy. Every kilogram of moisture needs about 2300 kJ of heat to vaporize and an additional 1500 kJ to be raised to a typical gasifier temperature of 700 °C. This large amount of energy (3800 kJ/kg) has to come from the energy released by the gasifier's exothermic reactions. Therefore, the lower the moisture, the higher the heat available in the product gas.

Outdoor storage may not work well because of rain and snow, but precipitation can have a beneficial effect on some herbaceous biomass, such as straw, since it leaches water-soluble agglomerates and corrosion-causing elements such as chlorine and potassium. The three types of moisture in a biomass gasifier are: (1) surface moisture, (2) chemical moisture, and (3) moisture in air or steam used for gasification.

While the chemical (also called inherent) moisture cannot be reduced, it is possible to reduce the surface moisture by drying, using the sensible heat in the gasifier product gas, the flue gas of the combustor, or heat from other external sources. Surface moisture less than 10 to 20% is desirable for most gasifier types (Cummers and Brown, 2002).

The temperature of the hot gas used for drying the biomass is a critical design parameter. Generally, it is in the range 50 to 60 °C. If much hotter gas is used, it can heat the biomass above 100 °C, and pyrolysis can set in on the outer surface of the biomass before the heat reaches the interior. Besides contributing to an energy loss though, such hot gas can cause volatile organic compounds to be released from the biomass that are potentially hazardous. They are detected by a "blue haze" in the exhaust gas (Cummers and Brown, 2002). The presence of excessive oxygen in the dryer can also lead to ignition of fuel dust in the dryer, resulting in a potential explosion. Therefore, oxygen concentration in the dryer should be kept below 10% to avoid this risk (Brammer and Bridgwater, 1999).

8.3.4 Conveying

The belt conveyor is the most common and perhaps the most reliable means for transportation of biomass. It allows a magnet hanging from the top to remove magnetic materials and other devices to remove oversized solids and other scrap materials as the biomass moves along the belt. The biomass that remains is fed into a silo for temporary storage.

8.3.5 Feeding

Feeding is the last step in the flow-handling stream. Many types of feeder are used depending on biomass type and other process parameters. This topic is discussed next.

8.4 BIOMASS FEEDERS

Based on the type of biomass, feeders can be divided into two broad groups: (1) those for harvested biomass and (2) those for nonharvested biomass.

Harvested fuels include long and slender plants like straw, grass, and bagasse, which carry considerable amounts of moisture. Examples of nonharvested fuels are wood chips, rice husk, shells, barks, and pruning. These fuels are not as long or as slender as harvested fuels, and some of them are actually granular in shape.

8.4.1 Feeding Systems for Harvested Fuel

Harvested biomass, such as straw and nonharvested hay, is pressed into bales in the field, and sometimes the bales are left in the field to dry (Figure 8.14).

FIGURE 8.14 Tall grass is cut in the field, baled, and left in the field for drying in Nova Scotia. *(Source: Photograph by the author.)*

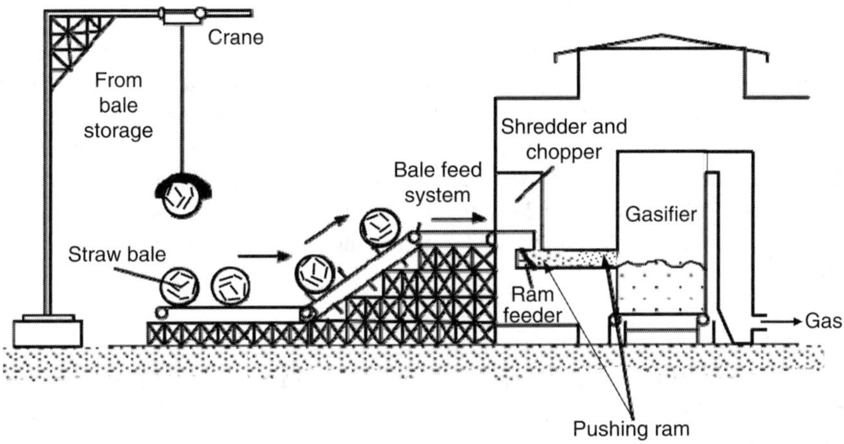

FIGURE 8.15 Typical handling system for straw bales.

Baling facilitates transportation and handling (Figure 8.15). Cranes are used to load the bales at a certain rate depending on the rate of fuel consumption. The bales are brought to the boiler house from storage by chain conveyors.

Whole bales are fed into a bale shredder and a rotary cutter chopper to reduce the straw to sizes adequate for feeding into a fluidized-bed gasifier or combustor. In the final leg, the chopped straw is fed into the furnace by one of several feeder types. Figure 8.15 shows a ram feeder, which pushes the straw into the furnace. In some cases, the straw falls into a double-screw stoker, which presses it into the furnace through a water-cooled tunnel.

8.4.2 Feeding Systems for Nonharvested Fuels

Wood and by-products from food-processing industries are generally granular in shape. Wood chips and bark may not be of the right size when delivered to the plant, so they need to be shredded to the desired size in a chopper. However, fuels like rice husk and coffee beans are of a fixed granular size and so do not need further chopping. Rice husk, a widely used biomass, is flaky and 2 to 10 mm × 1 to 3 mm in size. As such, it can be fed as it comes from the source, but it can be easily entrained in a fluidized bed. For this reason one can press it into pellets using either heat or a nominal binder in a press.

Feeders for nonharvested fuels are similar to those for conventional fuels like coal. Speed-controlled feeders take the fuel from the silo and drop measured amounts of it into several conveyors. Each conveyor takes the fuel to an air-swept spout that feeds it into the furnace. If the moisture in the fuel is too high, augers are used to push the fuel into the furnace.

8.4.3 Feeder Types

The six main feeder types for biomass are: (1) gravity chute, (2) screw conveyor, (3) pneumatic injection, (4) rotary spreader, (5) moving-hole feeder, and (6) belt feeder. These are broadly classified as traction, nontraction, and others as shown in Figure 8.16. In the traction type, there is linear motion of the surface carrying the fuel, as with a belt feeder or a moving-hole or drag-chain feeder. In the nontraction type, the motion is rotating and oscillatory screw feeders and rotary feeders belong to this group. Oscillatory feeders are of the vibratory or ram type. Other feeder types move the fuel by gravity or air pressure.

Gravity Chute

A gravity chute is a simple device in which fuel particles are dropped into the bed with the help of gravity. The pressure in the furnace needs to be at least slightly lower than the atmospheric pressure; otherwise, hot gas will blow back into the chute, creating operational hazards and possible choking of the feeder due to coking near its mouth.

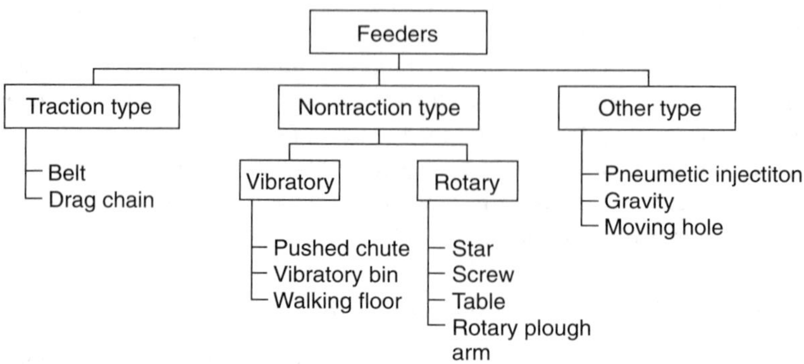

FIGURE 8.16 Types of feeders used in biomass plants.

In spite of the excellent mixing capabilities of a fluidized bed, a fuel-rich zone is often created near the outlet of a chute feeder that is subjected to severe corrosion. Since the fuel is not well dispersed in gravity chute feeding, much of the volatile matter is released near the feeder outlet, which causes a reducing environment. To reduce this problem, the chute can be extended into the furnace. However, the extension needs insulation and some cooling air to avoid premature devolatilization of the feed passing through it. Additionally, a pressure surge might blow fine fuel particles back into the chute while reducing conditions might encourage corrosion. An air jet can help disperse the fine particles away from the fuel-rich zone.

A gravity feeder is not a metering device. It can neither control nor measure the feed rate of the fuel. For this, a separate metering device such as a screw feeder is required upstream of the chute.

Screw Feeder

A screw feeder is a positive-displacement device. Not only can it move solid particles from a low-pressure zone to a high-pressure zone with a pressure seal; it can also measure the amount of fuel fed into the bed. By varying the speed of its drive, a screw feeder can easily control the feed rate. As with a gravity chute, the fuel coming out of a screw does not have any means for dispersion. An air dispersion jet employed under the screw feeder can serve this purpose.

Plugging of the screw is a common problem. Solids in the screw flights are compressed as they move downstream; sometimes they are packed so hard that they do not fall off the screw. Compaction against the sealed end of the trough carrying the screw is even worse, often leading to jamming of the screw. Plugging and jamming can be avoided by one of the following:

- Variable-pitch screw (Figure 8.17a)
- Variable diameter to avoid compression of fuels toward the feeder's discharge end (Figure 8.17b)
- Wire screw
- Multiple screws (Figure 8.18)

A wire screw is suitable for a highly fibrous biomass. It is made of a helical springlike wire with no central shaft or blades. Because there is minimum metal–feed contact, there is less chance of feed buildup even if the feed is cohesive.

Multiple screws are effective especially for large-biomass fuels. Figure 8.18 shows a feeder with two screws. Some feed systems use three, four, or more.

The hopper outlet, to which the inlet of a feeder is connected, needs careful design. Figure 8.9 showed two designs. The first (refer to Figure 8.9a) has a tapered wall hopper. It develops a large stagnant layer on the hopper's downstream wall. The second (refer to Figure 8.9b) is a vertical hopper wall toward the discharge end. This is superior to the traditional inclined wall because it develops a smaller stagnant layer and thus avoids formation of rat holes.

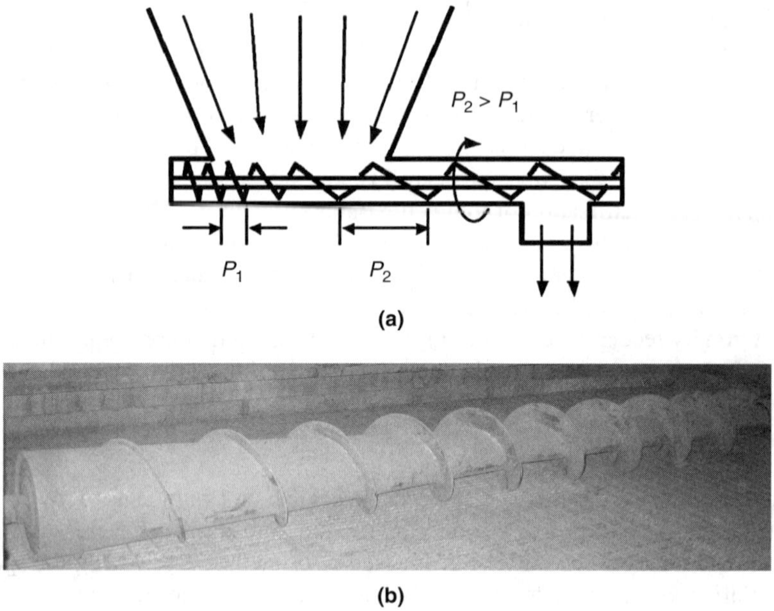

FIGURE 8.17 Two types of screw used for trouble-free feeding of biomass: (a) variable pith; (b) variable diameter. *(Source: Photograph by the author.)*

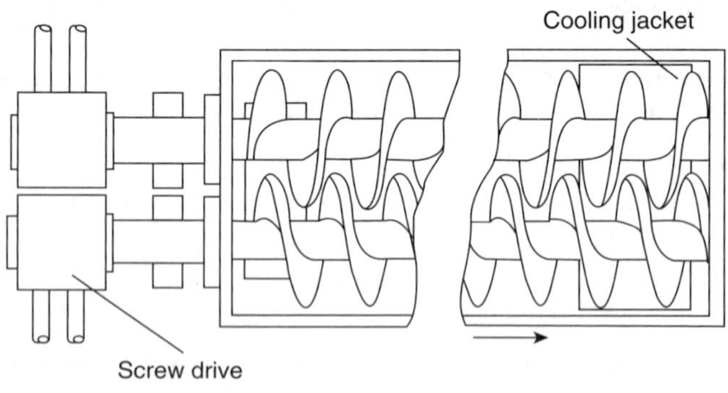

FIGURE 8.18 Double-screw feeders help uniform flow of biomass.

A screw feeder typically serves 3 m^2 or less area of a bubbling fluidized bed, so several feeders are needed for a large bed. A major and very common operational problem arises when the fuel contains high moisture. It has to be dried first before it enters the screw conveyor to avoid plugging.

Dai and Grace (2008) developed a model of the mechanism of solid flow through a screw feeder. They noted that the torque required by the screw is proportional to the vertical stress exerted on the hopper outlet by the bulk material in the hopper; it also depends strongly on screw diameter. The choke section

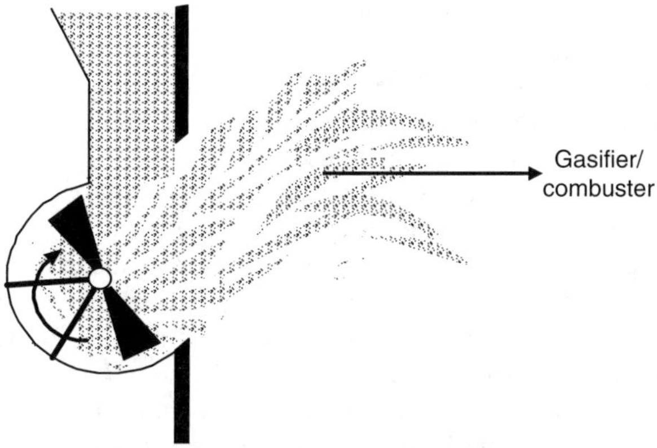

FIGURE 8.19 Rotary spreader for spreading the fuel over a large bed area.

(the part of the screw extending beyond the hopper exit) accounts for more than half of the total torque required to feed the biomass, especially with compressible particles. The torque, T, required by a screw of diameter, D_0, rotating in a shaft of diameter, D_c, is given as

$$T = K_i \sigma_v D_0^3 \tag{8.9}$$

where σ_v is the vertical stress for the flow and D_0 is the screw diameter. The constant, K_i, depends on the ratios P/D_0 and D_c/D_0 (normal stress/axial stress) and the wall friction, where D_c is the shaft diameter and P is the pitch of the screw.

Spreader

For a wide dispersion of fuel over the bed, spreader wheels are used (Figure 8.19). The spreader throws the fuel received from a screw or other type of metering feeder over a large area of the bed surface. Typically it comprises a pair of blades rotating at high speed; slightly opposite orientation of the blades helps throw the fuel over a larger lateral area. This is not a metering device. A major problem with the spreader is that it encourages segregation of particles in the bed.

Pneumatic Injection Feeder

A pneumatic injection feeder is not a metering device; rather, it helps feed already metered biomass into the reactor. This works well for gravity feeding, and it is especially suitable for fine solids. Pneumatic injection is preferred for less reactive fuels, which must reside in a gasifier bed longer for complete conversion. It transports dry fuel particles in an air stream at a velocity higher than their settling velocity. The fuel is typically fed from underneath a bubbling fluidized bed. The maximum velocity of air in the fuel transport lines may not exceed 11~15 m/s to avoid line erosion. The air for transporting constitutes part of the air for gasification.

Splitting of the fuel–air mixture into multiple fuel lines is a major problem with pneumatic injection. A specially designed feed splitter, like the one discussed in Basu (2006, p. 355), can be used.

In a underbed pneumatic system, air jets that carry solid particles with high momentum to enter into the fluidized bed, forming a plume that could punch through the bed. To avoid this, a cap sits at the top of the exit of each feeder nozzle. This cap reduces the momentum of the jets breaking into the freeboard of a bubbling fluidized bed. A highly erosive zone may be formed near each outlet nozzle of the feeders, which might corrode the tubes nearby.

Another innovative, but one that is less common, feed system uses pulsed air. Controlled-air pulses push the biomass into the gasifier, avoiding pyrolysis of feed in the gasifier feed line. A very small amount of air minimizes dilution of the product gas with nitrogen. The University of Western Ontario in Canada applied this design with success in a commercial mobile pyrolyzer.

Moving-Hole Feeder

A moving-hole feeder is particularly useful for fluffy biomass or solids with flakes, which are not free-flowing. Such solids can cause excessive packing in the hopper and screw feeder. Unlike other types, moving-hole feeders do not draw solids from one particular point in the silo.

A moving-hole feeder essentially consists of slots that traverse back and forth with no friction between the stored material and the feeder deck. At a desired rate, a moving hole or aperture slides under the hopper. The solids drop by gravity into the trough or belt that carry the feed at that rate.

With a moving-hole feeder there is no compaction of solids typically seen in screw feeders. Rat holes are also avoided by using vertical instead of sloped walls in the hopper.

Fuel Auger

A metering device such as a screw is used to meter biomass like hog fuel and feed it onto the main fuel belt. The belt carries the fuel to the gasifier front, where the fuel stream is divided into several 50%-capacity fuel trains. Each train consists of a surge bin with a metering bottom and a fuel auger to deliver the fuel into the furnace. The auger is cantilevered and driven at a constant speed through a gear reducer. The bearing of the auger shaft is located away from the heat of the gasifier. Cooling air is provided to cool the auger's inner trough as well as to propel the fuel toward the bed.

Ram Feeder for RDF

A ram feeder is essentially a hydraulic pusher. Refuse-derived fuel (RDF) is at times too fibrous or sticky to be handled by any of the aforementioned feeders.

In this case, a ram-type feeder can be effective in forcing them into the gasifier. A fuel auger can convey the solids into a hopper, at the bottom of which is the ram feeder. The ram pushes the RDF onto a sloped apron-type feeder that feeds the fuel chute (refer to Figure 8.15). From the fuel chute, the RDF drops into the fuel spout, where sweep air transports it into the furnace. The air also prevents any backflow of hot gases. The RDF stored in the inlet hopper provides a seal against positive furnace pressure. The apron feeders are driven by a variable-speed drive for controlling the amount of fuel going into the system.

Belt Feeder

Belt feeders are very effective for feeding non-free-flowing biomass that is cohesive, fibrous, friable, coarse, elastic, sticky, or bulky. However, they are not recommended for fine or granular solids. Typically a moving belt is located directly under the outlet chute of the fuel hopper. The belt is supported on rollers that can be mounted on load cells to directly measure the fuel feed rate. Such feeders are referred to as *belt-weigh feeders*.

The width and speed of the belt depend on the density and size of the feed material. A narrow belt with a high design speed may be the most economical, but it is limited by other considerations such as dust generation and hopper width. Most manufacturers provide data on available belt width, permissible speed, feed density, and recommended spacing of idlers supporting the belt. Such data can be used for design of the belt feeder and the feed system.

8.4.4 Fuel Feed in Fluidized Beds

Biomass feed in fluidized-bed gasifiers needs special considerations, which are discussed in the following sections. For bubbling fluidized beds, we have the choice of two types of feed systems: (1) overbed (Figure. 8.20a) and (2) underbed (Figure 8.20b).

Gasification is a relatively slow process compared to combustion, so the rapid mixing of fuels is not as critical as it is in a combustor. Table 8.1 compares the characteristics of the two types of feeder as used in a boiler. Such a comparison may be valid for fluidized-bed gasifiers but only on a qualitative basis.

Overbed feeders can handle coarse particles; underbed feeders need fine sizes with less moisture. An underbed system consists of crushers, bunkers, gravimetric feeders, air pumps, a splitter, and small fuel-transporting lines. An overbed feed system, on the other hand, consists of crushers, bunkers, gravimetric feeders, small storage bins, a belt conveyor, and spreaders.

Overbed System

The overbed system (Figure 8.20a) is simple, reliable, and economical, but it causes a loss of fine biomass particles through entrainment. In this system, the top size of the fuel particles is coarser than that in an underbed system, making

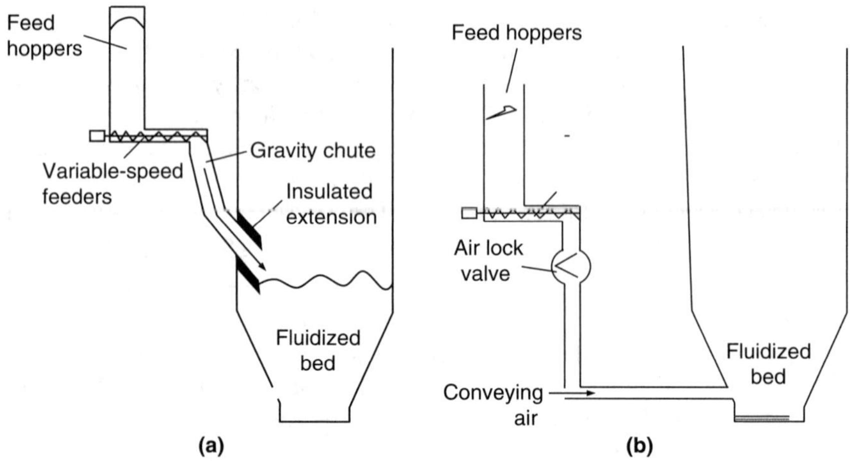

FIGURE 8.20 Position of feeders in a bubbling fluidized bed: (a) overbed feeding and (b) underbed feeding.

fuel preparation simpler and less expensive. However, the feed can contain a large amount of fines with a terminal velocity that is higher than the superficial velocity in the freeboard. When the terminal velocity is lower than the superficial velocity of the fluidized bed, the particles are elutriated before they completely gasify, resulting in a large carbon loss. This represents most of the carbon loss in a fluidized-bed gasifier.

In an overbed feed system, biomass particles are crushed to sizes less than 20 mm, which is usually coarser than the particle size used in the underbed system. In a typical setup, the fuel passes through bunkers, gravimetric feeders, and a belt conveyor, and is then dropped into a feed hopper.

Fewer feed points is an important characteristic of an overbed feed system. A rotary spreader throws the fuel particles over the bed surface. The coarser particles travel deeper into the gasifier while the finer particles drop closer to the feeder. The bed thus receives particles of a nonuniform size distribution. The maximum throwing distance of a typical spreader is around 4 to 5 m. The location of the spreaders is dependent on the dimensions of the bubbling bed. When the width is less than the depth, the spreaders are located on the side walls; when the depth is less than the width, they are located on the front wall. When both width and depth are greater than 4.5 m, the spreaders can be located on both side walls. Sometimes air is used to assist the throw of fuel by spreaders.

Underbed System

In an underbed feed system (Figure 8.20b), fuel particles are crushed into sizes smaller than 8 to 10 mm. Introduced in Section 8.4.2 as pneumatic feeding, this system is relatively expensive, complicated, and less reliable than the overbed

TABLE 8.1 Feed Points for Some Commercial Bubbling Fluidized-Bed Boilers

Boilers	Boiler Rating (MW_e)	Bed Area (m^2)	Feed Points	MWth per Feed	m^2 per Feeder	Feed Type	Fuel Type	HHV of Fuel (MJ/kg)
Shell	43 (MW_{th})	23.6	2	21.7	11.8	OB	Bituminous	
Black dog	130	93.44	12	31.0	7.8	OB	Bituminous	19.5–34.9
TVA	160	234	120	3.8	2.0	UB	Bituminous	24–25
Wakamatsu	50	99	86	1.7	1.2	UB	Bituminous	25.8
Stork	90	61	36	2.8	1.7	UB	Lignite	25

Note: OB = overbed spreader feeder; UB = underbed pneumatic feed.

system (especially with moist fuels), but it does achieve high char conversion efficiency.

Fuel entering at a feed point disperses over a much smaller area than it does in overbed feeding, so the feed points are more numerous and more closely spaced. Spacing greatly affects gasification. Because a deeper bed allows wider dispersion of the fuel and hence works with wider feed-point spacing, increased spacing with no sacrifice of char conversion efficiency can be achieved, but only if it is compensated by a corresponding increase in bed height. A decrease in bed height must be matched by increased feed-point spacing; otherwise, the conversion efficiency can drop. Coarser particles take longer to gasify and are less prone to entrainment. Therefore, wider spacing is preferred for them; finer particles require closer spacing.

The freeboard can provide room for further reaction of particles entrained from the bed. Freeboard design is important, especially when wide feed-point spacing is used.

Feed-Point Allocation

The excellent solids–solids mixing in a fluidized bed helps disperse the fuel over a small bed area of 1 to 2 m^2. A single feeder is adequate for a small bed having a cross-sectional size of less than 2 m^2. Larger beds need multiple feeders. The number required for a given bed depends on factors such as quality of fuel, type of feeding system, amount of fuel input, and bed area. Highly reactive fuels with high volatiles need a larger number of feed injection points because they react relatively fast; less reactive fuels require fewer feed points.

Industrial designs often call for redundancy. For example, if a reactor needs two overbed feeders, designers will provide three, each with a capacity that is at least 50% of the design feed rate. In this way, if one feeder is out of service, the plant can still maintain full output on the other two. The number of redundant feeders depends on the capacity and reliability required of the plant.

Symbols and Nomenclature

A = area of the cross-sectional area of silo (m^2)
B = parameter, depending on the silo (m)
C = parameter, depending on the silo (–)
d_p = particle size (m)
D = diameter of the silo (m)
D_0 = diameter of the screw (m)
D_c = shaft diameter (m)
dh = height of a differential element in the silo (m)
g = acceleration due to gravity (9.81 m/s^2)
H = height of the silo (m)
k_f = wall friction coefficient (–)
K = Janssen coefficient (–)
K_i = a constant, depending on D_0/D_0 or P/D_0 (–)

m = mass-flow rate (kg/s)
P = pitch of the screw (m)
P_w = normal pressure on the wall in the silo (Pa)
P_v = vertical pressure on the biomass in the silo (Pa)
P_0 = pressure at the base of the silo (Pa)
T = torque of the screw (Nm)
V_0 = average solid velocity through outlet (m/s)
τ = wall friction (Pa)
ρ, ρ_p = density of solids (kg/m^3)
ρ_a = density of air (kg/m^3)
σ_v = vertical stress for the flow (Pa)
μ = viscosity of air (kg/m.s)
θ = semi-included angle of hopper (degree)

Chapter 9

Production of Synthetic Fuels and Chemicals from Biomass

9.1 INTRODUCTION

Earlier chapters discussed methods of converting solid feedstock into gases and their use in energy production. Besides energy production, gasification and pyrolysis have important application in the production of chemicals and transport fuels. Many of our daily necessities like plastic, resin, and fertilizer can potentially come from biomass.

Syngas, a mixture of H_2 and CO, is an important product of gasification. It is a fuel as well as a basic building block for many hydrocarbons. Transport fuel and a large number of chemicals are produced from different syntheses of CO and H_2. These products can be divided into three broad groups: (1) energy feedstock (e.g., methane, carbon monoxide); (2) transportation fuels (e.g., biodiesel, biogas); and (3) chemical feedstock (e.g., methanol, ammonia). Presently, syngas is produced primarily from natural gas, but it can also be produced from

- Biomass
- Solid fossil fuels (e.g., coal, petcoke)
- Liquid fuels (e.g., refinery wastes)

Interest in biomass as a chemical feedstock is rising given that it is renewable and carbon-neutral. There is a growing shift toward "green chemicals" and "green fuels," which are derived from carbon-neutral biomass. Gasification and pyrolysis are effective and powerful ways to convert the biomass (or another fuel) into energy, chemicals, and transport fuels. This chapter discusses different ways to convert biomass-derived syngas into such useful products.

9.2 SYNGAS

The following sections discuss syngas—its physical properties and uses, as well as its production and its cleaning and conversion.

Biomass Gasification and Pyrolysis. DOI: 10.1016/B978-0-12-374988-8.00009-X

9.2.1 What Is Syngas?

As mentioned earlier, *syngas* is a mixture of hydrogen and carbon monoxide gases. It should not be confused with *SNG* (the abbreviation for "synthetic (or substitute) natural gas"), which is primarily made of methane gas. Syngas is an important feedstock for the chemical and energy industries. A large number of hydrocarbons traditionally produced from petroleum oil can also be produced from syngas.

Syngas may be produced from many hydrocarbons, including coal and petroleum coke, as well as from biomass. To distinguish syngas generated from biomass from that produced from fossil fuel, the former is sometimes called *biosyngas*. Here, *syngas* implies that derived from biomass unless specified otherwise.

One of the major applications of syngas is the production of liquid transport fuel. South African Synthetic Oil Limited (SASOL) has for many years been producing a large amount of liquid fuel from coal using Fischer-Tropsch synthesis of syngas produced from the gasification of coal. The same liquid fuel may be produced from biomass-derived syngas.

The typical product gas of biomass gasification contains hydrogen, moisture, carbon monoxide, carbon dioxide, methane, aliphatic hydrocarbons, benzene, and toluene, as well as small amounts of ammonia, hydrochloric acid, and hydrogen sulfide. From this mixture, the carbon monoxide and hydrogen must be separated to produce syngas.

9.2.2 Applications for Syngas

As mentioned, syngas is an important source of valuable chemicals. These include

- Hydrogen, produced in refineries
- Diesel gasoline, using Fischer-Tropsch synthesis
- Fertilizer, through ammonia
- Methanol, for the chemical industry
- Electricity, generated through combustion

It should be noted that a major fraction of the ammonia used for fertilizer production comes from syngas and nitrogen (see Section 9.4.3).

9.2.3 Production of Syngas

Gasification is the preferred route for the production of syngas from coal or biomass. The low price of natural gas is currently encouraging syngas production from it, but the situation may change when the price rises. A steam reformation reaction is used to produce syngas from natural gas that is mainly CH_4. This reaction is also the most widely used commercial method for bulk production of hydrogen, which is one of the two components of syngas.

In the steam reforming method, natural gas (CH_4) reacts with steam at high temperatures (700–1100 °C) in the presence of a metal-based catalyst (nickel).

$$CH_4 + H_2O \xrightarrow{\text{Catalyst}} CO + 3H_2 + 206\ \text{kJ/mol} \tag{9.1}$$

If hydrogen production is the main goal, the carbon monoxide produced is further subjected to a shift reaction (see Eq. 9.2), as described in the next section, to produce additional hydrogen and carbon dioxide.

The ratio of hydrogen and carbon monoxide in the gasification product gas is a critical parameter in the synthesis of the reactant gases into desired products such as gasoline, methanol, and methane. The product desired determines that ratio. For example, gasoline may need the H_2/CO ratio to be 0.5 to 1.0, while methanol may need it to be ~2.0 (Probstein and Hicks, 2006, p. 124). In a commercial gasifier the H_2/CO ratio of the product gas is typically less than 1.0, so the shift reaction is necessary to increase this ratio by increasing the hydrogen content at the expense of CO. The shift reaction often takes place in a separate reactor, as the temperature and other conditions in the main gasifier may not be conducive to it.

9.2.4 Gasification for Syngas Production

The two main routes for production of syngas from biomass or fossil fuel are low-temperature (~<1000 °C) and high-temperature gasification (~>1200 °C).

Low-temperature gasification is typically carried out at temperatures below 1000 °C. In most low-temperature gasifiers, the gasifying medium is air, which introduces undesired nitrogen in the gas. To avoid this, gasification can be carried out indirectly by one of the following means:

- An oxygen carrier (metal oxide) is used to transfer the oxygen from an air oxidizer to another reactor, where gasification takes place using the oxygen from the metal oxide.
- A combustion reaction in air is carried out in one reactor and heat-carrier solids carry the heat to a second reactor, where this heat is then used in gasification.
- Dilution of the product gas by nitrogen is avoided by the use of steam or oxygen as the gasifying medium.

Low-temperature gasification produces a number of heavier hydrocarbons along with carbon monoxide and hydrogen. These heavier hydrocarbons are further cracked, separated, and used for other applications. High-temperature gasification is carried out at temperatures above 1200 °C, where biomass is converted mainly into hydrogen and carbon monoxide. Primary gasification is often followed by the shift reaction, as described in the next section, to adjust the hydrogen-to-carbon monoxide ratio to suit the downstream application.

In any case, the product gas must be cleaned before it is used for synthesis reactions. Special attention must be paid to clean the syngas of tar and other catalyst-poisoning elements before it is used for Fischer-Tropsch synthesis, which uses iron- or cobalt-based catalysts.

Shift Reaction

For a reaction like Fischer-Tropsch synthesis that produces various gaseous and liquid hydrocarbons, a definite molar ratio of CO and H_2 in the syngas is necessary. This is done through the shift reaction that converts excess carbon monoxide into hydrogen:

$$CO + H_2O \xrightarrow{\text{Catalyst}} CO_2 + H_2 - 41.1\,\text{kJ/mol} \tag{9.2}$$

The reaction can be carried out either at higher temperatures (400–500 °C) or at lower temperatures (200–400 °C). For high temperatures, the shift reaction is often catalyzed using oxides of iron and chromium; it is equilibrium limited. At low temperatures, the shift reaction is kinetically limited; the catalyst is composed of copper, zinc oxide, and alumina, which help reduce the CO concentration down to about 1%.

9.2.5 Cleaning and Conditioning of Syngas

For synthesis reaction, a high degree of gas purity is needed, so the gas must be cleaned of particulates and other contaminating gases. The raw syngas may contain three principal types of impurity: (1) solid particulates (unconverted char, ash); (2) inorganic impurities (halides, alkali, sulfur compounds, nitrogen); and (3) organic impurities (tar, aromatics, carbon dioxide).

At high temperatures, the equilibrium shifts toward hydrogen-producing hydrogen-rich gas. The ash in the biomass appears as slag. At low temperatures, the ash remains in the product gas as dry ash. Cleaning has two aspects: removing undesired impurities and conditioning the gas to get the right ratio of H_2 and CO for the intended use. This use determines the level of cleaning and conditioning. Table 9.1 presents examples of product-gas specifications for different end uses.

Cleanup Options

For cleaning the gas of dust or particulates, there are four options: (1) cyclone, (2) fabric or other barrier filter, (3) electrostatic filter, and (4) solvent scrubber. Among organic impurities, tar is the most undesirable. The three main options for tar removal are

- Scrubbing with an organic liquid (e.g., methyl ester)
- Catalytic cracking by nickel-based catalysts or olivine sand
- High-temperature cracking

Inorganic impurities are best removed in sequence because some removal processes produce other components that need to be removed as well. First,

TABLE 9.1 Product-Gas Specifications for Various Applications

Specification	Hydrogen or Refinery Use	Ammonia Production	Methanol Synthesis	Fischer-Tropsch Synthesis
Hydrogen content	>98%	75%	71%	60%
Carbon monoxide content	<10–50 ppm(v)	[$CO + CO_2$] <20 ppm(v)	19%	30%
Carbon dioxide content	<10–50 ppm(v)		4–8%	
Nitrogen content	<2%	25%		
Other gases	N_2, Ar, CH_4	Ar, CH_4	N_2, Ar, CH_4	N_2, Ar, CH_4, CO_2
Balance		As low as possible	As low as possible	Low
H_2/N_2 ratio		~3		
H_2/CO ratio				0.6–2.0
$H_2/[2CO + 3CO_2]$ ratio			1.3–1.4	
Process temperature		350–550 °C	300–400 °C	200–350 °C
Process pressure	>50 bar	100–250 bar	50–300 bar	15–60 bar

Source: Adapted from Knoef, 2005, p. 224.

water quenching removes char and ash particles. Next, hydrolysis removes COS and HCN by converting them into H_2S and NH_3. The ammonia and halides can be washed with water, followed by adsorption of H_2S, which can be removed with the wash water. Solid or liquid adsorbents are used to remove carbon dioxide from the product gas.

9.3 BIO-OIL

Bio-oil (or biofuel) is any liquid fuel derived from a recently living organism, such as plants and their residues or animal extracts. In view of its importance, a detailed discussion of bio-oil is presented next.

9.3.1 What Is Bio-Oil?

Bio-oil is the liquid fraction of the pyrolysis product of biomass. For example, a fast pyrolyzer typically produces 75% bio-oil, 12% char, and 13% gas. Bio-oil is a highly oxygenated, free-flowing, dark-brown (nearly black) organic liquid (Figure 9.1) that contains a large amount of water (~25%) that is partly the

FIGURE 9.1 Bio-oil is a thick, black, tarry liquid.

original moisture in the biomass and partly the reaction product. The composition of bio-oil depends on the biomass it is made from as well as on the process used.

Table 9.2 presents the composition of a typical bio-oil. It shows that water, lignin fragments, carboxylic acids, and carbohydrates constitute its major components. When it comes from the liquid yield of pyrolysis, bio-oil is called *pyrolysis oil*. Several other terms are often used to describe bio-oil or are associated with it, including:

- Tar or pyroligneous tar
- Bio-oil
- Biocrude
- Wood liquid or liquid wood
- Liquid smoke
- Biofuel oil
- Wood distillates
- Pyrolysis oil
- Pyroligneous acids

Note that there is an important difference between pyrolysis oil and biocrude. The former is obtained via pyrolysis; the latter can be obtained via other methods such as supercritical gasification.

TABLE 9.2 Composition of Bio-Oil

Major Group	Compounds	Mass (%)
Water		20–30
Lignin fragments	Insoluble pyrolytic lignin	15–30
Aldehydes	Formaldehyde, acetaldehyde, hydroxyacetaldehyde, glyoxal, methylglyoxal	10–20
Carboxylic acids	Formic, acetic, propionic, butyric, pentanoic, hexanoic, glycolic	10–15
Carbohydrates	Cellobiosan, α-D-levoglucosan, oligosaccharides, 1.6 anhydroglucofuranose	5–10
Phenols	Phenol, cresols, guaiacols, syringols	2–5
Furfurals		1–4
Alcohols	Methanol, ethanol	2–5
Ketones	Acetol (1-hydroxy-2-propanone), cyclopentanone	1–5

Source: Adapted from Bridgwater et al., 2001, p. 989.

Bio-oil may be seen as a two-phase microemulsion. In the continuous phase are the decomposition products of hollocellulose; in the discontinuous phase are the pyrolytic lignin macromolecules. Hollocellulose is the fibrous residue that remains after the extractives, lignin, and ash-forming elements have been removed from the biomass. The same as crude petroleum oil, which is extracted from the ground, pyrolysis liquid and biocrude contain tar as their heaviest component.

Bio-oil is a class-3 substance falling under the flammable liquid designation in the UN regulations for transport of dangerous goods (Peacocke and Bridgwater et al., 2001, p. 1485).

9.3.2 Physical Properties of Bio-Oil

As we can sense from Figure 9.1, bio-oil is free-flowing. Its low viscosity is due to its high water content. Also, it has an acrid, smoky smell that can irritate eyes with long-term exposure. With a specific gravity of ~1.2, bio-oil is heavier than water or any oil derived from petroleum. A comparison of its physical and chemical properties with those of conventional fossil fuels is given in Table 9.3.

An important feature of bio-oil not reflected in Table 9.3 is that some of its properties change with time. For example, its viscosity increases and its volatility decreases (Mohan et al., 2006) with time. Some phase separation and

TABLE 9.3 Comparison of Physical and Chemical Properties of Bio-Oil and Three Liquid Fuels*

Property	Bio-Oil	Heating Oil	Gasoline	Diesel
Heating value (MJ/kg)	18–20	45.5	44[1]	42
Density at 15 °C (kg/m^3)	1200	865	737[1]	820–950[1]
Flash point (°C)	48–55	38	40[1]	42[1]
Pour point (°C)	–15	–6	–60	–29[5]
Viscosity at 40 °C (cP)	40–100 (25% water)[3]	1.8–3.4 cSt	0.37–0.44[3]	2.4[3]
pH	2.0–3.0	–		
Solids (% wt)[4]	0.2–1.0	–	0	0
Elemental Analysis (% weight)				
Carbon	42–47	86.4	84.9	87.4[2]
Hydrogen	6.0–8.0	12.7	14.76	12.1[2]
Nitrogen	<0.1	0.006	0.08	392 ppm[2]
Sulfur	<0.02	0.2–0.7		1.39[2]
Oxygen	46–51	0.04		
Ash	<0.02	<0.01		

*Except as indicated, all values are excerpted from *www.dynamotive.com*.

Note: cP—centipoise; cSt—centipoise. Values for gasoline and diesel are for a representative sample and can vary.

Sources:
[1]*http://www.engineeringtoolbox.com*.
[2]Hughey and Henerickson, 2001.
[3]Bridgwater et al., 2001, p. 990.
[4]Mohan et al., 2006.
[5]Maples, 2000.

deposition of gums may also occur with time, primarily because of polymerization, condensation, esterification, and etherification. This feature distinguishes bio-oil from mineral oils, the properties of which do not change with time.

Bio-oil is not soluble in water, although it contains a substantial amount of water. However, it is miscible in polar solvents, such as methanol and acetone, but immiscible with petroleum-derived oils. Bio-oil can accept water up to a maximum limit of 50% (total moisture). Any more water results in phase separation. Table 9.3 shows that bio-oil has a heating value nearly half that of conventional liquid fuels but has comparable flash and pour points.

9.3.3 Applications for Bio-Oil

Bio-oil is renewable and cleaner than nonrenewable mineral oil extracted from the ground (petroleum). Thus, it offers a "green" alternative in many applications where petro-oil is used. Bio-oil is mainly an energy source, but it may also be used as a feedstock for the production of "green chemicals."

Energy Production

Bio-oil may be fired in boilers and furnaces as a substitute for furnace oil in energy production. This allows a rapid and easy switchover from fossil fuels to biofuels, as it does not call for complete replacement or any major renovation of the firing system as would be needed if raw biomass were to be fired in a furnace or boiler designed for furnace oil. The combustion performance of a bio-oil–fired furnace should be studied carefully before such a switchover is made, because furnace oil and bio-oil have varying combustion characteristics, including significant differences in ignition, viscosity, energy content, stability, pH, and emission level. In many cases we can overcome these differences through proper design (Wagenaar et al., 2009).

Chemical Feedstock Production

Bio-oil is a hydrocarbon similar to petrocrude except that the former has more oxygen. Thus, most of the chemicals produced from petroleum can be produced from bio-oil. These include:

- Resins
- Food flavorings
- Agro-chemicals
- Fertilizers
- Levoglucosan
- Adhesives
- Preservatives
- Acetic acid
- Hydroxyacetaldehyde

Transport Fuel Production

Bio-oil contains less hydrogen per carbon (H/C) atom than do conventional transport fuels like diesel and gasoline, but it can be hydrogenated (hydrogen added) to make up for this deficiency and thereby produce transport fuels with a high H/C ratio. The hydrogen required for the hydrogenation reaction normally comes from an external source, but it can also be supplied by reforming a part of the bio-oil into syngas. This method is practiced by Dynamotive, a Canadian company.

9.3.4 Production of Bio-Oil

Several options for the production of bio-oil are available. They are either thermochemical or biochemical.

- Gasification of biomass and the synthesis of the product gases into liquid (thermochemical)
- Production of biocrude using fast pyrolysis of biomass (thermochemical)
- Production of bio-diesel (fatty acid methyl ester, or FAME) from vegetable oil or fats through transesterification (biochemical)
- Production of ethanol from grains and cellulosic materials (biochemical)

The important steps in the production of bio-oil from biomass are as follows:

1. Receipt at the plant and storage
2. Drying and sizing
3. Reaction (pyrolysis, gasification, fermentation, hydrolysis, etc.)
4. Separation of products into solids, vapor (liquid), and gases
5. Collection of the vapor and its condensation into liquid
6. Upgrading of the liquid to transport fuel or extraction of chemicals from it

9.4 CONVERSION OF SYNGAS INTO CHEMICALS

As mentioned earlier, syngas is an important building block for a host of hydrocarbons. Commercially it finds use in two major areas: (1) alcohols (e.g., methanol, higher alcohols) and (2) chemicals (e.g., glycerol, fumaric acid, ammonia). The following section briefly describes the production of some of these products.

9.4.1 Methanol Production

Methanol (CH_3OH) is an important feedstock for the production of transport fuels and many chemicals. The production of gasoline from methanol is an established commercial process. Methanol is produced through the synthesis of syngas (CO and H_2) in the presence of catalysts (Figure 9.2) (see Higman and van der Burgt, 2008, p. 266):

$$CO + 2\,H_2 \xrightarrow{\text{Catalyst}} CH_3OH - 91\ \text{kJ/mol} \tag{9.3}$$

Methanol synthesis is an exothermic reaction influenced by both temperature and pressure. The equilibrium concentration of methanol in this reaction increases with pressure (in the 50–300 atm range), but reduces with temperature (in the 240–400 °C range). In the absence of a suitable catalyst, the actual yield is very low, so catalysts based on Zn, Cu, Al, and Cr are used.

Syngas, which is the feedstock for methanol production, can be produced from biomass through either thermal or hydrothermal gasification. One of the most commonly used commercial methods involves natural gas (CH_4). This process uses the steam reforming of methane as shown in the next equation:

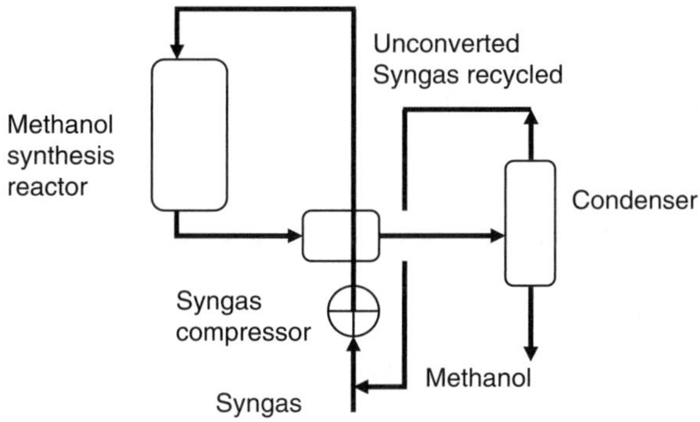

FIGURE 9.2 Methanol production process.

$$CH_4 + H_2O \rightarrow CO + 3\,H_2 + 206\ \mathrm{kJ/mol} \tag{9.4}$$

We note from this reforming reaction that for every mol of CO produced, three moles of H_2 are produced, but the methanol synthesis reaction (Eq. 9.3) requires only two moles of hydrogen for every mole of carbon monoxide. Thus, there is an extra hydrogen molecule for every mol of methanol. In such a situation, carbon dioxide, if available, may be used in the following reaction to produce an additional methanol molecule utilizing the excess hydrogen:

$$CO_2 + 3\,H_2 \rightarrow CH_3OH + H_2O - 50\ \mathrm{kJ/mol} \tag{9.5}$$

Two major routes for methanol synthesis reaction (Reed, 2002, p. III-225) are: (1) high pressure (~30 MPa, 300–400 °C) and (2) low pressure (5–10 MPa, 220–350 °C).

In the high-pressure process, the syngas is first compressed. The pressurized syngas is passed through and then fed into either a fixed- or a fluidized-bed reactor for synthesis in the presence of a catalyst at 300 to 350 atm and 300 to 400 °C. A fluidized bed has the advantage of continuous catalyst regeneration and efficient removal of the generated heat. The catalyst used is an oxide of Zn and Cr.

The product is next cooled to condense the methanol. Since the conversion is generally small, the unconverted syngas is recycled to the reactor to be further converted. Today, the most widely used catalyst is a mixture of copper, zinc oxide, and alumina.

The low-pressure process is similar to the high-pressure process, but of course it uses low pressure and low temperature. In one of several variations, a fixed bed of Cu/Zn/Al catalyst is used at 5 to 10 MPa and 220 to 290 °C (Reed, 2002, p. II-225).

Liquid-phase synthesis is another option, but it is not yet proven. However, it can give a much higher level of conversion (~90%) compared to 20% for the high-pressure process (Chu et al., 2002). Here, the syngas is fed into a slurry of the catalysts in an appropriate solvent. The compressed syngas is mixed with recycled gas and then heated in a heat exchanger to the desired reactor inlet temperature, usually about 220 to 230 °C. In a cold-quench operation, only about two-thirds of the feed gas is preheated; the rest is used to cool the product gas between the individual catalyst layers.

Example 9.1

The production of methanol from syngas is given by the reaction

$$CO + 2\,H_2 \rightarrow CH_3OH \quad \text{(i)}$$

The reaction heat at 25 °C is –90.7 kJ/mol. Using the van Hoff equation,

$$\frac{d\ln K}{dT} = \frac{\Delta H_T^0}{RT^2}$$

Calculate the equilibrium constant, K. Using this constant, find the fraction of the hydrogen in the syngas that will be converted into methanol at 1 atm, 50 atm, and 300 atm at that temperature.

Solution

Let us assume that the reaction started with 1 mol of CO and 2 mol of H_2. If in the equilibrium state only x moles of CO have been converted, it will have consumed $2x$ moles of H_2 and produced x moles of CH_3OH (as per Eq. i), leaving $(1 - x)$ moles of unreacted CO and $2(1 - x)$ moles of H_2. The total number of moles will comprise unreacted moles and the methanol produced. Hence, the total moles will be $1 - x + 2(1 - x) + x = 3 - 2x$.

Noting that partial pressure is proportional to mole fraction, the equilibrium constant is defined as

$$K = \frac{P_{CH_3OH}}{P_{CO}(P_{H_2})^2} = \frac{xP}{(3-2x)} \times \frac{(3-2x)}{(1-x)P} \times \left[\frac{3-2x}{2(1-x)P}\right]^2 \quad \text{(ii)}$$

The equilibrium constant, K, is calculated from the Gibbs free energy using Eq. (6.3).

$$K = \exp\left(\frac{-\Delta G^o{}_T}{RT}\right) \quad \text{(iii)}$$

So, for $T = 25 + 273 = 298$ K, we take the value of $\Delta G^\circ{}_T$ for methanol from Table 6.5:

$$\Delta G^o{}_{298} = -161.6 \text{ kJ/mol}$$

The universal gas constant, R, is known to be 0.008314 kJ/mol K. Substituting these values in Eq. (iii) we get

$$K = \exp(161.6/(0.008314 \times 298)) = 2.12 \times 10^{28}$$

Using this value in Eq. (i), we get a quadratic equation of x. Now, solving x, we get the following:

$$x = 1.0$$

So the equilibrium concentration of the product is

$$CO = 1 - 1 = 0;\ H_2 = 2(1 - 1) = 2\ \text{mol; and } CH_3OH = 1\ \text{mol}$$

At 25 °C, the reaction will produce 2 moles of hydrogen and 1 mole of methanol.

9.4.2 Fischer-Tropsch Synthesis

Fischer-Tropsch (FT) synthesis, developed in the 1920s, is a highly successful method for the production of liquid hydrocarbons from syngas. The FT process can produce high-quality diesel oil from biomass or coal with no aromatics and with a high Cetane number (>70). The composition of this product is very similar to that of petrodiesel. Thus, it can be blended with petrodiesel or used directly in an engine.

Biomass conversion with the FT process may have several additional advantages because its gasification product typically contains $H_2 : CO$ ratio of about unity, which is ideal for iron catalysts. Furthermore, biomass gasification products contain CO_2, which is beneficial for the production of liquid products (Reed, 2002, p. 242). The absence of sulfur in biomass also helps most catalysts.

The most successful and well-known use of FT synthesis is for the production of liquid fuel from coal by SASOL in South Africa, where syngas produced from coal is converted into petroleum products. The FT process is also useful for conversion of biomass into liquid fuels and chemicals.

The FT reaction, which is typically carried out in the range 200 to 350 °C and 20 to 300 atm (Reed, 2002, p. II–238), may be written as

$$n\text{CO} + 2n\text{H}_2 \xrightarrow{\text{Catalyst}} (\text{CH}_2)_n + n\text{H}_2\text{O} + \text{Heat} \tag{9.6}$$

where the hydrocarbon product is represented by the generic formula $(CH_2)_n$.

Fisher-Tropsch reactions produce a wide spectrum of oxygenated compounds, including alcohols and aliphatic hydrocarbons ranging in carbon numbers from $C_1 - C_3$ (gases) to C_{35} + (solid waxes). For synthetic fuels the desired products are olefinic hydrocarbons in the C_5 to C_{10} range (Probstein and Hicks, 2006, p. 128). The upper end of the range favors a gasoline product. Although iron-based catalysts are most favored, cobalt- and nickel-based catalysts have also been used with varying selectivity.

9.4.3 Ammonia Synthesis

Ammonia (NH_3) is an important chemical used for a large number of applications, including production of fertilizers, disinfectants, nitric acid, and refrigerants. It is produced by passing hydrogen and nitrogen over a bed of catalyst at high pressure but moderate temperature. The hydrogen for this reaction can come from biomass gasification.

$$N_2 + 3H_2 \xrightarrow{\text{Catalyst}} 2NH_3 \tag{9.7}$$

Catalysts play an important role in this reaction. Iron catalysts (FeO, Fe_2O_3) with added promoters like oxides of aluminium, calcium, potassium, silicon, and magnesium are used (Reed, 2002, p. III-250).

Because the gasification of biomass yields syngas, which contains both CO and H_2, for production of ammonia, the syngas must first be stripped of its CO through the shift reaction (Eq. 9.2). As mentioned earlier, the shift conversion is aided by commercial catalysts, such as iron oxide and chromium oxide, that work in a high-temperature range (350–475 °C); zinc oxide–copper oxide catalysts work well in a low-temperature range (200–250 °C).

In a typical ammonia synthesis process, the syngas is first passed through the shift reactor, where CO is converted into H_2 and CO_2 following the shift reaction. Then the gas is passed through a CO_2 scrubber, where a scrubbing liquid absorbs the CO_2; this liquid is passed to a regenerator for regeneration by stripping the CO_2 from it. The cleaned gas then goes through a methanation reactor to remove any residual CO or CO_2 by converting it into CH_4. The pure mixture of hydrogen obtained is mixed with pure nitrogen and is then compressed to the required high pressure of the ammonia synthesis. The product, a blend of ammonia and unconverted gas, is condensed, and the unconverted syngas is recycled to the ammonia converter.

9.4.4 Glycerol Synthesis

Biodiesel from fat or oil produces a large amount (about 10%) of glycerol ($HOCH_2CH[OH]CH_2OH$) as a by-product. Large-scale commercial production of biodiesel can therefore bring a huge amount of glycerol into the market. For example, for every kg of biodiesel, 0.1 kg of glycerol is produced (86% FAME, 9% glycerol, 4% alcohol, and 1% fertilizer) (*www.biodiesel.org*). If produced in the required purity (>99%), glycerol may be sold for cosmetic and pharmaceutical production, but that market is not large enough to absorb it all. Therefore, alternative commercial uses need to be explored. These include:

- Catalytic conversion of glycerol into biogas (C_8–C_{16} range) (Hoang et al., 2007)
- Liquid-phase or gas-phase reforming to produce hydrogen (Xu et al., 1996)

A large number of other chemicals may potentially come from glycerol. Zhou et al. (2008) reviewed several approaches for a range of chemicals and fuels. Through processes like oxidation, transesterification, esterification, hydrogenolysis, carboxylation, catalytic dehydration, pyrolysis, and gasification, many value-added chemicals can be produced from glycerol.

9.5 TRANSPORT FUELS FROM BIOMASS

Biodiesel, ethanol, and biogas are transport fuels produced from biomass that are used in the transportation industry. The composition of biodiesel and biogas may not be exactly the same as their equivalence from petroleum, but they perform the same task. Ethanol derived from biomass is either used as the sole fuel or mixed with gasoline in spark-ignition engines.

There are two thermochemical routes available for the production of diesel and gasoline from syngas: (1) gasoline, through the methanol-to-gasoline (MTG) process; and (2) diesel, through the FT process. The two biochemical means for production of ethanol and diesel are

- Diesel, through the transesterification of fatty acids
- Ethanol, through the fermentation of sugar

It may be noted that in both schemes part of the syngas's energy content (30–50%) is lost during conversion into liquid transport fuel. It is apparent from Table 9.4 that this loss in conversion from biomass to methanol or ethanol can be as high as 50%, and further loss can occur when the methanol is converted into a transport fuel like gasoline. For this reason, when we consider the overall energy conversion efficiency of a car run on biogas and compare that with an electric car, the former shows a rather low fuel-to-wheel energy ratio.

9.5.1 Biochemical Ethanol Production

Ethanol is the most extensively used biofuel in the transportation industry. Ethanol can be mixed with gasoline (petroleum) or used alone for operating spark-ignition engines, just as biodiesel can be mixed with petrodiesel for operating compression-ignition engines. In most cases engine modifications may not be needed for substitution of mineral oil with bio-oil–derived fuels. Ethanol is produced mainly from food crops, but, less commonly, it can also be produced from nonfood ligno-cellulosic biomass.

TABLE 9.4 Energy Losses in Methanol Production

Conversion Process	Energy Loss (%)
Biomass to methanol	30–47
Coal to methanol	41–75

Source: Data compiled from Reed, 2002, p. III-226.

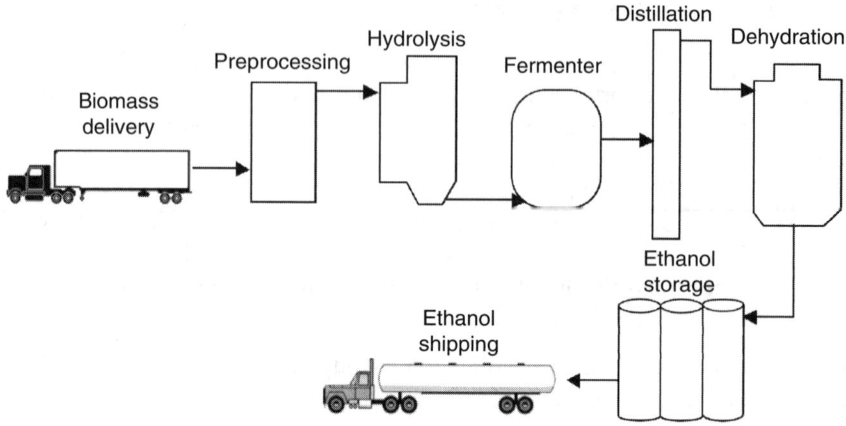

FIGURE 9.3 Ethanol production from food cereal.

Ethanol from Food Sources

Ethanol (C_2H_6O) is presently produced primarily from glucose from grain (corn, maize, etc.), sugar (sugarcane), and energy crops using the fermentation-based biochemical process. A typical process, as shown in Figure 9.3, comprises the following major steps:

Milling: Corn is ground to a fine powder called cornmeal.
Liquefying: A large amount of water is added to make the cornmeal into a solution.
Hydrolysis: Enzymes are added to the solution to break large carbohydrate molecules into shorter glucose molecules.
Fermentation: The glucose mixture is taken to the fermentation batch reactor, where yeast is added. The yeast converts the glucose into ethanol and carbon dioxide as represented by the equation

$$C_6H_{12}O_6\,(\text{glucose}) + \text{yeast} \xrightarrow{\text{Fermentation}} 2C_2H_6O\,(\text{ethanol}) + 2CO_2 \quad (9.8)$$

Distillation: The product of fermentation contains a large amount of water and some solids, so the water is removed through distillation. Distillation purifies ethanol to about 95 to 96% purity. The solids are pumped out and discarded as a protein-rich stock, which may be used only for animal feed.
Dehydration: The ethanol produced is good enough for car engines in countries like Brazil, but further purification is needed if it has to be blended with mineral gasoline for ordinary cars. In this stage, a molecular sieve is used for dehydration. Small beads with pores large enough for water but not for ethanol absorb the water.

A large amount of energy is consumed in distillation and other steps in this process. By one estimate, for the production of 1 liter of purified ethanol, about

12,350 kJ of energy is needed for processing, especially for dehydration. An additional 7440 kJ/L of energy consumed in harvesting the corn is required (Wang and Pantini, 2000). Although a liter of ethanol releases 21,200 kJ of energy when burnt, the farming and processing of the corn consumes about 19,790 kJ of energy. The net energy production is therefore a meager 1410 kJ (21,200 – 19,790) per liter of ethanol.

The shortcoming of this process is that it uses a valuable food source—indeed, a staple food in many countries. The search for an alternative is therefore ongoing. Though not fully commercial yet, some methods are available using either the biochemical or the thermochemical process.

Ethanol from Nonfood Sources

The conventional means of producing ethanol from food sources like corn and sugarcane is, commercially, highly successful. In contrast, the production of ethanol from nonfood biomass (ligno-cellulose), although feasible in principle, is not widely used. More processing is required to make the sugar monomers in ligno-cellulose feedstock available to the microorganisms that produce ethanol by fermentation. However, production from food sources, even though it strains the food supply and is wasteful, is widespread.

Consider that only 50% of the dry kernel mass is transformed into ethanol, while the remaining kernel and the entire stock of the corn plant, regardless that it is grown using cultivation energy and incurs expenses, remains unutilized. It is difficult to ferment this part, which contains ligno-cellulose mass, so it is discarded as waste. Alternative methods are being developed to convert the cellulosic components of biomass into ethanol so that they can also be utilized for transport fuel. This option is discussed further in Section 9.5.4.

9.5.2 Gasoline

Petrogasoline is a mixture of hydrocarbons having a carbon number (i.e., the number of carbon-per-hydrocarbon molecules) primarily in the range of 5 to 11. These hydrocarbons belong to the following groups:

- Paraffins or alkanes
- Aromatics
- Olefins or alkenes
- Cycloalkanes or naphthenes

Gasoline Production from Methanol

Methanol may be converted into gasoline using several processes. One of these, Exxon Mobil's methanol-to-gasoline (MTG) process, is well known (Figure 9.4). Methanol is converted into hydrocarbons consisting of mainly (>75%) gasoline-grade materials (C_5–C_{12}) with a small amount of liquefied petroleum gas (C_3–C_4) and fuel gas (C_1–C_2). Mobil uses both fixed beds and fluidized beds

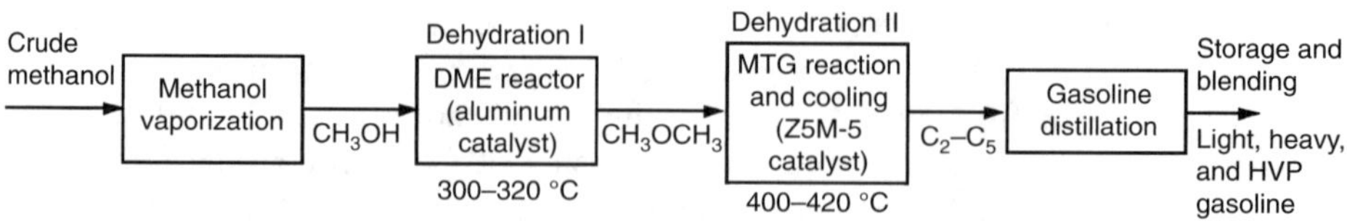

FIGURE 9.4 Production of gasoline from methanol through the MTG process.

of proprietary catalysts for this conversion. The reaction is carried out in two stages: the first stage is dehydration to produce dimethyl ether intermediate; the second stage is also dehydration, this time over a zeolite catalyst, ZSM-5, to give gasoline.

$$2CH_3OH \rightarrow (-H_2O) \xrightarrow[\text{Alumina catalyst}]{300\text{–}320\ ^\circ\text{C}} CH_3OCH_3 \rightarrow (-H_2O) \xrightarrow[\text{ZSM catalyst}]{400\text{–}420\ ^\circ\text{C}} C_2\text{–}C_5$$
$$\rightarrow \text{paraffins} + \text{aromatics} + \text{cycloparaffins} \quad (9.9)$$

where ($-H_2O$) represents the dehydration step.

The typical composition of the gasoline in weight percentage (see *nzic.org.nz/ChemProcesses/energy/7D.pdf*) is as follows:

- Highly branched alkanes: 53%
- Highly branched alkenes: 12%
- Napthenes: 7%
- Aromatics: 28%

The dehydration process produces a large amount of water. For example, from 1000 kg of methanol, 387 kg of gasoline, 46 kg of liquefied petroleum gas, 7 kg of fuel gas, and 560 kg of water are produced (Adrian et al., 2007). Figure 9.4 shows a simplified scheme for the production of gasoline from methanol. This gasoline, sometimes referred to as *MTG gasoline*, is completely compatible with petrogasoline.

9.5.3 Diesel

Generally, the oil burnt in a diesel (compression-ignition) engine is called *diesel*. If produced from petroleum, it is called *petrodiesel*, and if produced from biomass, it is called *biodiesel*. Mineral diesel (or petrodiesel) is made of a large number of saturated and aromatic hydrocarbons. The average chemical formula can be $C_{12}H_{23}$. Petrodiesel (also called fossil diesel) is produced from the fractional distillation of crude oil between 200 °C and 350 °C at atmospheric pressure, resulting in a mixture of carbon chains that typically contain between 8 and 21 carbon atoms per molecule (Collins, 2007).

According to the American Society for Testing and Materials (ASTM), biodiesel (B100) is defined as "a fuel comprised of mono-alkyl (methyl) esters of long chain fatty acids derived from vegetable oils or animal fats, and meeting the requirements of ASTM D 6751." Its characteristics are similar to those of petrodiesel, but not identical. Biodiesel, which can be mixed with petrodiesel for burning in diesel engines, has several positive features for use in engines, as listed in the following:

- Petrodiesel contains up to 20% polyaromatic hydrocarbon, while biodiesel contains none, making it safer for storage.
- Biodiesel has a higher flash point, making it safer to handle.
- Being oxygenated, biodiesel is a better lubricant than petrodiesel is and therefore gives longer engine life.
- Its higher oxygen content allows biodiesel to burn more completely.

Biodiesel Production from Methanol

Biodiesel is generally produced from vegetable oil and/or from animal fats with major constituents that are triglycerides. It is produced by transesterification of vegetable oil or fat in the presence of a catalyst. Biodiesel carries the name *fatty acid methyl (or ethyl) ester*, commonly abbreviated as FAME. A popular production method involves mixing waste vegetable oil or fat with the catalyst and methanol (or ethanol) in appropriate proportion. A typical proportion is 87% oil, 1% NaOH catalyst, and 12% alcohol. Both acid and base catalysts can be used, but the base catalyst NaOH is the most common. Because NaOH is not recyclable, a "nongreen" feed is required to produce "green" biodiesel. Efforts are being made to produce recyclable catalysts and thereby make the product pure "green."

Figure 9.5 shows the reaction for the conversion of triglyceride into biodiesel (FAME) and its by-product, glycerol. Glycerol cannot be used as a transport fuel, and its disposal is a major issue.

An alternative noncatalytic conversion route for biodiesel is under development in which transesterification of triglycerides is by supercritical methanol (above 293 °C, 8.1 MPa) without a catalyst (Kusdiana et al., 2006). The methanol can be recycled and reused, but the process for this must be carried out at high temperatures and pressures. Efforts are also being made to use woody biomass (ligno-cellulose) instead of fats or oil to produce biodiesel using the supercritical method (Minami and Saka, 2006). The reaction is carried out in

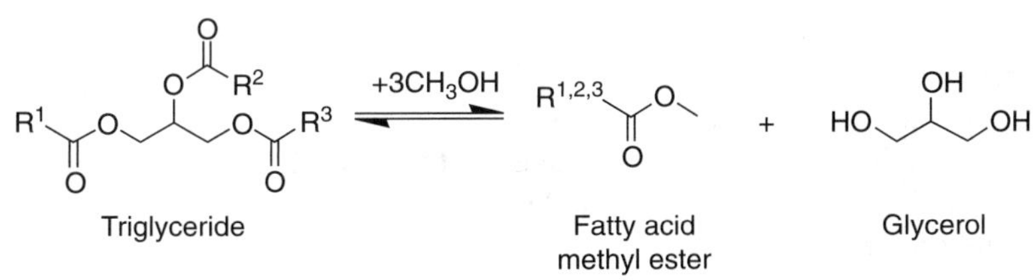

FIGURE 9.5 Diesel (fatty acid methyl ester) production from triglyceride.

a fixed or fluidized bed. The fluidized bed has the advantage of continuous catalyst regeneration and efficient removal of the heat of reaction.

9.5.4 Transport Fuel Production from Nonfood Biomass

Use of food cereals, such as wheat and corn for the production of biodiesel or ethanol, has been commercially successful; however, it has had a major impact on the world's food market, driving up prices and creating shortages. Alternative sources of biodiesel are being researched. Instead of sugar beets or rape seed, cellulosic biomass like wood may be used as the feedstock. With cellulosic materials, the industry can significantly increase the yield of fuel per unit of cultivated area.

There are two options for production of ethanol or gasoline from nonfood sources: thermal and biochemical.

Thermal Process

In the thermal process, cellulosic feedstock is subjected to fast pyrolysis (Chapter 3). The liquid produced is refined and upgraded to gasoline or ethanol. Since cellulose is the feedstock, the ethanol from it is often referred to as *cellulosic ethanol.* An alternative thermal process involves gasification of the biomass to produce syngas and synthesis of the syngas into diesel oil using the FT process. This process was described in Section 9.4.2 and is illustrated in Figure 9.6.

Biochemical Process

Figure 9.7 illustrates the biochemical process for production of ethanol from nonfood ligno-cellulosic biomass. To produce alcohol, the long-chain sugar molecules in the cellulose must be broken down into free sugar molecules. Only then can the sugar be fermented into alcohol (ethanol), as in the food-based process (refer to Figure 9.3). This extra step of breaking down to free sugar molecules is not necessary in the latter process because there the feedstock (corn, sugarcane, etc.) is already in sugar form.

The breakdown of cellulose into sugar can be carried out by either (1) acid hydrolysis or (2) enzymatic hydrolysis. The production of cellulosic ethanol typically takes five steps:

1. Feed preparation (i.e., mechanical cleaning and sizing, physico-chemical preparation)
2. Hydrolysis (conversion into sugar)
3. Fermentation (conversion of the sugar into ethanol)
4. Distillation (removal of water and solids)
5. Dehydration (final drying)

The second step is different for the two biochemical processes. All other steps are the same.

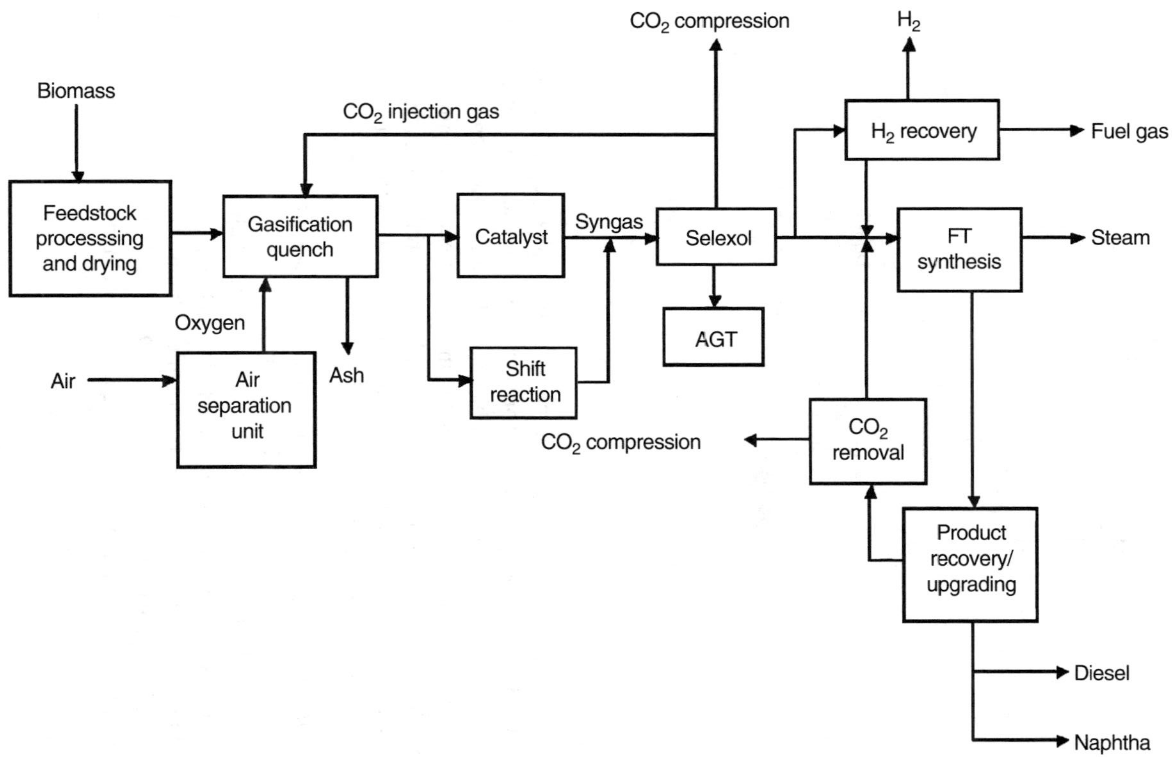

FIGURE 9.6 Transport fuel production from coal and biomass using Fisher-Tropsch synthesis. *(Source: Adapted from White et al., 2007.)*

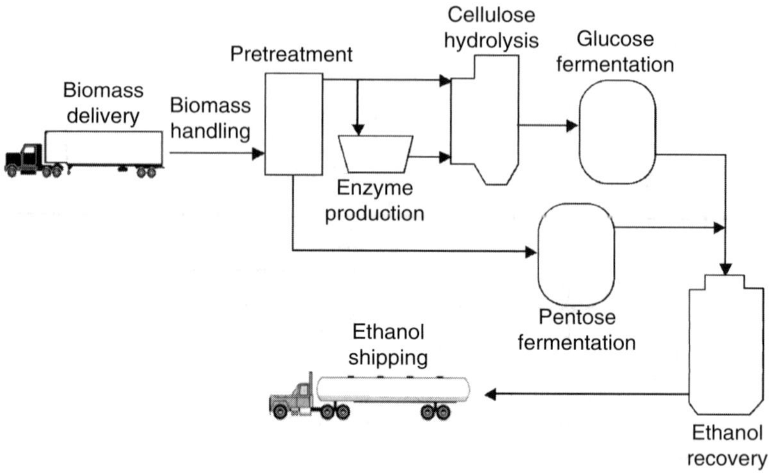

FIGURE 9.7 Biochemical process for producing ethanol from ligno-cellulosic feed.

Feed Preparation

This step prepares the biomass for processing. It involves cleaning and then pretreatment. Unlike food grain (e.g., corn, wheat), ligno-cellulose often come mixed with dirt and debris. These must be cleaned from the delivered biomass, which is then shredded into small particles. In pretreatment, the hemicellulose/lignin sheath that surrounds the cellulose in plant material is disrupted. Physical, chemical, or biological pretreatment, as follows, makes the cellulose more accessible to the hydrolysis process.

Physical methods: grinding, milling, shearing (energy intensive) and steam explosion (to produce some inhibitory compounds)
Chemical methods: treatment with acid (for pH neutralization and recovery of chemicals), treatment with alkalis (for pH adjustment and recycling of chemicals), and treatment with organic solvents (solvent removal and recycling is expensive)
Biological method: enzymatic treatment of the cellulose (time consuming)

Hydrolyis

Acid hydrolysis uses dilute acid at high temperature and pressure. Concentrated acid at lower temperature and pressure may be used, but this produces a toxic by-product that inhibits fermentation and so must be removed.

In enzymatic hydrolysis, cellulose chains are broken into glucose molecules by cellulose enzymes, in a process similar to what occurs in the stomach of a cow to convert grass or fodder cellulose into sugar. Xylanose and hemicellulose enzymes can convert many cellulosic agricultural residues into fermentable sugars. These residues include corn stover, distiller grains, wheat straw, and sugarcane bagasse, as well as energy crops such as switch grass. Lignin is difficult to convert into sugar, so it is discarded as waste. Figure 9.7 shows a process based on cellulose hydrolysis.

Fermentation of Hemicellulosic Sugars

Through a series of biochemical reactions, bacteria convert xylose and other hemicellulose and cellulose sugars into ethanol. The yeast or other microorganisms for the fermentation of cellulose and that for hemicellulose are not necessarily the same. In any case, they consume sugar molecules and produce ethanol and carbon dioxide.

Distillation and Dehydration

Dilute ethanol broth produced during the fermentation of hemicellulose and cellulose sugars is distilled to remove water and concentrate the ethanol. Solid residues containing lignin and microbial cells can be burned to produce heat or used to generate electricity consumed by ethanol production. Alternately, the solids can be converted to co-products, such as animal feed and nutrients for crops). The last step in the process involves removal of the remaining water from the distilled ethanol.

Appendix A

Definition of Biomass

In the United States, the definition of *biomass* has been hotly debated. Currently, the generally accepted definition can be found in the American Clean Energy and Security Act of 2009, HR 2454, excerpted as follows.

The term "renewable biomass" means any of the following:

(A) Plant material, including waste material, harvested or collected from actively managed agricultural land that was in cultivation, cleared, or fallow and non-forested on the date of enactment of this section;

(B) Plant material, including waste material, harvested or collected from pastureland that was non-forested on such date of enactment;

(C) Nonhazardous vegetative matter derived from waste, including separated yard waste, landscape right-of-way trimmings, construction and demolition debris or food waste (but not municipal solid waste, recyclable waste paper, painted, treated or pressurized wood, or wood contaminated with plastic or metals);

(D) Animal waste or animal byproducts, including products of animal waste digesters;

(E) Algae;

(F) Trees, brush, slash, residues, or any other vegetative matter removed from within 600 feet of any building, campground, or route designated for evacuation by a public official with responsibility for emergency preparedness, or from within 300 feet of a paved road, electric transmission line, utility tower, or water supply line;

(G) Residues from or byproducts of milled logs;

(H) Any of the following removed from forested land that is not Federal and is not high conservation priority land:

- *(i) Trees, brush, slash, residues, inter-planted energy crops, or any other vegetative matter removed from an actively managed tree plantation established—*
  - *(I) Prior to the date of enactment of this section; or*
  - *(II) On land that, as of the date of enactment of this section, was cultivated or fallow and non-forested.*
- *(ii) Trees, logging residue, thinnings, cull trees, pulpwood, and brush removed from naturally regenerated forests or other non-plantation forests, including for the purposes of hazardous fuel reduction or preventative treatment for reducing or containing insect or disease infestation.*

Biomass Gasification and Pyrolysis. DOI: 10.1016/B978-0-12-374988-8.00015-5

(iii) Logging residue, thinnings, cull trees, pulpwood, brush and species that are non-native and noxious, from stands that were planted and managed after the date of enactment of this section to restore or maintain native forest types.

(iv) Dead or severely damaged trees removed within 5 years of fire, blow down, or other natural disaster, and badly infested trees:

(I) Materials, pre-commercial thinnings, or removed invasive species from National Forest System land and public lands (as defined in section 103 of the Federal Land Policy and Management Act of 1976 (43 U.S.C. 1702)), including those that are byproducts of preventive treatments (such as trees, wood, brush, thinnings, chips, and slash), that are removed as part of a federally recognized timber sale, or that are removed to reduce hazardous fuels, to reduce or contain disease or insect infestation, or to restore ecosystem health, and that are—

(i) Not from components of the National Wilderness Preservation System, Wilderness Study Areas, inventoried road-less areas, old growth or mature forest stands, components of the National Landscape Conservation System, National Monuments, National Conservation Areas, Designated Primitive Areas, or Wild and Scenic Rivers corridors;

(ii) Harvested in environmentally sustainable quantities, as determined by the appropriate Federal land manager; and

(iii) Harvested in accordance with Federal and State law and applicable land management plans.

Another accepted definition is that of the Ontario Corporations Tax Act, excerpted as follows:

The term "biomass resource" means

(a) organic matter that is derived from a plant and available on a renewable basis, including organic matter derived from dedicated energy crops, dedicated trees, agricultural food and feed crops, or

(b) waste organic material from harvesting or processing agricultural products, including animal waste and rendered animal fat, forestry products, including wood waste, and sewage.

Physical Constants

Atmospheric pressure

$101.325\ N/m^2$

101.325 kPa

1.013 bar

Avogadro's number

6.022×10^{23}/mol

Boltzmann's constant

1.380×10^{-23} J/K

Gravitational acceleration (sea level), g

$9.807\ m/s^2$

Planck's constant

6.625×10^{-34} J.s

Speed of light in vacuum

2.998×10^{8} m/s

Stefan-Boltzmann constant

$5.670 \times 10^{-8}\ W/m^2.K^4$

Universal gas constant, R

$8.205 \times 10^{-2}\ m^3$.atm/kmol.K $= 8.314 \times 10^{-2}\ m^3$.bar/kmol.K

= 8.314 kJ/kmol.K

= 282 N.m/kg.K

= 8.314 kPa.m^3/kmol.K

= 1.98 kCal/kmol.K

Biomass Gasification and Pyrolysis. DOI: 10.1016/B978-0-12-374988-8.00017-9

Appendix C

Selected Design Data Tables

Table C.1 Fusibility of Biomass Ash

	Temperature (°C)			
Type	*Initial Deformation*	*Softening*	*Hemispherical*	*Fluid*
Corn cob[1]	900		1020	
Corn stalk[1]	820		1091	
Grape pruning (oxidizing)[2]	1313	1368	1374	1424
Grape pruning (reducing)[2]	1310	1360	1371	1382
Olive pit[1]	850		1480	
RDF pellet[1]	890		1130	
RDF (oxidizing)[3]	1065	1092	1131	1193
RDF (reducing)[3]	1024	1063	1097	1182
Rice hulls	1439		>1650	
Rice straw[1]	1060		1250	
Walnut shell[1]	820		1225	

[1]*Source:* Osman and Goss, 1983.
[2]*Source:* Rossi, 1984, pp. 69–99.
[3]*Source:* Alter and Campbell, 1979, pp. 127–142.

Biomass Gasification and Pyrolysis. DOI: 10.1016/B978-0-12-374988-8.00018-0

TABLE C.2 Standard Heating Value of Constituents of Typical Product Gas from Biomass Gasification

Gases	H_2	CO	CO_2	CH_4	C_2H_6	C_2H_4	C_2H_2	C_3H_8
HHV (MJ/Nm^3)[2]	12.74	12.63		39.82	70.29	63.41	58.06	101.24
LHV (MJ/Nm^3)[2]	10.78	12.63		35.88	64.34	59.45	56.07	99.09
Viscosity[1] (μP)	90	182	150	112	94	103	104	82
Thermal conductivity[1] (W/m.K)	0.1820	0.0251	0.0166	0.0343	0.0218	0.0214	0.0213	0.0183
Specific heat[1] (kJ/kg.K)	3.467	1.05	0.85	2.226	1.926	1.691	1.775	1.708
Gases	C_3H_6	i-C_4H_8	i-C_4H_{10}	n-C_4H_{10}	C_6H_6	N_2	NH_3	H_2S
HHV (MJ/Nm^3)[2]	93.57	125.08	133.12	134.06	142.89		13.07	25.10
LHV (MJ/Nm^3)[2]	87.57	116.93	122.91	123.81	141.41		10.13	23.14
Viscosity[1] (μP)						180		
Thermal conductivity[1] (W/m.K)						0.026		
Specific heat[1] (kJ/kg.K)						1.05		

[1]*Source*: Data compiled from Jenkins, in Kitani and Hall, 1989, p. 887.
[2]*Source*: Data compiled from Waldheim and Nilsson, 2001.

TABLE C.3 Composition of Standard Air at Atmospheric Pressure

Gas	By Volume	By Weight	Molecular Weight
Nitrogen	78.09	75.47	28.02
Oxygen	20.95	23.2	32
Argon	0.933	1.28	39.94
Carbon dioxide	0.03	0.046	44.01

Table C.4 Specific Heat of Gases

Gas	Molecular Weight	Specific Heat* at Temperature, T (K)	Range of Validity (K)
H_2S	34	30.139 + 0.015*T	300–600
H_2O_{steam}	18	34.4 + 0.000628*T + $0.0000052T^2$	300–2500
H_2	2	27.71 + 0.0034*T	273–2500
CH_4	16	22.35 + 0.048*T	273–1200
CO	28	27.62 + 0.005T	273–2500
CO_2	44	43.28 + 0.0114*T-818363/T^2	273–1200
O_2	32	34.62 + 0.00108T-785712/T^2	300–5000
N_2	28	27.21 + 0.0042T	300–5000

*Amounts in kJ/kmol.K.

Source: Adapted from Perry and Green, 1997.

Table C.5 Physical Properties of Air at Various Temperatures

Temperature (K)	Density (kg/m^3)	Dynamic Viscosity, $\mu.10^7$ ($N.s/m^2$)	Kinematic Viscosity, $\gamma.10^6$ (m^2/s)	Thermal Conductivity, $K_g.10^3$ (W/m.K)	Thermal Diffusivity, $\alpha.10^6$ (m^2/s)	Prandtl Number
100	3.5562	71.1	2.00	9.34	2.54	0.786
150	2.3364	103.4	4.426	13.8	5.84	0.758
200	1.7458	132.5	7.590	18.1	10.3	0.737
250	1.3947	159.6	11.44	22.3	15.9	0.720
300	1.1614	184.6	15.89	26.3	22.5	0.707
350	0.9950	208.2	20.92	30.0	29.9	0.700
400	0.8711	230.1	26.41	33.8	38.3	0.690
450	0.7740	250.7	32.39	37.3	47.2	0.686
500	0.6964	270.1	38.79	40.7	56.7	0.684
550	0.6329	288.4	45.57	43.9	66.7	0.683
600	0.5804	305.8	52.69	46.9	76.9	0.685
650	0.5356	322.5	60.21	49.7	87.3	0.690
700	0.4975	338.8	68.10	52.4	98.0	0.695
750	0.4643	354.6	796.37	54.9	109	0.702
800	0.4354	369.8	84.93	57.3	120	0.709
850	0.4097	384.3	93.80	59.6	131	0.716
900	0.3868	398.1	102.9	62.0	143	0.720
950	0.3666	411.3	112.2	64.3	155	0.723
1000	0.3482	424.4	121.9	66.7	168	0.726
1100	0.3166	449.0	141.8	71.5	195	0.728
1200	0.2902	473.0	162.9	76.3	224	0.728
1300	0.2679	496.0	185.1	82	238	0.719
1400	0.2488	530	213	91	303	0.703
1500	0.2322	557	240	100	350	0.685
1600	0.2177	584	268	106	390	0.688

TABLE C.6 Heat of Formation of Various Elements and Compounds at Standard Condition, 25 °C, and 1 Bar Pressure

Substance	$\Delta H_f°$ (kJ/mol)	$S°$ (J/K.mol)	$\Delta G_f°$ (kJ/mol)	Substance	$\Delta H_f°$ (kJ/mol)	$S°$ (J/K.mol)	$\Delta G_f°$ (kJ/mol)
$Al_{(s)}$	0	28.3	0	$NH_{3\,(g)}$	–46.1	192.5	–16.5
$Al_2O_{3\,(s)}$	–1675.7	50.9	–1582.3	$N_2H_{4\,(l)}$	50.6	121.2	149.3
$Br_{2\,(l)}$	0	151.6	0	$NH_4Cl_{(s)}$	–314.4	94.6	–202.9
$HBr_{(g)}$	–36.4	198.7	–53.5	$NH_4NO_{3\,(s)}$	–365.6	151.1	–183.9
$Ca_{(s)}$	0	41.4	0	$NO_{(g)}$	90.3	210.8	86.6
$CaCO_{3\,(s)}$ (calcite)	–1206.9	92.9	–1128.8	$NO_{2\,(g)}$	33.2	240.1	51.3
$CaCl_{2\,(s)}$	–795.8	104.6	–748.1	$N_2O_{(g)}$	82.1	219.9	104.2
$C_{(s)}$ (graphite)	0	5.7	0	$N_2O_{4\,(g)}$	9.2	304.3	97.9
				$HNO_{3\,(l)}$	–174.1	155.6	–80.7
$CCl_{4\,(l)}$	–135.4	216.4	–65.2	$O_{(g)}$	249.2	161.1	231.7
$CCl_{4\,(g)}$	–96.0	309.9	–60.6	$O_{2\,(g)}$	0	205.1	0
$CHCl_{3\,(l)}$	–134.5	201.7	–73.7	$O_{3\,(g)}$	142.7	238.9	163.2
$CH_{4\,(g)}$	–74.8	186.3	–50.7				
$C_2H_{2\,(g)}$	226.7	200.9	209.2				
$C_2H_{4\,(g)}$	52.3	219.6	68.2				
$C_2H_{6\,(g)}$	–84.7	229.6	–32.8				
$C_3H_{8\,(g)}$	–103.8	269.9	–23.5				
$C_6H_{6\,(l)}$	49.0	172.8	124.5				
$CH_3OH_{(l)}$	–238.7	126.8	–166.3				
$C_2H_5OH_{(l)}$	–277.7	160.7	–178.8	$K_{(s)}$	0	64.2	0
$CH_3CO_2H_{(l)}$	–484.5	159.8	–389.9	$KCl_{(s)}$	–436.7	82.6	–409.1
$CO_{(g)}$	–110.5	197.7	–137.2	$KClO_{3\,(s)}$	–397.7	143.1	–296.3
$CO_{2\,(g)}$	393.5	213.7	–394.4	$KOH_{(s)}$	–428.8	78.9	–379.1
$COCl_{2\,(g)}$	–218.8	283.5	–204.6				
$CS_{2\,(g)}$	+117.4	237.8	67.1				
$Cl_{2\,(g)}$	0	223.1	0				

Continues

TABLE C.6 Heat of Formation of Various Elements and Compounds at Standard Condition, 25 °C, and 1 Bar Pressure *Continued*

Substance	$\Delta H_f°$ (kJ/mol)	$S°$ (J/K.mol)	$\Delta G_f°$ (kJ/mol)	Substance	$\Delta H_f°$ (kJ/mol)	$S°$ (J/K.mol)	$\Delta G_f°$ (kJ/mol)
$HCl_{(g)}$	−92.3	186.9	−95.3	$Na_{(s)}$	0	51.2	0
$CrCl_{3\,(s)}$	−556.5	123.0	−486.1	$NaCl_{(s)}$	−411.2	72.1	−384.1
$Cu_{(s)}$	0	33.2	0	$NaOH_{(s)}$	−425.6	64.5	−379.5
$CuO_{(s)}$	−157.3	42.6	−129.7	$Na_2CO_{3\,(s)}$	−1130.7	135.0	−1044.0
CuCl	−137.2	86.2	−119.9				
$CuCl_{2\,(s)}$	−220.1	108.1	−175.7	$S_{(g)}$	278.8	167.8	238.3
$F_{2\,(g)}$	0	202.8	0	$SF_{6\,(g)}$	−1209.0	291.8	−1105.3
$HF_{(g)}$	−271.1	173.8	−273.2	$H_2S_{(g)}$	−20.6	205.8	−33.6
$He_{(g)}$	0	126.0	0	$SO_{2\,(g)}$	−296.8	248.2	−300.2
$H_{2\,(g)}$	0	130.7	0	$SO_{3\,(g)}$	−395.7	256.8	−371.1
$H_2O_{(l)}$	−285.8	69.9	−237.1	$H_2SO_{4\,(l)}$	−814.0	156.9	−690.0
$H_2O_{(g)}$	−241.8	188.8	−228.6				
$H_2O_{2\,(l)}$	−187.8	109.6	−120.4				
$Fe_{(s)}$	0	27.8	0				
$FeO_{(s)}$	−272.0	57.6	245.1				
$Fe_2O_{3\,(s)}$	−824.2	87.4	−742.2				
$Fe_3O_{4\,(s)}$	−1118.4	146.4	−1015.4				
$FeCl_{2\,(s)}$	−341.8	118.0	−302.3				
$FeCl_{3\,(s)}$	−399.5	142.3	−344.0				
$FeS_{2\,(s)}$	−178.2	52.9	−166.9				
$Pb_{(s)}$	0	64.8	0				
$Ne_{(g)}$	0	146.2	0				
$N_{2\,(g)}$	0	191.6	0				

Source: Data compiled from the University of Saskatoon, chemistry department, *http://www.saskschools.ca/curr_content/chem30_05/.*

Note: Subscripts in parentheses indicate the state: solid (s), liquid (l), or gaseous (g).

Table C.7 Equilibrium Constants for the Water–Gas, Boudouard, and Methane Formation Reactions (JANAF Thermochemical Tables)

Temperature (K)	$K_{PW}\left(=\frac{P_{H_2}P_{CO}}{P_{H_2O}}\right)$	$K_{PB}\left(=\frac{P_{CO}^2}{P_{CO_2}}\right)$	$K_{PM}\left(=\frac{P_{CH_4}}{P_{H_2}^2}\right)$
400	7.709×10^{-11}	5.225×10^{14}	9.481×10^{4}
600	5.058×10^{-5}	1.870×10^{-6}	8.291×10^{2}
800	4.406×10^{-2}	1.090×10^{-2}	5.246×10^{0}
1000	2.617×10^{0}	1.900×10^{0}	2.727×10^{-2}
1500	6.081×10^{2}	1.622×10^{3}	3.762×10^{-8}

Note: See Chapter 5 for definitions.

Table C.8 Specific Heat of Biomass and Related Materials

Type	Specific Heat (kJ/kg.K)	Temperature (K)
Carbon[1]	0.70	299–349
	1.60	329–1723
Cellulose[2]	1.34	
Graphite	0.84[3]	273–373
	1.62[1]	329–1723
Wood (Oven dry, avg. 20 species)	1.37[4]	273–379
Wood charcoal	0.84[3]	273–273

[1]*Source:* Perry, Green, and Maloney, 1984.
[2]*Source:* Kollman and Cote, 1968.
[3]*Source:* Baumeister, 1967.
[4]*Source:* Dunlap, 1912.

Glossary

Carbohydrates: Organic compounds of carbon, hydrogen, and oxygen, with the general formula $C_m(H_2O)_n$. The carbon and hydrogen are in the 2 : 1 atom ratio. Carbohydrates can be viewed as hydrates of carbon. They include sugars, starches, cellulose, and other cellular products.

Cellulose: The main constituent of cell walls, with the generic formula ($C_6H_{10}O_5$).

Cracking: The breaking up of large complex organic molecules into smaller molecules using pressure and temperature with or without a catalyst. The product of cracking depends on temperature, pressure, and catalyst used. No single unique reaction takes place in the cracker. The hydrocarbon molecules are broken up fairly randomly to produce mixtures of smaller hydrocarbons.

Depolymerization: The decomposition of a polymer into smaller fragments, or the breakdown of macromolecular compounds into relatively simple compounds.

Esterification: The chemical process for making esters, which are compounds of the chemical structure R-COOR′ in which R and R′ are either alkyl or aryl groups. The most common method for preparing esters is to heat a carboxylic acid, R-CO-OH, with an alcohol, R′-OH, and remove the water that is formed.

Esters: Any chemical compounds derived by reacting an oxoacid (it contains an oxo group, X = O) with a hydroxyl compound such as an alcohol or a phenol.

Ethanol: A popular alcohol (C_2H_5OH) used in spark-ignition engines, either alone or blended with petroleum-derived gasoline.

Gasoline: In the United States and Canada, the petroleum-derived oil that runs normal spark-ignition car engines is called gasoline. Many other places it is called petrol. Gasoline is a mixture of a large number of hydrocarbons containing 4 to 12 carbon atoms per molecule in proportions that can vary depending on the crude oil and the user's specification. In the United States, gasoline is usually a blend of straight-run gasoline, reformate, alkylate, and some butane. The approximate composition is 15% C4 to C8 straight-chain alkanes, 25 to 40% C4 to C10 branched alkanes, 10% cycloalkanes, <25% aromatics (<1.0% benzene), and 10% straight-chain and cyclic alkenes (ACS, 2005). The average heating value of gasoline is 44.4 MJ/kg and its specific gravity is 0.67 to 0.77. Its average molecular weight is ~108 (Ritter, 2005).

Hemicellulose: An important component of plant cell walls that can be any of several heteropolymers present in almost all plant cell walls along with cellulose.

Hydrolysis: The breaking of hydrogen bonds in long-chained organic molecules.

Biomass Gasification and Pyrolysis. DOI: 10.1016/B978-0-12-374988-8.00016-7

Hydrocracking: A cracking process that uses a catalyst and occurs at high hydrogen partial pressure to "crack" the fractions into smaller molecules, to produce high-octane gasoline and other good-quality, stable distillates.

Hydrotreating: A process used in a refinery in which the feedstock is treated with hydrogen at elevated temperature and pressure in the presence of appropriate catalysts to remove contaminants such as sulfur, nitrogen, metals, and condensed-ring aromatics or metals.

Lignin: A component of the cell wall of wood. It is a complex polymer that binds cellulose cells in biomass.

Methanol: An important alcohol (CH_3OH) that serves as a feedstock for a host of chemicals and liquid transportation fuels.

Polymerization: The building up of larger molecules by combining smaller molecules that are necessarily similar.

Producer gas: Primarily a mixture of carbon monoxide, hydrogen, and nitrogen produced by blowing air and steam through the fuel bed.

Protein: Any organic compound made up of amino acids arranged in a linear chain and folded into a globular form. These high-molecular-weight compounds of carbon, hydrogen, oxygen, and nitrogen are synthesized by plants and animals.

Reforming: The structural manipulation of a molecule to improve its product quality. The process does not always involve major change in the molar mass. The steam reforming of methane is a widely used method of producing hydrogen.

Starch: A polysaccharide carbohydrate consisting of a large number of glucose units joined together by glycosidic bonds, with the generic formula $(C_6H_{10}O_5)_n$. All green plants produce starch for energy storage.

Steam reforming: A method for producing hydrogen from methane. Steam reacts with methane to produce hydrogen and carbon monoxide when heated to very high temperatures in the presence of a metal-based catalyst.

Substituent: An atom or a group of bonded atoms that can be considered to have replaced a hydrogen atom in a parent molecular entity.

Sugar: Any monosaccharide or disaccharide used especially by organisms to store energy. Glucose ($C_6H_{12}O_6$), a monosaccharide, is the simple sugar that stores chemical energy that biological cells convert to other types of energy.

Syngas: A mixture of carbon monoxide and hydrogen.

Synthesis: The building up of larger molecules from smaller ones. The molecules being synthesized need not be similar.

Synthetic natural gas (SNG): A methane gas artificially produced from other gaseous, solid, or liquid fuels using various methods.

Triglycerides: A glyceride in which the glycerol is esterified with three fatty acids; the main constituent of vegetable oil and animal fats. The chemical formula is $RCOO—CH_2CH(—OOCR')\ CH_2—OOCR''$, where R, R′, and R″ are longer alkyl chains.

References

Abdulsalam, P., 2005. A comparative study of hydrodynamics and gasification performance of two types of spouted bed reactor designs. PhD Thesis, Asian Institute of Technology, School of Environment, Resources and Development, Bangkok.

Acharya, B., Dutta, A., Basu, P., 2009. Chemical looping gasification of biomass for hydrogen enriched gas production with in-process carbon-dioxide capture. Energy & Fuels 23 (10), 5077–5083.

Adrian, M.H., Brandl, A., Hooper, M., Zhao, X., Tabak, S.A., 2007. An Alternative Route for Coal to Liquids: Methanol-to-Gasoline (MTG) Technology. Gasification Technologies Conference, San Francisco.

Altafini, C.R., Wander, P.R., Barreto, R.M., 2003. Prediction of the working parameters of a wood waste gasifier through an equilibrium model. Energy Conversion and Management 44 (17), 2763–2777.

Alter, H., Campbell, J.A., 1979. The preparation and properties of densified refuse-derived fuel. In: Jones, J.L., Radding, S.B. (Eds.), Thermochemical Conversion of Solid Waste. ACS Symposium Series. 130, Washington, DC, pp. 127–142.

Amin, S., Reid, R.C., Modell, M., 1975. Reforming and decomposition of glucose in an aqueous phase. Intersociety Conference on Environmental System, San Francisco, ASME 75-ENAs-21.

Antal, M.J., 2003. The Art, Science, and Technology of Charcoal Production. Industrial Engineering Chemistry Research 42, 1619–1640.

Antal, M.J., Allen, S.G., Schulman, D., Xu, X., Divilio, R.J., 2000. Biomass gasification in supercritical water. Industrial Engineering Chemistry 39, 4040–4053.

Antal, M.J., Vgrhegyi, G., 1995. Cellulose pyrolysis kinetics: The current state of knowledge. Ind. Eng. Chem. Res. 34, 703–717.

Appel, B.S., Adams, T.N., Roberts, M.J., Lange, W.F., Freiss, J.H., Einfeldt, C.T., Carnesi, M.C., 2004. Process for conversion of organic, waste, or low-value materials into useful products. US Patent 2004/0192980 A1, September.

Arthur, J.R., 1951. Reactions between carbon and oxygen. Transactions of the Faraday Society 47, 164–178.

ASTM, 2000. Standard Test Method for Shear Testing of Bulk Solids Using the Jenike Shear Cell. American Society for Testing and Materials, D-6128–00.

Aznar, M.P., Delgado, J., Corella, J., Borque, J.A., Campos, I.J., 1997. Steam gasification in fluidized bed of a synthetic refuse containing chlorine with a catalytic gas cleaning at high temperature. In: Bridgwater, A.V., Boocock, D.G.B. (Eds.), Developments in Thermochemical Biomass Conversion. Blackie Academic and Professional, pp. 1194–1208.

Babu, B.V., Chaurasia, A.S., 2004a. Parametric study of thermal and thermodynamic properties on pyrolysis of biomass in thermally thick regime. Energy Conversion and Management 45 (1), 53–72.

Biomass Gasification and Pyrolysis. DOI: 10.1016/B978-0-12-374988-8.00019-2

Babu, B.V., Chaurasia, A.S., 2004b. Pyrolysis of biomass: improved models for simultaneous kinetics and transport of heat, mass and momentum. Energy Conversion and Management 45 (9–10), 1297–1327.

Barea, A.G., 2009. Personal communication.

Barrio, M., Hustad, J.E., 2001. CO_2 gasification of birch char and the effect of CO inhibition on the calculation of chemical kinetics. In: Bridgwater, A.V. (Ed.), Progress in Thermochemical Biomass Conversion, vol. 1. Blackwell Science, pp. 47–60.

Barrio, M., Gøbel, B., Risnes, H., Henriksen, U., Hustad, J.E., Sørensen, L.H., 2001. Steam gasification of wood char and the effect of hydrogen inhibition on the chemical kinetics. In: Bridgwater, A.V. (Ed.), Progress in Thermochemical Biomass Conversion, vol. 1. Blackwell Science, pp. 32–46.

Basu, P., 2006. Combustion and Gasification in Fluidized Beds. Taylor & Francis, pp. 355–357.

Basu, P., 1977. Burning rate of carbon in fluidized beds. Fuel 56 (4), 390–392.

Baumeister, T. (Ed.), 1967. Standard Handbook for Mechanical Engineers. McGraw-Hill.

Behrendt, F., Neubauer, Y., Oevermann, M., Wilmes, B., Zobel, N., 2008. Direct liquifaction and biomass. Chemical Engineering Technology 31 (5), 667–677.

Bergman, P.C.A., 2005. Combined torrefaction and pelletisation. Energy Research Center of The Netherlands, ECN Report, ECN-C-05-073.

Bergman, P.C.A., Kiel, J.H.A., 2005. Torrefaction for biomass upgrading. 14th European Biomass Conference & Exhibition, Paris, ECN Report ECN-RX-05-180.

Bergman, P.C.A., Boersma, A.R., Kiel, J.H.A., Prins, M.J., Ptasinski, K.J., Janssen, F.J.J.G., 2005a. Torrefaction for entrained-flow gasification of biomass. Energy Research Centre of The Netherlands, ECN Report ECN-C-05-067.

Bergman, P.C.A., Prins, M.J., Boersma, A.R., Ptasinski, K.J., Kiel, J.H.A., Janssen F.J.J.G., 2005b. Combined Torrefaction and Pellestisation. Energy Research Centre of The Netherlands, ECN Report ECN-05-073.

Bingyan, X., Zengfan, L., Chungzhi, W., Haitao, H., Xiguang, Z., 1994. Circulating fluidized bed gasifier for biomass. In: Nan, L., Best, G., Coelho, C., de Carvalho, O. (Eds.), Integrated Energy Systems in China—The Cold Northeastern Region Experience. FAO, The United Nations, Rome.

Biomass Energy Centre, *www.biomassenergycentre.org.uk*. Accessed July 2009.

Blasi, C.D., 2009. Combustion and gasification rates of lignocellulosic chars. Progress in Energy and Combustion Science 35 (2), 121–140.

Blasi, C.D., 1993. Modeling and simulation of combustion processes of charring and non-charring solid fuels. Progress in Energy and Combustion Science 19 (1), 71–104.

Blasi C.D., Branca C., Galgano A., Meier D., Brodzinski I., Malmros, O., 2007. Supercritical gasification of wastewater from updraft wood gasifiers. Biomass and Bioenergy 31 (11–12), 802–811.

Boerrigter, H., Rauch, R., 2005. Syngas production and utilization (Chapter 10). In: Knoef, H.A.M. (Ed.), Biomass Gasification Handbook. Biomass Technology Group (BTG), Enschede.

Boley, C.C., Landers, W.S., 1969. Entrainment drying and carbonization of wood waste. Bureau of Mines, Report of Investigations 7282, Washington, DC.

Boukis, I.P., Grammelis, P., Bezergianni, S., Bridgwater, A.V., 2007. CFB air-blown flash pyrolysis. Part I: Engineering design and cold model performance. Fuel 86, 1372–1386.

Boukis, N., Franz, G., Habicht, W., Dinjus, E., 2001. Corrosion resistant materials for SCWO—applications: experimental results from long time experiments, CORROSION 2001, March, Paper 01353.

Boukis N., Galla U., D'Jesus P., Muller H., Dinjus E., 2005. Gasification of wet biomass in supercritical water: results of pilot plant experiments, 14th European Biomass Conference, Paris, pp. 964–967.

Boutin, O., Lédé, J., 2001. Use of a concentrated radiation for the determination of cellulose thermal decomposition mechanism. In: Bridgwater, A.V. (Ed.), Progress in Thermochemical Biomass Conversion. Blackwell Science, pp. 1034–1045.

Bowerman, F.R., 1969. Introduction. In: Corey, R.C. (Ed.), Principles and Practices of Incineration. John Wiley.

Bradbury, A.G.W., Sakai, Y., Shafizadeh, F., 1979. A kinetic model for pyrolysis of cellulose. Journal of Applied Polymer Science 23 (11), 3271–3280.

Bramer, E.A., Holthuis, M.R., Brem, G., 2006. A novel thermogravimetric vortex reactor for the determination of the primary pyrolysis rate of biomass. In: Bridgwater, A.V., Boocock, D.G.B. (Eds.), Science in Thermal and Chemical Biomass Conversion, vol. 2. CPL Press, pp. 1115–1124.

Brammer, J.G., Bridgwater, A.V., 1999. Drying technologies for an integrated gasification bioenergy plant. Renewable and Sustainable Energy Reviews 3, 243–289.

Bridgwater, A.V., 2002. Fast Pyrolysis of Biomass: A Handbook, vol. 2. CPL Press.

Bridgwater, A.V., 1999. Principles and practice of biomass fast pyrolysis processes for liquids. Journal of Analytical and Applied Physics 51 (1–2), 3–22.

Bridgwater, A.V., Czernik, S., Piskorz, J., 2001. An overview of fast pyrolysis. In: Bridgwater, A.V. (Ed.), Progress in Thermochemical Biomass Conversion. Blackwell Science, pp. 977–997.

Bui, T., Loof, R., Bhattacharya, S.C., 1994. Multi-stage reactor for thermal gasification of wood. Energy 19 (4), 397–404.

Callaghan, C.A., 2006. Kinetics and catalysis of the water-gas-shift reaction: a microkinetic and graph theoretic approach. Ph.D. thesis, Chemical Engineering Department, Worcester Polytechnic Institute, p. 116.

Carleton, A.J., 1972. The effect of fluid-drag forces on the discharge of free-flowing solids from hoppers. Powder Technology 6 (2), 91–96.

Carlos, L., 2005. High temperature air/steam gasification of biomass in an updraft fixed batch type gasifier. Ph.D. thesis, Royal Institute of Technology, Energy Furnace and Technology, Stockholm.

Cetin, E., Moghtaderi, B., Gupta, R., Wall, T.F., 2005. Biomass gasification kinetics: influence of pressure and char structure. Combustion Science and Technology 177 (4), 765–791.

Chakraverty, A., Mujumdar, A.S., Raghavan, G.S.V., Ramaswamy, H.S., 2003. Handbook of Post-Harvest Technology. Taylor & Francis.

Channiwala, S.A., Parikh, P.P., 2002. A unified correlation for estimating HHV of solid, liquid and gaseous fuels. Fuel 81, 1051–1063.

Chase, M.W., 1998. NIST JANAF Thermochemical Tables. American Chemical Society.

Chembukulam, S.K., Dandge, A.S., Kovilur, N.L., Seshagiri, R.K., Vaidyeswaran, R., 1981. Smokeless fuel from carbonized sawdust. Industrial and Engineering Chemistry, Product Research and Development 20 (4), 714–719.

Chen, P., Zhao, Z., Wu, C., Zhu, J., Chen, Y., 2005. Biomass gasification in circulating fluidized bed. In: Cen, K. (Ed.), Circulating Fluidized Bed Technology, vol 8. International Academic Publishers, pp. 507–514.

Chowdhury, R., Bhattacharya, P., Chakravarty, M., 1994. Modeling and simulation of a down draft rice husk gasifier. International Journal of Energy Research 18 (6), 581–594.

Chu, W., Zhang, T., He, C., Wu, Y., 2002. Low-temperature methanol synthesis in liquid phase on novel copper based catalysts. Catalyst Letters 79 (1–4), 129–132.

Ciferno, J.P., Marano, J.J., 2002. Benchmarking biomass gasification technologies for fuels, chemicals and hydrogen production. US Department of Energy, National Energy Technology Laboratory.

Collins, C.D., 2007. Implementing Phytoremediation of Petroleum Hydrocarbons. Methods in Biotechnology, vol. 23. Humana Press, pp. 99–108.

Corella, J., Sanz, A., 2005. Modeling circulating fluidized bed biomass gasifiers: a pseudo-rigorous model for stationary state. Fuel Processing Technology 86 (9), 1021–1053.

Corella, J., Toledo, J.M., Molina, G., 2007. A Review on Dual Fluidized Bed Biomass Gasifiers. Industrial Engineering Chemistry and Research 46 (21), 6831–6839.

Cummers, K.R., Brown, R.C., 2002. Ancillary equipment for biomass gasification. Biomass and Bioenergy 23 (2), 113–128.

Dai, J., Grace, J.R., 2008. A model for biomass screw feeding. Powder Technology 186 (1), 40–55.

Darton, R.C., LaNauze, R.D., Davidson, J.F., Harrison, D., 1977. Bubble growth due to coalescence in fluidized beds. Transactions of the Institution of Chemical Engineers 55 (4), 274–280.

Dayton, D., 2002. A Review of the Literature on Catalytic Biomass Tar Destruction. National Renewable Energy Laboratory, NREL Report TP-510–32815.

Debdoubi, A., el Amarti, A., Colacio, E., Blesa, M.J., Hajjaj, L.H., 2006. The effect of heating rate on yields and compositions of oil products from esparto pyrolysis. International Journal of Energy Research 30 (15), 1243–1250.

DeGroot, W.F., Shafizadeh, F., 1984. Kinetics of gasification of Douglas fir and cottonweed chars by carbon dioxide. Fuel 63, 210.

Delgado, J., Aznar, M.P., Corella, J., 1996. Calcined dolomite, magnesite, and calcite for cleaning hot gas from a fluidized bed biomass gasifier with steam: life and usefulness. Industrial and Engineering Chemistry Research 35 (10), 3637–3643.

Demirbas, A., 2009. Biorefineries: Current activities and future developments. Energy Conversion and Management 50, 2782–2801.

Demirbas A., 2001. Biomass resources facilities and biomass conversion processing for fuels and chemicals. Energy Conversion and Management 42 (11), 1357–1378.

Demirbas, A., 2000. Mechanisms of liquefaction and pyrolysis reactions of biomass. Energy Conversion and Management 41 (6), 633–646.

Department of Energy, 1978. Coal conversion systems technical data book. USDOE, contract no. EX-76-C-0102286, HCP/T2286-01, pp. IIIA-52–53.

Desch, H.E., Dinwoodie, J.M., 1981. Timber: its structure, properties and utilization, 6th ed. Macmillan Press.

Devi, L., Ptasinski, K.J., Janssen, F.J.J.G., 2003. A review of the primary measures for tar elimination in biomass gasification processes. Biomass and Bioenergy 24 (2), 125–140.

Di Blasi, C., 2008. Modeling chemical and physical processes of wood and biomass pyrolysis. Progress in Energy and Combustion Science 34 (1), 47–90.

Diebold, J.P., Bridgwater, A.V., 1997. Overview of fast pyrolysis of biomass for the production of liquid fuels. In: Bridgwater, A.V., Boocock, D.G.B. (Eds.), Developments in Thermochemical Biomass Conversion. Blackie Academic & Professional, pp. 5–27.

Diebold, J.P., Power, A., 1988. Engineering aspects of the vortex pyrolysis reactor to produce primary pyrolysis oil vapors for use in resins and adhesives. In: Bridgwater, A.V., Kuester, J.L. (Eds.), Research in Thermochemcial Biomass Conversion. Elsevier, Applied Science, pp. 609–628.

Diebold, J.P., Milne, T.A., Czernik, S., Osamaa, A., Bridgwater, A.V., Cuevas, A., Gust, S., Hufman, D., Piskorz, J., 1997. Proposed specification for various grades of pyrolysis oils. In:

Bridgwater, A.V., Boocock, D.G.B. (Eds.), Developments in Thermochemical Biomass Conversion. Blackie Academic & Professional.

Dinjus, E., Kruse, A., 2004. Hot compressed water—a suitable and sustainable solvent and reaction medium. Journal of Physics of Condensed Matter 16, S1161–S1169.

Dryer, F.L., Glassman, I., 1973. High temperature oxidation of CO and CH_4. 14th Symposium of Combustion. The Combustion Institute, Pittsburgh, pp. 987–1003.

Dunlap, F., 1912. Specific heat of wood, USDA Forest Service Bulletin 110, Washington, DC.

Ebasco Services Inc., 1981. Supplemental Studies for Anthracite Coal Gasification to Produce, Fuels and Chemicals, Nepgas Project, Volume II. Supplemental Tasks. Energy Development and Resource Corporation, Nanticoke, PA, contract number FG01–79RA20221.

Encinar, J.M., Gonzalez, J.F., Rodriguez, J.J., Ramiro, M.J., 2001. Catalysed and uncatalysed steam gasification of eucalyptus char: influence of variables and kinetic study. Fuel 80 (14), 2025–2036.

Enden P.J. van den, Lora, E.S., 2004. Design approach for a biomass fed fluidized bed gasifier using the simulation software CSFB. Biomass and Bioenergy 26, 281–287.

Evans, R.J., Milne, T.A., 1997. Chemistry of Tar Formation and Maturation in the Thermochemical Conversion of Biomass. In: Bridgwater, A.V., Boocock, D.G.B. (Eds.), Developments in Thermochemical Biomass Conversion, vol. 2. Blackie Academic & Professional, pp. 803–816.

Feng W., Hedzer J., Kooi V.D., Arons J.S., 2004a. Biomass conversions in subcritical and supercritical water: driving force, phase equilibria, and thermodynamic analysis. Chemical Engineering Journal 43, 1459–1467.

Feng W., Hedzer J., Kooi V.D., Arons J.S., 2004b. Phase equilibria for biomass conversion processes in subcritical and supercritical water. Chemical Engineering Journal 98, 105–113.

Florin, N.H., Harris, A.T., 2008. Enhanced hydrogen production from biomass with in situ carbon dioxide capture using calcium oxide sorbents. Chemical Engineering Science 63 (2), 287–316.

Freda, C., Canneto, G., Mariotti, P., Fanelli, E., Molino, A., Braccio, G., 2008. Cold model testing of an internal circulating fluid bed gasifier. Gasification Workshop, Marmara Research Centre, Turkey.

Friedrich, C., Kritzer, P.N., Boukis, N., Franz, G., Dinjus, E., 1999. The corrosion of tantalum in oxidizing sub- and supercritical aqueous solutions of HCl, H2SO4 and H3PO4. Journal of Material Science 34 (13), 3137–3141.

Gao, K., Wu, J., Wang, Y., Zhang, D., 2006. Bubble dynamics and its effect on the performance of a jet fluidized bed gasifier simulated using CFD. Fuel 85 (9), 1221–1231.

García, X.A., Alarcón, N.A., Gordon, A.L., 1999. Steam gasification of tars using a CaO catalyst source. Fuel Processing Technology 58 (2), 83–102.

Gasafi, E., Reinecke, M., Kruse, A., Schebek, L., 2008. Economic analysis of sewage sludge gasification in supercritical water for hydrogen production. Biomass and Bioenergy 32, 1085–1096.

Gaur, S., Reed, T.B., 1995. An atlas of thermal data for biomass and other fuels. National Renewable Energy Laboratory, NREL no. NREL/TP-433-7965.

Gesner, A., 1861. A practical treatise on coal petroleum and other distilled oils (n.p.).

Gil, J., Corella, J., Aznar, M.P., Caballero, M.A., 1999. Biomass gasification in atmospheric and bubbling fluidized bed: Effect of the type of gasifying agent on the product distribution. Biomass and Bioenergy 17 (5), 389–403.

Gomez-Barea, A., Leckner, B., Campoy, M., 2008. Conversion of char in CFB gasifiers. In: Werther, J., Nowak, W., Wirth, K., Hartge, E. (Eds.), Circulating Fluidized Bed Technology, vol. 10. Tu-Tech Innovation GmbH, Hamburg, pp. 727–732.

Graham, R.G., Bain, R., 1993. Biomass gasification: hot gas clean up. International Energy Agency, Biomass Gasification Working Group.
Grezin, A.K., Zakharov, N.D., 1988. Thermodynamic analysis of air separation equipment with a throttling refrigerating cycle. Chemical and Petroleum Engineering (translated from Russian) Khimicheskoe i Neftyanoe Mashinostroenie 24 (5), 223–227.
Grotkjær, T., Da-Johansen, K., Jensen, A.D., Glarborg, P., 2003. An experimental study of biomass ignition. Fuel 82 (1), 825–833.
Guo, B., Li, D., Cheng, C., Lu, Z., Shen, Y., 2001. Simulation of biomass gasification with a hybrid neural network model. Bioresource Technology 76 (2), 77–83.
Guo, B., Shen, Y., Li, D., Zhao, F., 1997. Modelling coal gasification with a hybrid neural network. Fuel 76 (12), 1159–1164.
Guo, L.J., Lu, Y.J., Zhang, X.M., Ji, C.M., Guan, Y., Pei, A.X., 2007. Hydrogen production by biomass gasification in supercritical water—a systematic experimental and analytical study. Catalysis Today 129 (3–4), 275–286.
Haar, L., Gallagher, J.S., Kell, G.S., 1984. NBS/NRC Steam Tables. Hemisphere Publishing.
Hajek, M., Judd, M.R., 1995. Use of neural networks in modeling the interactions between gas analysers at coal gasifiers. Fuel 74 (9), 1347–1351.
Hao, X.H., Guo, L.J., Mao, X., Zhang, X.M., Chen, X.J., 2003. Hydrogen production from glucose used as a model compound of biomass gasified in supercritical water. International Journal of Hydrogen Energy 28, 55–64.
Hasler, P., Bühler, R., Nussbaumer, T., 1997. Evaluation of gas cleaning technologies for small-scale biomass gasifiers. Swiss Federal Office of Energy and Swiss Federal Office for Education and Science.
Hemati, M., Laguerie, C., 1988. Determination of the kinetics of the wood sawdust steam-gasification of charcoal in a thermobalance. Entropie 142, 29–40.
Hemati, M., Laguerie, C., 1987. The kinetic study of the pyrolysis of sawdust in a thermobalance: kinetic approach to the pyrolysis of oak sawdust. The Chemical Engineering Journal 35 (3), 147–156.
Herguido, J., Corella, J., Gonzalez-Saiz, J., 1992. Steam gasification of lignocellulosic residues in a fluidized bed at a small pilot scale: effect of the type of feedstock. Industrial and Engineering Chemistry Research 31 (5), 1274–1282.
Higman, C., van der Burgt, M., 2008. Gasification, 2nd ed. Gulf Professional Publishing/ Elsevier.
Hoang, T., Ballantyne, G., Danuthai, T., Lobban, L.L., Resasco, D., Mallinson, R.G., 2007. Glycerol to Gasoline Conversion. Spring National Meeting, AIChE, 2007.
Hodge, B.K., 2010. Biomass (Chapter 12). Alternative Energy Systems & Applications. Wiley, pp. 296–329.
Hofbauer, H., 2002. Biomass CHP–Plant Güssing: a success story. Paper presented at Pyrolysis and Gasification of Biomass and Waste Expert Meeting, Strasbourg.
Holgate, H.R., Meyer, J.C., Tester, W.J., 1995. Glucose hydrolysis and oxidation in supercritical water. AIChE Journal 41, 637–648.
Hossfeld, R.J., Barnum, R.A., 2007. How to avoid flow stoppages during storage and handling. Power Engineering 111 (10), 42–50.
Hsu, I.C., 1979. Heat transfer to isobutene flowing inside a horizontal tube at supercritical pressure. PhD thesis, University of California, Lawrence Berkeley Laboratory.
Hughey, C.A., Henerickson, C.L., 2001. Elemental composition analysis of processed and unprocessed diesel fuel by electrospray ionization Fourier transform ion cyclotron resonance mass spectrometry, Energy Fuels 15 (5), 1186–1193.

Hulet, C., Briens, C., Berruti, F., Chan, E.W., 2005. A review of short residence time cracking processes. International Journal of Chemical Reactor Engineering 3, 1–71.

IFP–Innovation Energy Environment, 2007. Potential biomass mobilization for biofuel production worldwide, in Europe and in France, Panaroma, *www.ifp.fr*.

Jacob, K., 2000. Bin and Hopper Design. Dow Chemical Company, *www.uakron.edu*.

Janssen, H.A., 1895. Versuche über Getreidedruck in Silozellen. Zeitschrift des Verein Deutscher Ingenieure 39, 1045–1049.

Jarungthammachote, S., Dutta, A., 2008. Equilibrium modeling of gasification: Gibbs free energy minimization approach and its application to spouted bed and spout-fluid bed gasifiers. Energy Conversion and Management 49 (6), 1345–1356.

Jenike, A.W., 1964. Storage and Flow of Solids. Bulletin of the University of Utah, 53 and 26, Bull. No. 123, University of Utah Engineering Experiment Station.

Jenkins, B.M., Jones, A.D., Turn, S.Q., Williams, R.B., 1996. Emission factors for polycyclic aromatic hydrocarbons from biomass burning. Environmental Science Technology 30 (8), 2462–2469.

Jenkins, B.M., 1989. Physical properties of biomass. In: Kitani, O., Hall, C.W. (Eds.), Biomass Handbook. Gordon & Breach Science Publishers, Amsterdam.

Jenkins, B.M., Ebeling J.M., 1985. Correlations of physical and chemical properties of terrestrial biomass with conversion. In: Proceedings of Energy from Biomass and Wastes. IX Institute of Gas Technology, Chicago, pp. 371–400.

Ji, P., Feng, W., Chen, B., Yuan, Q., 2006. Finding appropriate operating conditions for hydrogen purification and recovery in supercritical water gasification of biomass. Chemical Engineering Journal 124, 7–13.

Johanson, J.R., 1965. Method of calculating rate of discharge from hoppers and bins. Transactions of Society of Mining Engineers 232 (3), 69–80.

Jones, J.M., Nawaz, M., Darvell, L.I., Ross, A.B., Pourkashanian, M., Williams, A., 2006. Towards biomass classification for energy applications. In: Bridgwater, A.V., Boocock, D.G.B. (Eds.), Science in Thermal and Chemical Biomass Conversion, vol. 1. CPL Press, pp. 331–339.

Kabyemela, B.M., Adschiri, T., Malaluan, R.M., Arai, K., 1997. Kinetics of Glucose Epimerization and Decomposition in Subcritical and Supercritical Water. Industrial Engineering Chemistry Research 36, 1552–1558.

Kalogirou, S.A., 2001. Artificial neural networks in renewable energy systems applications: a review. Renewable and Sustainable Energy Reviews 5 (4), 373–401.

Kalogirou, S.A., Panteliou, S., Dentsoras, A., 1999. Artificial neural networks used for the performance prediction of a thermosiphon solar water heater. Renewable Energy 18 (1), 87–99.

Kaushal, P., Pröll, T., Hofbauer, H., 2008. Model for biomass char combustion in the riser of a dual fluidized bed gasification unit: Part 1—Model development and sensitivity analysis. Fuel Processing Technology 89 (7), 651–659.

Kersten, S.R.A., 2002. Biomass gasification in circulating fluidized beds. Dissertation, Twente University, Twente University Press.

Kinoshita, C.M., Wang, Y., Zhou, J., 1994. Tar formation under different biomass gasification conditions. Journal of Analytical and Applied Pyrolysis 29 (2), 169–181.

Kitani, O., Hall, C.W., 1989. Biomass Handbook. Gordon & Breach Science Publishers.

Klass, D.L., 1998. Biomass for Renewable Energy, Fuels, and Chemicals. Academic Press, pp. 30, 276–277, 233, 239.

Klose, W., Wolki, M., 2005. On the intrinsic reaction rate of biomass char gasification with carbon dioxide and steam. Fuel 84 (7–8), 885–892.

Knight, R.A., 2000. Experience with raw gas analysis from pressurized gasification of biomass. Biomass and Bioenergy 18 (1), 67–77.

Knoef, H.A.M. (Ed.), 2005. Handbook Biomass Gasification. BTG Publisher, Enschede, The Netherlands, pp. 32, 239–241.

Knoef, H.A.M., 2000. Inventory of Biomass Gasifier Manufacturers and Installations. Final report to the European Commission, Contract DIS/1734/98–NL, Biomass Technology Group B.V., University of Twente, Enschede, The Netherlands.

Kollman, F.F.P, Cote, W.A. 1968. Principle of wood science and technology, vol. I, Solid Wood. Springer-Verlag.

Koltz, I.M., Rosenberg, R.M., 2008. Chemical Thermodynamics, 7th ed. Wiley, p. 237.

Krammer P., Mittelstadt S., Vogel H., 1999. Investigating the synthesis potential in supercritical water. Chemical Engineering Technology 22, 126–130.

Kritz, P., 2004. Corrosion in high-temperature and supercritical water and aqueous solutions: a review. Journal of Supercritical Fluids 29, 1–29.

Kruse, A., Abel, J., Dinjus, E., Kluth, M., Petrich, G., Schacht, M., Sadri, H., Schmieder, H., 1999. Gasification of biomass and model compounds in hot compressed water. International Meeting of the GVC-Fachausschub Hochdruckverfahrenstechnik, Karlsruhe, Germany.

Kruse, A., Henningsen, T., Sinag, A., Pfeiffer, J.J., 2003. Biomass gasification in supercritical water: influence of the dry matter content and the formation of phenols. Industrial and Engineering Chemistry Research 42 (16), 3711–3717.

Kudo, K., Yoshida, E., 1957. The decomposition process of wood constituents in the course of carbonization I: the decomposition of carbohydrate and lignin in Mizunara. Journal of the Japan Wood Research Society 3 (4), 125–127.

Kumar, J.V., Pratt, B.C., 1996. Compositional analysis of some renewable biofuels. American Laboratory 28 (8), 15–20.

Kunii, D., Levenspiel, O., 1991. Fluidization Engineering. Butterworth-Heinemann.

Kusdiana, D., Minami, E., Saka, S., 2006. Non-catalytic biodiesel fuel production by supercritical methanol pre-treatment. In: Bridgwater, A.V., Boocock, D.G.B. (Eds.), Science in Thermal and Chemical Biomass Conversion. CPL Press, pp. 424–435.

Lammars, G., Beenackers, A.A.C.M., Corella, J., 1997. Catalytic tar removal from biomass producer gas with secondary air. In: Bridgwater, A.V., Boocock, D.G.B. (Eds.), Developments in Thermochemical Biomass Conversion. Blackie Academic & Professional.

Lee, I., Kim, M.S., Ihm, S.K., 2002. Gasification of glucose in supercritical water. Industrial and Engineering Chemistry Research 41, 1182–1188.

Lewellen, W.S., Segur, H., Varma, A.K., 1977. Modeling two-phase flow in a swirl combustor. ARAP report No. 310.

Li, B., Kado, S., Mukainakano, Y., Miyazawa, T., Miyao, T., Naito, S., Okimura, K., Kunimori, K., Tomishige, K., 2007. Surface modification of Ni catalysts with trace Pt for oxidative steam reforming of methane. Journal of Catalysis 245, 144–155.

Li, X.T., Grace, J.R., Watkinson, A.P., Lim, C.J., Ergüdenler, A., 2001. Equilibrium modeling of gasification: a free energy minimization approach and its application to a circulating fluidized bed coal gasifier. Fuel 80 (2), 195–207.

Liden, A.G., Berruti, F., Scott, D.S., 1988. A kinetic model for the production of liquids from the flash pyrolysis of biomass. Chemical Engineering Communication 65, 207–221.

Liliedahl, T., Sjostrom, K., 1997. Modeling of char-gas reaction kinetics. Fuel 76 (1), 29–37.

Lim, M.T., Alimuddin, Z., 2008. Bubbling fluidized bed gasification—performance, process findings and energy analysis. Renewable Energy 33 (10), 2339–2343.

Littlewood, K., 1977. Gasification: theory and application. Progress in Energy and Combustion Science 3 (1), 35–71.

Liu, H., Li, S., Zhang, S., Chen, L., Zhou, G., Wang, J., Wang, X., 2008. Catalytic performance of monolithic foam Ni/SiC catalyst in carbon dioxide reforming of methane to synthesis gas. Catalysis Letters 120 (1–2), 111–115.

Loppinet-Serani, A., Aymonier, C., Cansell, F., 2008. Current and foreseeable applications of supercritical water for energy and the environment. ChemSusChem 1, 486–503.

Lu, Y.J., Guo, L.J., Ji, C.M., Zhang, X.M., Hao, X.H., Yan, Q.H., 2006. Hydrogen production by biomass gasification in supercritical water—a parametric study. International journal of Hydrogen Energy 31, 822–831.

MacLean J.D., 1941. Thermal conductivity of wood. Transactions of American Society of Heating and Ventilating Engineers 47, 323–354.

Malhotra, A., 2006. Thermodynamic Properties of Supercritical Steam, *www.steamcenter.com*, Jaipur, *www.lulu.com/items/volume_13/254000/254766/1/preview/preview-thpb.doc*.

Maniatis, K., 2001. Progress in biomass gasification: an overview. In: Bridgwater, A.S. (Ed.), Progress in Thermochemical Biomass Conversion, vol. I. Blackwell Science, pp. 1–31.

Maples, R.E., 2000. Petroleum Refinery Process Economics, 2nd ed. PennWell.

Marrone, P.A., Hong, G.T., 2008. Corrosion control methods in supercritical water oxidation and gasification processes, NACE Conference, New Orleans.

Marrone P.A., Hong G.T., Spritzer, M.H., 2007. Developments in supercritical water as a medium for oxidation, reforming, and synthesis. Journal of Advanced Oxidation Technologies 10 (1), 157–168.

Mastellone, M.L., Arena, U., 2008. Olivine as a tar removal catalyst during fluidized bed gasification of plastic waste. AIChE Journal 54 (6), 1656–1667.

Matsurmura, Y., 2002. Evaluation of supercritical water gasification and biomethanation for wet biomass utilization in Japan. Energy Conversion and Management 43, 1301–1310.

Matsumura, Y., Minowa, T., 2004. Fundamental design of a continuous biomass gasification process using a supercritical water fluidized bed. International Journal of Hydrogen Energy 29, 701–707.

Matsumura, Y., Minowa, T., Potic, B., Kerstein, S.R.A., Prins, W., van Swaiij, W.P.M., van de Beld, B., Elliott, D.C., Neuenschwander, G.G., Kruse, A., Antal, M.J., 2005. Biomass gasification in near and supercritical water-status and prospect. Biomass and Bioenergy 29 (4), 269–292.

Matsurmura, Y., Minowa, T., Xu, X., Nuessle, F.Q., Adschiri, T., Antal, M.J., 1997. High pressure CO2 removal in supercritical water gasification of biomass. In: Bridgewater, A.V., Boocock, D.G.B. (Eds.), Developments in Thermochemical Biomass Conversion. Blackie Academic & Professional, pp. 864–877.

McKendry, P., 2002. Energy production from biomass, part 1—overview of biomass. Bioresource Technology 83 (1), 37–46.

Mettanant, V., Basu, P., Butler, J., 2009. Agglomeration of biomass fed fluidized bed gasifier and combustor. Canadian Journal of Chemical Engineering 87 (Oct), 656–684.

Mettanant, V., Basu, P., Leon, M.A., 2009. Gasification of rice husk in supercritical water, 8th World Conference on Chemical Engineering, Montreal, August, paper #971.

Miao, Q., Zhu, J., Yin, X.L., Wu, C.Z., 2008. Modeling of biomass gasification in circulating fluidized beds. In: Werther, J., Nowak, W., Wirth, K., Hartge, E. (Eds.), Circulating Fluidized Bed Technology, vol. 9. Tu-Tech Innovation GmbH, Hamburg, pp. 685–690.

Miller, R.B., 1999. Structure of wood. In: Wood Handbook—Wood as an Engineering Material, USDA Forest Service, Forest Products Laboratory, Technical report FPL-FTR-113, Madison, WI.

Milne, T., 2002. Pyrolysis: the thermal behaviour of biomass below 600 °C (Chapter 5). In: Reed, T.B. (Ed.), Encyclopedia of Biomass Thermal Conversion, 3rd ed. Biomass Energy Foundation Press, pp. II-96–131.

Milne, T.A., Evans, R.J., Abatzoglou, N., 1998. Biomass Gasifier Tars: Their Nature, Formation, and Conversion, NREL/TP-570-25357.

Minami, E., Saka, S., 2006. Chemical conversion of woody biomass and supercritical methanol to liquid fuels and chemicals. In: Bridgwater, A.V., Boocock, D.G.B. (Eds.), Science in Thermal and Chemical Biomass Conversion. CPL Press, pp. 1028–1037.

Minchener, A.J., 2005. Coal gasification for advanced power generation. Fuel 23 (17), 2222–2235.

Minowa T., Zhen F., Ogi T., 1998. Cellulose decomposition in hot-compressed water with alkali or nickel catalyst. Journal of Supercritical Fluids 13, 253–259.

Mohan, D., Pittman, C.U., Steele, P.H., 2006. Pyrolysis of wood/biomass for bio-oil: a critical review. Energy and Fuels 20 (3), 848–889.

Mozaffarian, M., Deurwaarder, E.P., Kerste, S.R.A., 2004. Green gas (SNG) production by supercritical gasification of biomass, The Netherlands Energy Research Foundation report ECN-C-04-081), *www.ecn.nl.*

Mühlen, H.J., Sowa, F., 1995. Factors influencing the ignition of coal particles–studies with a pressurized heated-grid apparatus. Fuel 74 (11), 1551–1554.

Narváez, I., Orío, A., Aznar, M.P., Corella, J., 1996. Biomass gasification with air in an atmospheric bubbling fluidized bed. Effect of six operational variables on the quality of the produced raw gas. Industrial Engineering Chemistry and Research 35 (7), 2110–2120.

Neeft, J.P.A., Knoef, H.A.M., Zielke, U., Sjöström, K., Hasler, P., Simell, P.A., Dorrington, M.A., Abatzoglou, N., Deutch, S., Greil, C., Buffinga, G.J., Brage, C., Soumalainen, M., 1999. Guideline for Sampling an Analysis of Tar and Particles in Biomass Producer Gas, Version 3.1; Energy project EEN5-1999-00507 (tar protocol).

Norton, T., Sun, D.W., Grant, J., Fallon, R., Dodd, V., 2007. Applications of computational fluid dynamics (CFD) in the modeling and design of ventilation systems in the agricultural industry: a review. Bioresource Technology 98 (12), 2386–2414.

Osman, E.A., Goss, J.R., 1983. Ash chemical composition, deformation and fusion temperatures for wood and agricultural resides. Paper no. 83-3549, American Society of Agricultural Engineers, St. Joseph, MI.

Overend, R.P., 2004. Thermo-chemical gasification technology and products, April, Presentation at Global Climate and Energy Project, Stanford University.

Overend, R.P., 1982. Wood gasification: review of recent Canadian experience. National Research Council of Canada, report no. 20094, Ottawa.

Overmann, M., Gerber, S., Behrendt, F., 2008. Euler-Euler and Euler-Lagrange modelling of wood gasification in fluidized beds. In: Werther, J., Nowak, W., Wirth, K., Hartge, E. (Eds.), Circulating Fluidized Bed Technology, vol. 9. Tu-Tech Innovation GmbH, Hamburg, pp. 733–738.

Peacocke, G.V.C., Bridgwater, A.V., 2001. Transport, handling and storage of biomass derived fast pyrolysis liquid. In: Bridgwater, A.V. (Ed.), Progress in Thermochemical Biomass Conversion, vol. 2. Blackwell Science, Oxford, pp. 1482–1499.

Perry, R.H., Green, D.W., 1997. Perry's Chemical Engineer's Handbook, 7th ed. McGraw-Hill, pp. 2-161–2-169.

Perry, R.H., Green, D.W., Maloney, J.O. (Eds.), 1984. Chemical Engineer's Handbook. McGraw-Hill.

Petersen, L., Werther, J., 2005. Experimental investigation and modeling of gasification of sewage sludge in the circulating fluidized bed. Chemical Engineering and Processing 44 (7), 717–736.

Peterson A.A., Vogel, F., Lachance, R.P., Froling, M., Antal, M.J., Tester, J.W., 2008. Thermochemical biofuel production in hydrothermal media: a review of sub- and supercritical water technologies. Energy Environment Science 1, 32–65.

Pfeifer, C.M., Pourmodjib, M., Rauch, R., Hofbauer, H., 2005. Bed material and additive variations for a dual fluidized bed steam gasifier in laboratory, pilot and demonstration scale. In: Cen, K. (Ed.), Circulating Fluidized Bed Technology VIII. International Academic Publishers, pp. 482–489.

Pimchua, A., Dutta, A., Basu, P., 2009. Torrefaction of agriculture residue to enhance combustible properties 8th world conference in Chemical Engineering, Montreal, paper 1815.

Piskorz, J., Scott, D.S., Radlien, D., 1988. Composition of oils obtained by fast pyrolysis of different woods (Chapter 16). In: Soltes, J., Milne, T.A. (Eds.), Pyrolysis Oils from Biomass: Producing, Analyzing and Upgrading. American Chemical Society, pp. 167–178.

Prins, M.J., Krzysztof, J.P., Janssen, F.J.J.G., 2006. More efficient biomass gasification via torrefaction. Energy 31 (15), 3458–3470.

Prins, W., Kersten, S.R.A., Pennington, J.M.L., van de Beld, L., 2005. Gasification of wet biomass in supercritical water. In: Knoeff, H.A.M. (Ed.), Handbook of Biomass Gasification. BTG Biomass Technology Group, Entschede, p. 234.

Probstein, R.F., Hicks, R.E., 2006. Synthetic Fuels. Dover Publications, pp. 63, 98–99.

Quaak, P., Knoef, H.A.M., Stassen, H., 1999. Energy from biomass: a review of combustion and gasification technologies. Technical paper no. 422, pp. 56–57. Energy Series, The World Bank.

Rao, M.S., Singha, S.P., Sodha, M.S., Dubey, A.K., Shyam, M., 2004. Stoichiometric, mass, energy and energy balance analysis of countercurrent fixed-bed gasification of post-consumer residues. Biomass and Bioenergy 27 (2), 155–171.

Rapagnà, S., Jand, N., Kiennemann, A., Foscolo, P.U., 2000. Steam-gasification of biomass in a fluidized-bed of olivine particles. Biomass and Bioenergy 19 (3), 187–197.

Razvigorova, M., Goranova, M., Minkova, V., Cerny, J., 1994. On the composition of volatiles evolved during the production of carbon adsorbents from vegetable wastes. Fuel 73 (11), 1718–1722.

Reed, T.B., 2002. Kinetics of char gasification reactions above 500 °C (Chapter 7). Encyclopedia of Biomass Thermal Conversion, 3rd ed. Biomass Energy Foundation Press, p. II–289.

Reed, T.B., Das, A., 1988. Handbook of biomass downdraft gasifier for engine–system. Solar Energy Research Institute, SERUSP-271-3022, NTIS.

Rezaiyan, J., Cheremisinoff, N.P., 2005. Gasification Technologies: A Primer for Engineers and Technologists. Taylor and Francis Group, CRC Press, p. 15.

Risnes, H., Sørensen, L.H., Hustad, J.E., 2001. CO_2 reactivity of chars from wheat, spruce, and coal. In: Bridgwater, A.V. (Ed.), Progress in Thermochemical Biomass Conversion, vol. 1. Blackwell Science, pp. 61–72.

Ritter, S., 2005. Science & Technology–Gasoline. Chemcial and Engineering News 83 (8), 37.

Rocha, J.D., Brown, S.D., Love, G.D., Snape, C.E., 1997. Hydropyrolysis: a versatile technique for solid fuel liquefaction, sulphur speciation and biomarker release. Journal of Analytical and Applied Pyrolysis 40–41, 91–103.

Ross, D., Noda, R., Adachi, M., Nishi, K., Tanaka, N., Horio, M., 2005. Biomass gasification with clay minerals using a multi-bed reactor system. In: Cen, K. (Ed.), Circulating Fluidized Bed Technology VIII. International Academic Publishers, pp. 468–473.

Rossi, A., 1984. Fuel characteristics of wood and nonwood biomass fuels. In: Tilman, D.A., Jahn, E.C. (Eds.), Progress in Biomass Conversion, vol. 5, Academic Press, pp. 69–99.

Sanner, W.S, Ortuglio, C., Walters, J.G., Wolfson, D.E., 1970. Conversion of municipal and industrial refuse into useful materials by pyrolysis. U.S. Bureau of Mines, RI 7428.

Scott, D.S., Piskorz, J., 1984. The Continuous Flash Pyrolysis of Biomass, Can. J. Chem. Eng. 62 (3), 404–412.

Schmieder, H., Abeln, J., Boukis, E., Dinjus, A., Kruse, M., Kluth, G., Petric, E., Sadri, E., Schacht, M., 2000. Hydrothermal gasification of biomass and organic wastes. Journal of Supercritical Fluids 17, 145–153.

Seebauer, V., Petek, J., Staudinger, G., 1997. Effects of particle size, heating rate and pressure on measurement of pyrolysis kinetics by thermogravimetric analysis. Fuel 76 (13), 1277–1282.

Sensoz, S., Angin, D., 2008. Pyrolysis of safflower seed press cake: part 1—the effect of pyrolysis parameters on the product yields. Bioresource Technology 99, 5492–5497.

Serani, A.L., Aymonier, C., Cansell, F., 2008. Current and Foreseeable Applications of Supercritical Water for Energy and the Environment, ChemSusChem, DOI: 10.1002/cssc.2007001671, pp. 486–503.

Shafizadeh, F., 1984. Chemistry of solid wood. Advances in Chemistry Series No. 207, American Chemical Society.

Shaw, R.W., Dahmen, N., 2000. Destruction of toxic organic materials using supercritical water oxidation: current state of the technology, In: Kiran, E., Debenedetti, P.G., Peters C.J., (Eds.), Supercritical Fluids—Fundamentals and Applications. NATO Science Series E, vol. 366. Kluwer Academic Publishers, pp. 425–438.

Shen, J., Wand, X., Perez, M., Mourant, D., Rhodes, M., Li, C., 2009. Effects of particle size on the fast pyrolysis of oil mallee woody biomass. Fuel 88 (10), 1810–1817.

Sinag A., Kruse A., Rathert J., 2004. Influence of the heating rate and the type of catalyst on the formation of key intermediates and on the generation of gases during hydropyrolysis of glucose in supercritical water in a batch reactor. Industrial and Engineering Chemistry Research 43, 502–508.

Sit, S.P., Grace, J.R., 1978. Interphase mass transfer in an aggregative fluidized bed. Chemical Engineering Science 33 (8), 1115–1122.

Smith, E.B., 2005. Basic Chemical Thermodynamics. Imperial College Press, pp. 49–57.

Smith, P.V., Davis, B.M., Liu, G., 2002. Operation of the PSDF transport gasifier, Gasification Technologies Session: New Development (KBR #1728), 19th Annual Conference, Pittsburgh, pp. 1–17.

Smoot, L.D., Smith, P.J., 1985. Coal combustion and gasification. Plenum Chemical Engineering Series, pp. 88–89.

Soltes, E.J., Elder, T.J., 1981. Pyrolysis. In: Goldstein, I.S. (Ed.), Organic Chemicals from Biomass. CRC Press.

Souza-Santos, M.L.D., 2004. Solid fuels combustion and gasification. CRC Press.

Ståhl, K., Neergaard, M., Nieminen, J., 2001. Final report: Värnamo demonstration programme. In: Bridgwater, A.V. (Ed.), Progress in Thermochemical Biomass Conversion. Blackwell, pp. 549–563.

Stevens, D.J., 2001. Hot gas conditioning: recent progress with larger scale biomass gasification systems. National Renewable Energy Laboratory, report NREL/SR-510-29952.

Stiegel, G.J., 2005. Overview of Gasification Technologies. Global Energy and Energy Project (GCEP) Advanced Coal Workshop, Provo, UT.

Sun, H., Song, B., Yang, Y., Kim, S., Li, H., Cheng, J., 2007. The characteristics of steam gasification of biomass and water filter carbon, Korean Journal of Chemical Engineering 24 (2), 341–346.

Susanto, H., Beenackers, A.A.C.M., 1996. Moving bed gasifier with internal recycle of pyrolysis gas. Fuel 75 (11), 1339–1347.

Sutton, D., Kelleher, B., Ross, J.R.H., 2001. Review of literature on catalysts for biomass gasification. Fuel Processing Technology 73 (3), 155–173.

Syred, N., Kurniawan, K., Griffith, T., Gralton, T., Ray, R., 2007. Development of fragmentation models for solid fuel combustion and gasification as subroutines for inclusion in CFD codes. Fuel 86 (14), 2221–2231.

Termuehlen, H., Emsperger, W., 2003. Clean and Efficient Coal-Fired Power Plants: Development Toward Advanced Technologies. ASME Press, p. 23.

Tester, J.W., Holgate, H.R., Armellini, F.A., Webley, P.A., Killilea, W.R., Hong, G.T., Barner, H.E., 1993. Supercritical water oxidation technology process development and fundamental research. In: Teddler, W.D., Pohland, F.G. (Eds.), Emerging Technologies for Hazardous Waste Management III, vol. 518. American Chemical Society Symposium Series, Washington, pp. 35–76.

Thunman, H., Leckner, B., 2002. Thermal conductivity of wood—models for different stages of combustion. Biomass and Bioenergy 23 (1), 47–54.

Thurner, F., Mann, U., 1981. Kinetic investigation of wood pyrolysis. Industrial Engineering Chemistry 20 (3), 482–488.

Tillman, D.A. 1978. Wood as an Energy Resource. Academic Press.

Uematsu, M., Franck, E.U., 1980. Static dielectric constant of water and steam. Journal of Physical and Chemical Reference Data 9 (4), 1291–1306.

UNFCCC, 2005. Clarifications of definition of biomass and consideration of changes in carbon pools due to a CDM project activity, EB-20, Appendix 8, July.

U.S. DOE, 1978. Coal Conversion Systems Technical Data Book.

Vamvuka, D., Woodburn, E.T., Senior, P.R., 1995. Modelling of an entrained flow coal gasifier 1. Development of the model and general predictions. Fuel 74 (10), 1452–1460.

van Heek, K.H., Muhlen, H.J., 1990. Chemical kinetics of carbon and char gasification. In: Lahaye, J., Ehrburger, P. (Eds.), Fundamental Issues in Control of Carbon Gasification Reactivity. Kluger Academic Publisher, pp. 1–34.

Van Loo, S., Koppejan, J., 2008. The Handbook of Biomass Combustion and Co-firing. Task 32. Earthscan, London.

Van Swaaij, W.P.M., 2003. Technical feasibility of biomass gasification in a fluidized bed with supercritical water, Report GP-01. University of Twente, p. 160.

Vilienskii, Hezmalian, D.M., 1978. Dynamics of the Combustion of Pulverized Fuel (Dinamika Gorenia Prilevidnovo Topliva). Energia, Moscow.

Wagenaar, B.M., Florijn, J.H., Gansekoele, E., Venderbosch, R.H., 2009. Bio-oil as natural gas substitute in a 350 MW Power Station. Biomass Technology Group. Accessed from *www.btgworld.com*.

Walker, P.L., Rusinko, F., Austin, L.G., 1959. Gas reactions of carbon. Advances in Catalysis 11, 133–221.

Wang, M., Pantini, D., 2000. Corn Based Ethanol Does Indeed Achieve Energy Benefits. Center for Transportation Research, Argonne National Laboratory, February.

Wang, S., Lu, G.Q., 1996. Carbon dioxide reforming of methane to produce synthesis gas over metal-supported catalysts: state of the art. Energy & Fuels 10, 896–904.

Wang, Y., Kinoshita, C.M., 1993. Kinetic model of biomass gasification. Solar Energy 51 (1), 19–25.

Wang, Y., Yan, L., 2008. CFD studies on biomass thermochemical conversion. International Journal of Molecular Sciences 9 (6), 1108–1130.

Watanabe M., Inomata H., Osada M., Sato T., Adschiri T., Arai K., 2003. Catalytic effects of NaOH and ZrO2 for partial oxidative gasification of n-hexadecane and lignin in supercritical water. Fuel 82, 545–552.

Weigner, P., Martens, F., Uhlenberg, J., Wolff, J., 2002. Increased flexibility of Shell gasification plant. IChemE conference on gasification: the clean choice for carbon management. Noordwijk, The Netherlands.

Wen, C.Y., Bailie, R.C., Lin, C.Y., O'Brien, W.S., 1974. Production of low BTU gas involving coal pyrolysis and gasification. Advances in Chemistry Series, vol. 131, American Chemical Society, Washington, DC.

Werpy, T., Petersen, G., 2004. Top value added chemicals from biomass, vol. 1. National Renewable Energy Laboratory, August, pp. 11–13.

Westbrook, C.K., Dryer, F.K., 1981. Simplified reaction mechanisms for the oxidation of hydrocarbon fuels in flames. Combustion Science and Technology 27, 31–43.

White, C., Gray, D., Tomlinson, G., 2007. Co-conversion of coal and biomass to produce Fisher-Tropsch transportation fuels. Gasification Technologies Conference, San Francisco.

www.absoluteastronomy.com.

www.biodiesel.org.

www.kamengo.com.

Xia, B., Sun, D.W., 2002. Applications of computational fluid dynamics (CFD) in the food industry: a review. Computers and Electronics in Agriculture 34 (1–3), 5–24.

Xiao, G., Ni, M.N., Chi, Y., Jin, B.S., Xiao, R., Zhong, Z.P., Huang, Y.J., 2009. Gasification characteristics of MSW and an ANN prediction model. Waste Management 29 (1), 240–244.

Xu, X., Matsumura, Y., Stenberg, J., Antal M.J., 1996. Carbon-catalyzed gasification of organic feed-stocks in supercritical water. Industrial Engineering Chemistry Research 35 (8), 2522–2530.

Yamagata, K., Nishikawa, K., Hasegawa, S., Fyjii, T., 1972. Forced convective heat transfer to supercritical water flowing in tubes. Int. J. Heat Mass Transfer 15, 2575–3593.

Yoshida, Y., Dowaki, K., Matsumura, Y., Matsuhashi, R., Li, D., Ishitani, H., Komiyama, H., 2003. Comprehensive comparison of efficiency and CO2 emissions between biomass energy conversion technologies—position of supercritical water gasification biomass technologies. Biomass & Bioenergy 25, 257–272.

Yu, D.M., Aihara, M., Antal, M.J., 1993. Hydrogen production by steam reforming of glucose in supercritical water. Energy & Fuels 7, 574–577.

Zeleznik, F.J., Gordon, S., 1968. Calculation of complex chemical equilibria. Industrial and Engineering Chemistry 60 (6), 27–57.

Zen, H.C. (Ed.), 2006. *Coal Gasification Technology* (in Chinese), Chemical Industry Press, Beijing, pp. 140, 154.

Zhou, C., Beltraminin, J.N., Fan, Y.X., Lu, F.Q., 2008. Chemoselective catalytic conversion of glycerol as a biorenewable source to valuable commodity chemicals. Chemical Society Review 37 (3), 527–554.

Zhu, X., Venderbosch, R., 2005. A correlation between stoichiometrical ratio of fuel and its higher heating value. Fuel 84, 1007–1010.

Zimont, V.I., Trushin, Y.M., 1969. Total combustion kinetics of hydrocarbon fuels. Combustion, Explosions and Shock Waves 5 (4), 567–573.

Zolin, A., Jensen, A., Jensen, P.A., Frandsen, F., Johansen, K.D., 2001. The influence of inorganic materials on the thermal deactivation of fuel chars. Energy and Fuels 15 (5), 1110–1122.

Index

Biomass Gasification and Pyrolysis. DOI: 10.1016/B978-0-12-374988-8.00021-0